CONSERVATION

CONSERVATION

Linking Ecology, Economics, and Culture

Monique Borgerhoff Mulder
and
Peter Coppolillo

PRINCETON UNIVERSITY PRESS
PRINCETON AND OXFORD

Library of Congress Cataloging-in-Publication Data

Borgerhoff Mulder, Monique.
Conservation : linking ecology, economics, and culture
Monique Borgerhoff Mulder and Peter Coppolillo.
p. cm.
Includes bibliographical references and index.
ISBN 0-691-04979-3 (cl : alk. paper) — ISBN 0-691-04980-7
(pbk. : alk. paper)
1. Nature conservation—Philosophy.   2. Nature
conservation—Economic aspects.   3. Nature
conservation—Social aspects.   I. Coppolillo, Peter, 1970–
II. Title.

QH75.M848   2005
333.72—dc22   2004044336

British Library Cataloging-in-Publication Data is available

This book has been composed in Sabon

Printed on acid-free paper. ∞

pup.princeton.edu

Printed in the United States of America

3 5 7 9 10 8 6 4 2

ISBN-13: 978-0-691-04980-9 (pbk.)

For Barnabas, Henry, and Nina
and everyone looking towards the future

Contents

Preface

We had just driven back to our campsite on the edge of the Katuma River after a hard day's work weighing and measuring children in a small village adjacent to Katavi National Park, Tanzania. As usual, we were cooking beans and rice on the fire, watching hippos lumber out of the water in search of grass as dusk fell.

"Mum, should we tell them?" ventured the little five-year-old boy.

He was referring to the excellent buffalo stew that we, the anthropologists in the team, had eaten for lunch in the village. He carried on:

"They'll be awfully cross, because that buffalo was poached, you know. But what else was there to eat? We were so hungry."

We were a group of biologists, ecologists, and anthropologists, all with research interests in conservation. We were starting a new project on biodiversity protection in a remote zone of western Tanzania where, as in so many other parts of the world, the interests of conservationists and the local communities are in conflict, at least in the short term. Almost daily we would engage this debate, variously from the viewpoints of philosophy, ecology, conservation biology, ethics, policy, and pragmatism. Tonight no doubt the buffalo stew incident would lead to ethics. Should the anthropologists turn a blind eye? Or should they report to the national park staff that two buffalo had been shot by a villager? The decision is not easy. It is certainly more polite just to keep quiet; anthropologists are, after all, guests and guests should not tell tales. Silence is also more pragmatic and self-serving, since a gabbing anthropologist will not long be welcome in the village. On the other hand, failing to report a poaching incident would undercut the shared goal of our project, to work out conservation solutions in and outside of Katavi National Park. Not only our scientific colleagues, but also many villagers, would be disappointed by the duplicity.

So how do we balance the rights and interests of human populations with our equally human obligation to preserve some measure of biodiversity and a healthy functioning planet for future generations? This is more than campfire chat. It is one of the key challenges we face at the dawn of the twenty-first century. There is, of course, no shortage of millennial challenges but, because of the escalating rate at which natural resources are being eroded, the biodiversity crisis seems to us the most pressing of all. And indeed it is drawing attention, not only in the worlds of academia and development, but also all over the media. So why another book on the topic?

For the very simple reason that there is no even-handed introduction to the key issues of the debate. In one part of the library you can find fiery material written by activists who eloquently represent the interests of local communities, works that are firmly endorsed by many in the social sciences, the humanities, and development circles. In a different part of the library, you can find equally vociferous and inspired writings from conservation biologists, struggling to give voice to the interests of the natural world (for want of a better term). Although various compromises and standoffs have been reached throughout the long and turbulent history of this debate, the vast majority of literature up to the present day belongs to one of two camps, "conservation first" or "people first." Those who embrace the former rarely accept the assumptions, arguments, data, and conclusions of the latter, and vice versa. Even writers who sincerely try to strive for balance end up stacking the deck one way or another, ultimately because of their adherence to a standard set of disciplinary tenets. If you try to make sense of the field, or teach a course in the area, you find the literature lacks easy bridging material. Though we have both met plenty of thoughtful people who in their day-to-day work negotiate successfully between competing sets of interests regarding "conservation" and "people," there is little for incoming students (or indeed the interested public) to read where the assumptions behind different ideological positions are made explicit, or where research is relevant to both camps. Only on this basis can reconciliatory solutions be reached that move beyond political agendas or personal ethics. In a very fundamental sense, then, we think of this book as an invitation to those who would like to keep their minds open to insights both from the natural and the social sciences.

The goal of this book is to delineate a toolkit of terms, concepts, models, ideas, and issues that the next generation will need if they are to contribute new solutions to age-old conservation problems. It is only through an expressly interdisciplinary training that thinking people interested in conservation will be able to make a difference. For too long, social and biological scientists have been talking past each other, like ships in the night. This has been not so much because they are uninterested in each other's objectives, but because they share little vocabulary with which to discuss their differences. As authors, we do not expect to change our readers' emotional or political

commitment to how such conflicts should be solved. Nor do we wish to suggest that there is always an acceptable compromise whereby the value of conserving biodiversity can be balanced against the survival and needs of a local community. We do, however, intend to sensitize our readers to burning issues on each side of the fence, and to clarify debates that are currently all fire and no light.

Our commitment to a multidisciplinary approach to studying the human dimensions of biodiversity conservation is inspired by E. O. Wilson's arguments in the opening chapters of *Consilience* (1998). He proposes that to reach honest, pragmatic, and effective solutions we need to move freely between ethics, environmental policy, biology, and the social sciences. Wilson develops his view by arguing persuasively for a new unity of knowledge, in fact, nothing short of a new Enlightenment. Indeed, conservation biologists facing the contemporary ecological crisis now, more than ever before, need to build on contributions from the social sciences, including economics, anthropology, sociology, and the political sciences to forge their solutions. Increasingly, their goal is to maintain biological diversity in the context of exploding human populations, changing livelihood patterns, global markets, and a shrinking base of natural resources and habitats.

In this book we turn to the practicalities of what such training might entail, by introducing readers to an intellectual toolkit with which social and ecological processes can be analyzed. Our concern with a multidisciplinary approach is more than just faddish, faddish though it is. If you believe (as we do) that rights and duties should come in a packet, in other words that a person is ineligible for rights if she or he fails to perform duties, and, conversely, that the performance of duties should confer rights, then you would also concur that it is indefensible to side single-heartedly with either "conservation people" or "people people." As humans we have the right to enjoy nature and the ecological services that nature provides, but we also have the duty to restrain our behavior in such as way as to ensure the continued enjoyment of this right. The question is how to do this, and the answers lie across many disciplines.

Unsurprisingly, this book gives no single solution to the puzzle of how to balance the rights of local communities with the obligations of conservationists. There is, of course, no one answer. We place no bets on any one horse. Rather, we analyze the rallying cries that are found throughout the literature—integrated conservation and development projects, direct payments, protectionism, community-based conservation, ecological economics, private reserves, co-management, and so forth, inviting the reader to consider both the merits and demerits of each and, more important, the suitability of each to the specific conflict at hand. Our goal then is to outline a checklist whereby the reader can determine the range of appropriate solutions to particular conflicts that arise from the need to protect biodiversity.

There are further limits that we need to set ourselves. What kind of conservation are we talking about, and who are the people at the focus of this book? First, to conservation. This word gets bandied around in all sorts of contexts and we cannot deal with all its uses. We consider only the conservation of natural and renewable resources; accordingly, we do not deal with historic buildings, nor with mineral deposits, nor even with water conservation in metropolises. Additionally, we focus only on those suites of natural resources that represent some measure of biodiversity. Thus, we do not consider how a farmer can be induced to adopt soil conservation measures in an intensive agricultural system, a topic that falls squarely in the fields of agronomy and development studies. Since biodiversity itself has many meanings, this is a somewhat fuzzy boundary, but in general we limit ourselves to discussion of the conservation of intact or at least functioning natural ecosystems. This does *not* mean we deal only with

challenges that arise from conserving reserves or parks. However, since the majority of threatened ecosystems that have been written about enjoy some level of protection (if only on paper) most of our discussion of the human dimensions of biodiversity conservation is drawn from sites in and adjacent to protected areas, broadly defined.

Who are the local people whose interests are infringed by conservationist policies? Much of the antecedent literature on the relationships between local people and protected areas deals with "traditional" or "native" peoples. This focus is awkward for two reasons. First, the labels *traditional* (or *native*) necessarily restrict focus to communities that are indigenous, a problematic category partly because it is rarely known how long a group of people have inhabited an area (or indeed whether they were the first to do so), and partly because the term *indigenous* often conveys the idea of a harmonious relationship between the human community and the environment, a view that is open to question. We pay equal attention to all local communities affected by conservation actions, irrespective of whether these communities are indigenous, acculturated to modern ways of life, or involved in commercial production. Accordingly, we use interchangeably the terms *indigenous, traditional, local,* and *resident peoples,* following the commonest usage for any particular case. Second, the principal threats to biodiversity come from habitat loss and over-exploitation, and people of all kinds (from local fishermen to logging operators) are behind these forces. To cast a spotlight only on traditional communities living at the forest edge would be illogical. For this reason we devote several chapters to the broader community, political, and global factors involved in biodiversity loss. Third, though we refer at certain instances to local or indigenous populations in developed nations, such as the Aborigines of Australia or Native American groups in North America, our attention is primarily on the developing world. This reflects the specific challenges faced by developing nations, their particular vulnerability to the forces of globalization, and the fact that much of the planet's remaining biodiversity lies in the developing world.

So let us turn finally to the structure of this book, which falls into three sections. Conservation, both its goals and its methods, has always been a moving target. In the first four chapters of this book we examine the changing practice of conservation and the development of its underlying science. "Why conserve biodiversity" is the focus of chapter 1, which presents the various ecological, social, and philosophical answers that have been given, and examines how these viewpoints have changed over time and in different parts of the world. In chapter 2 we trace the roots and political evolution of modern conservation thinking, particularly with respect to the links between protectionism, policy, and development aid. Chapter 3 gives a historical overview to principal shifts in ecological thinking that have shaped conservation objectives and practices, and introduces the new crisis discipline of conservation biology. In chapter 4 various anthropological perspectives on the role of local communities in protecting biodiversity within their territories are critically examined.

The five chapters following provide the intellectual disciplinary core of the book, offering a series of ever more encompassing frameworks within which the human dimensions of biodiversity conservation can be studied. Having already dealt with the dynamics of natural resources and human communities (the disciplines of ecology and anthropology), we move from the study of the individual actor (behavioral ecology and anthropology, chapter 5), through to the institutions that frame individual actions (neo-institutional economics, political science, sociology, and anthropology, chapter 6), to the national and global context (political ecology, ecological economics, and international policy studies, chapters 7 to 9). Thus, chapter 5 explores behavioral ecological approaches used by anthropologists to

examine where, when, and how humans might behave as conservationists. Chapter 6 broadens the focus up to the community, and examines how local social institutions affect resource use and management. Political ecology is the centerpiece of chapter 7, a field that emphasizes the links between individual behavior, local institutions, and the state, while chapter 8 injects some real politics into the global dimensions to these debates by focusing on topics such as local peoples' alliances with international nongovernmental organizations (NGOs), bioprospecting, and the marketing of rainforest products. Finally, chapter 9 turns to the international sphere, exploring the roles of business and international treaties in addressing global conservation challenges.

Specific strategies for dealing with conservation challenges in the developing world are the focus of the final two chapters. Using ideas developed earlier, chapter 10 analyzes the successes and failures of different methods, focusing on case studies embracing some of the more conventional strategies—protectionism, outreach, conservation education, ecotourism, integrated conservation and development, and extractive reserves, and also looks at evaluation and monitoring. The final chapter (chapter 11) deals with some novel integrative solutions, initiatives where the conventional barriers between the social and biological sciences are breaking down, and draws together many of the conclusions of earlier chapters by emphasizing how important it is both to define clearly the goals of a conservation project and to distinguish these from the means of reaching that end. It also points to how human aspects of biodiversity conservation should be considered in the future.

We have many acknowledgments. MBM would like to thank extraordinarily enthusiastic students in a course (Conservation & People) offered at the University of California at Davis over several years for the inspiration to write a book of this breadth. Their ideas and discussion made it painfully clear that narrower treatments are incomplete, failing to alert students to the brittle seams along the social and natural sciences where innovation is most needed. Thanks also to Steve Brush, Kelly Stewart, Joe Spaeder, Truman Young, and others who have offered engaging lectures over the years, and to Nancy McLaughlin and the fabulous staff in Anthropology. PC would like to acknowledge his colleagues at the Wildlife Conservation Society who have stimulated his thinking, challenged his ideas, and offered countless insights.

Huge amounts of reading went into this project, as evidenced by the gargantuan bibliography. Almost all of it was rewarding, but there were a few scholars who strongly shaped our understanding, to whom we are both grateful for their clarity of exposition and apologetic if we have unwittingly lifted some of their ideas without due citation at every juncture (so diffuse and wide-ranging were their influences). In this respect we are particularly thankful (for insights as well as in some cases further suggestions, comments, and lively debate) to Arun Agrawal, Michael Alvard, Andy Balmford, Kelly Bannister, Fikret Berkes, Katerina Brandon, Steve Brush, Tim Caro, Ronald Carroll, Marcus Colchester, Madhav Gadgil, Ricardo Godoy, Richard Groves, Robert Hitchcock, Peter Little, Katherine McAfee, David Maybury-Lewis, Gary Meffe, John Merson, Ben Orlove, Elinor Ostrom, Nancy Peluso, Dustin Penn, Carlos Peres, Kent Redford, John Robinson, Satoshi Sarkar, Kelly Stewart, Bill Sutherland, Timothy Swanson, Eric Alden Smith, Clem Tisdell, Bruce Winter-halder, David Wilkie, and Truman Young. Andy Balmford and three other anonymous reviewers read through the complete manuscript, and to them we are particularly grateful for all their suggestions, corrections, and encouragement.

We have also benefited from a plethora of discussion, advice, help, and suggestions from current and erstwhile graduate students at UC Davis: in particular, Jeremy Brooks, Maggie Franzen, Christopher Holmes, Sarah Otterstrom, Patricia Pinho, Aron Pretty, Lore Ruttan, Ryan Sensenig, and Joe Spaeder ("40 Below Joe" who, perhaps wisely, jumped ship from authorship when there was still time to leap). And we thank Corine Graham for all her help, her organizational skills with the pictures, and her constant good cheer. Tim Caro diligently proofread the whole book on trains and ferries in Scandinavia—a million thanks.

Particularly illuminating have been numerous discussions with thoughtful Tanzanians involved with wildlife and other policy issues, most memorably, The Honorable Mizengo Pinda, the late Ally Kyambile, Alfred Kikoti, Joseph Chua, Leo Mashauri, Antony Sijimbi, and the many villagers of Mpimbwe who deal with the consequences of conservation policy on a daily basis.

Last but not least, families bear the brunt of books, without whom none of this could have been possible. Huge thanks to Tim and Chris, who could always see the wood from the trees, and to Barnabas, Henry, and Nina, who aren't out of the woods yet.

Commonly Used Abbreviations

ADMADE   Administrative Design for Game Management Areas

BSP   Biodiversity Support Program

CAMPFIRE   Community Area Management Programme for Indigenous Resources

CBC   community-based conservation

CBD   Convention on Biological Diversity

CBNRM   community-based natural resources management

CC   community conservation

CI   Conservation International

CITES   Convention on International Trade in Endangered Species of Wild Fauna and Flora

CS   Cultural Survival

GATT   General Agreement on Tariffs and Trade

GEF   Global Environmental Facility

IBT   island biogeography theory

ICD   integrated conservation and development

IIED   International Institute for Environment and Development

INBio   National Biodiversity Institute

IPR   intellectual property rights

IUCN   International Union for the Conservation of Nature, formerly; now World Conservation Union

IWGIA   International Work Group for Indigenous Affairs

MAB   Man and the Biosphere Program

MSY   maximum sustainable yield

NGO   nongovernmental organization

NTFP   nontimber forest product

PAR   Participatory Action Research

RAN   Rainforest Action Network

SI   Survival International

TEK   traditional ecological knowledge

TNC   The Nature Conservancy

UN   United Nations

UNCED   United Nations Conference on Environment and Development

UNESCO   United Nations Educational, Scientific, and Cultural Organization

USAID	U.S. Agency for International Development	WHS	World Heritage Site
WCS	Wildlife Conservation Society	WRI	World Resources Institute
		WTO	World Trade Organization
WHC	World Heritage Convention	WWF	WorldWide Fund for Nature/World Wildlife Fund

CHAPTER 1

The Many Roads to Conservation

1.1 Introduction

It is customary to start a book about conservation with a doom-laden outline of the impending biological and ecological crises that face the planet. In fact, most readers will have more than a passing awareness of the gravity of the current situation. A few images that depict the skeletal white hand of logging roads etched into an Amazonian satellite view, the scarred red earth of Madagascar blowing in the dust, or an immense lump of fallen rhino with the horn hacked away are enough to focus attention on biodiversity loss. At one level, the problem posed by the erosion of biodiversity is simple—it is happening and it needs to be slowed, better, stopped. But once you start examining proposed conservation strategies, or more specifically exploring the disciplines where such strategies are being crafted, only more questions arise. What is biodiversity? Why does it need conserving? What are the goals of biodiversity conservation? Who is involved in the practice of conservation, and what are their objectives? Such questions must be faced directly, since they frame any so-called solutions to the biodiversity crisis. Inevitably people will give wildly divergent answers to these kinds of questions, depending on when and where they live, their social, economic, and political circumstances, their personal experiences, and (at least among academics) their intellectual training or disciplinary affiliation. Yet answers to

these questions take us a long way toward understanding changes in conservation thinking through time and help unravel the unique problems associated with the contemporary challenges discussed in this book.

In this chapter we look at some of these questions. Philosophers, artists, naturalists, ecologists, and activists have all written on these matters, and we cannot claim to offer anything other than a coarse overview of their positions. We start out with a summary of the principal threats to biodiversity (1.2) and an intellectual map of the reasons biodiversity should be saved (1.3 and 1.4). The core of the chapter lies in an examination of the historical evolution of conservation thinking. By focusing on this story in the United States, we demonstrate how tightly interwoven changing social and environmental philosophies are with conservation practice, in particular, protectionism (1.5), management (1.6), and multiple use (1.7). We also draw on a wider range of related issues from other parts of the world in the boxes. With this background we conclude by showing how, at the dawn of the modern conservation era, purely utilitarian objectives had largely succumbed to the highly preservationist goal of comprehensive biodiversity protection. We also point to the importance of distinguishing conservation goals and means, and to the fact that conservation is always about choices based on ethical values (1.8).

1.2 Principal Threats to Biodiversity

Biodiversity refers broadly to the full set of species, genetic variation within species, the variety of ecosystems that contain the species, and the natural abundance in which these items occur (OTA, 1987); in other words, it is an umbrella term for all of nature's variety (McNeely, 1990). Leaving historical (chapter 2), ecological (chapter 3), political (chapter 7), and global (chapter 9) aspects of the concept of biodiversity for later consideration, we consider here its major threats.

It barely needs repeating, the world is a very different place from what it was 10,000 years ago, and even 100 years ago. The changes of the last century radically restructured humankind's economic activities, political relations, and social and demographic profile. One prominent feature of this change is the accelerating scale of human impact on the Earth's natural biophysical systems—climate, stratospheric ozone, terrestrial and marine ecosystems, and the great cycles of water, nitrogen, and sulfur, all of which sustain the conditions on which life depends (McMichael et al., 1999; Vitousek et al., 1997). Clearly, the world our children inherit will be far more crowded, more polluted, and less habitable than the one we occupy today (Meffe and Carroll, 1997).

Closely associated with these broad changes in the global environment is the erosion of biodiversity, in particular, the loss of species, populations, and their habitats. The intricate relationships between levels of biodiversity, productivity, ecological complexity, stability, and environmental health are not well understood (May, 1999), but a pop rivet analogy nicely captures many biologists' thinking about loss of this diversity, given all the uncertainties. How many rivets can a financially strapped airline operator remove from its aircraft without impairing safety in flight? As with the removal of species and habitats from our planet, no one really wants to do the critical test. In the absence of precise details on the tolerance of human welfare to the current wave of extinctions, there is nevertheless increasing recognition that biological diversity is a key environmental asset under threat. In part, this is because of the evidence that certain ecosystems, such as speciose forests and highly productive marshes, play essential roles in maintaining a healthy, functioning ecosystem, and in part because many people hold strong ethical beliefs that nature deserves respect, often for purely intrinsic reasons (see below). The erosion of biodiversity is reflected in extinction rates (Wilson, 1988). Though extinction is a natural evolutionary process, its rate over the last century ranged from 100 to 10,000 species per year (Pimm et al., 1995) compared to a one to ten species a year background rate, based on paleontologists' estimates. This has been called variously an "extinction crisis" (Soulé, 1986) or an "extinction spasm" (Myers, 1987).

Mass extinctions have occurred before; there have been five, at the end of the Ordovician, Devonian, Permian, Triassic, and Cretaceous, spanning geologic time from 430 mya to 65 mya, when the dinosaurs met their end. These were all "natural events," and usually occurred over many hundreds or thousands of years. The present extinction spasm is considered to be unnatural (and unique) insofar as it is driven by a single species—humans. In contrast to other extinctions, which were probably driven by large-scale climatic changes that affected many or all species, our current crisis affects species in a more systematic way: large-bodied, economically significant, and habitat-sensitive species are being extirpated and replaced by smaller, generalist species that thrive in human-dominated places. The current situation is notable also for its extreme rapidity, and the fact that between one-third and two-thirds of all species on earth will disappear within the foreseeable future if the present trend continues unchecked (Pimm et al., 1995; Wilson, 1992). With species loss goes the eradication of locally adapted populations, habitats, and the evolutionary and ecological processes whereby these species coevolve and coexist.

Biodiversity is strongly patterned. Its greatest storehouse lies in areas that are warm and humid, especially tropical rainforests which, although they occupy less than an estimated 7 percent of the Earth's surface, are thought to contain at least 50 percent of the world's species (Wilson, 1988). Although species diversity is also associated with altitude, area size, successional stage, and other factors (considered in chapter 3), the latitudinal effect is strong. For example, of the twenty-five biodiversity hotspots, areas highlighted for the prioritization of global conservation efforts, sixteen lie in the tropics (Figure 1.1). This means that remaining biodiversity is found largely in developing countries, where conservation resources are scarcest, where habitat conversion is most rapid (Dobson et al., 1997), and where the threat to biodiversity is greatest. Notably, too, species richness is concentrated in areas of high human density, both at a global scale (Cincotta et al., 2000; see chapter 8) and within continents (at least as shown for Africa; Balmford et al., 2001), even when latitudinal differences are controlled. Given that the population projections for 2050 are highest for many of the world's poorest countries (Bongaarts and Bulatao, 2000), the challenge for conservationists is all the more acute. None of this means that blame is to be laid at the door of developing rather than developed nations. In fact, almost the converse: a good proportion of the extraction of natural resources supports businesses owned by Westerners or Westernized national elites (chapter 7). With so much of the Western and developed world already environmentally degraded, a guilty focus necessarily settles on areas of remaining diversity (chapter 9).

The direct and indirect causes of biodiversity declines are extremely complex, and rarely if ever exclusively local. Virtually everything we do affects species diversity. Sometimes these outcomes are positive: for instance, there are activities that potentially enhance species diversity and habitat protection, such as programs that establish wildlife reserves, mount species survival programs, or manage botanical and zoological parks, as well as many traditional natural resource management practices that may be quite effective (chapters 4 and 5). However, the majority of human activities, particularly in recent years, are detrimental to diversity. The broad anthropogenic processes of deforestation, desertification, pollution, agricultural expansion, and urban sprawl drive species extinction, spearheaded by an "evil quartet" of mechanisms (habitat loss, overexploitation, introduced species, and pollution; e.g., Purvis et al., 2000). Furthermore, though the 6.5 billion (and growing) world population is a key ingredient of biodiversity loss, equally important factors are where these people live and the inequities in what they consume. In the developing world many people are forced to live in fragile areas not well suited to human habitation, whereas in industrial countries (and among elites in the developing world) the wealthy consume a disproportionate share of nature's products, whether this be water, wood, or wilderness experience (Figure 1.2). In fact, standing back from the local sites of biodiversity erosion, it becomes clear that the most distal causes of biodiversity decline are probably the most important: the steady narrowing of traded products from agriculture, forestry, and fisheries, promoting monoculture and genetic loss; deficiencies in knowledge and its applications; and legal and institutional systems that promote unsustainable exploitation (WRI, 1992). These global processes are examined further in chapters 7 and 9.

1.3 Why Conserve Nature? Instrumental Values

Why is there a need to conserve biodiversity? The reasons are neither obvious nor widely agreed upon (Norton, 2000). Envi-

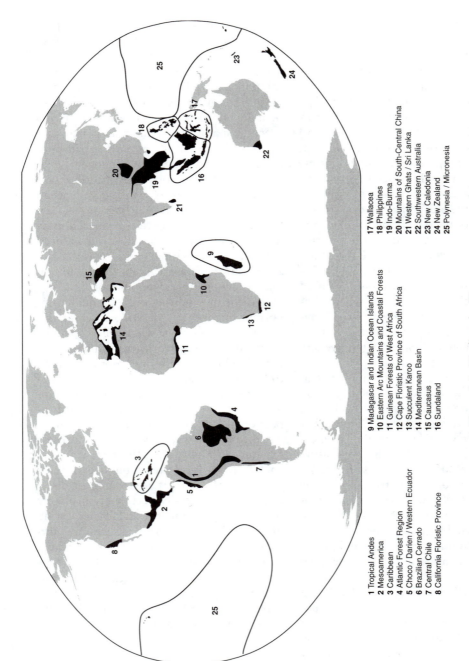

1 Tropical Andes
2 Mesoamerica
3 Caribbean
4 Atlantic Forest Region
5 Chocó / Darién / Western Ecuador
6 Brazilian Cerrado
7 Central Chile
8 California Floristic Province

9 Madagascar and Indian Ocean Islands
10 Eastern Arc Mountains and Coastal Forests
11 Guinean Forests of West Africa
12 Cape Floristic Province of South Africa
13 Succulent Karoo
14 Mediterranean Basin
15 Caucasus
16 Sundaland

17 Wallacea
18 Philippines
19 Indo-Burma
20 Mountains of South-Central China
21 Western Ghats / Sri Lanka
22 Southwestern Australia
23 New Caledonia
24 New Zealand
25 Polynesia / Micronesia

Figure 1.1. Biodiversity Hotspots.

Map shows twenty-five areas (shaded dark gray) containing biodiversity hotspots; hotspots comprise 3 to 30 percent of each shaded area. Hotspots are defined by two criteria: species endemism and degree of threat. To qualify as a hotspot an area must contain at least 0.5 percent of the world's 300,000 plant species, and have lost more than 70 percent of its primary vegetation (Myers et al., 2000). These hotspots (an aggregate expanse of 800,767 km^2) contain the sole remaining habitats of 44 percent of the Earth's plant species and 35 percent of its vertebrate species; 38 percent of this area is currently protected, legally if not effectively, in parks or reserves. Reprinted by permission from *Nature* 403, 853–858 (2000) Macmillan Publishers Ltd.

ronmental philosophers identify two very different sets of arguments, based on the utilitarian (or instrumental) versus the intrinsic (or inherent) value of nature. The utilitarian value of nature refers to the product or function that nature can provide, whereas intrinsic value inheres in the natural object or system itself, irrespective of whether it has any use. Arguments for conserving biodiversity that are based on the utilitarian value are often labeled anthropocentric (human-centered), whereas the arguments predicated on intrinsic value are often called biocentric (or ecocentric) since the value exists independent of its use to human beings. Though philosophers are somewhat troubled by the tautology entailed in the anthropocentric position, pointing out that enjoyment or use of an object is ultimately based on some inherent value (Sagoff, 1988), the intrinsic/utilitarian distinction remains enormously helpful for thinking about the arguments for conserving biodiversity.

The utilitarian value of biodiversity may be divided into four basic categories: goods, services, information, and spiritualism (Table 1.1). As regards nature's *goods*, people need food fuel, fiber, and medicine, items they can obtain both through collection in the wild and through cultivation. The utility of the vast majority of species is still unknown, with respect to both undiscovered medicinal properties (see chapter 8) and genetic diversity. As regards the latter, although more than 20,000 edible plants are known, and perhaps 3,000 have been used by humankind throughout history, the world's food supply is dominated by many fewer (Prescott-Allen and Prescott-Allen, 1990) and according to some analyses a mere handful of crops (Vietmeyer, 1986). The genetic diversity that lies in the wild ancestors of wheat, oats, and barley that still occur in the arid hills of Galilee could, for example, save conventional food resources from incurable disease or uncontrollable pests. Nature's *services* (often referred

to as ecosystem services) are a product of such a vast, invisible natural economy that they were, until quite recently, overlooked! For example, green plants replenish oxygen and remove carbon dioxide from the atmosphere, fungal and microbial organisms decompose dead organic material and recycle plant nutrients, and rhizobial bacteria turn atmospheric nitrogen into usable nitrate fertilizer for plants. Fear about the stress to which ecosystems can safely be subjected when several of their component species go extinct was popularized by Commoner (1971), and this question has become central to the field of conservation biology (see chapter 3). The *information* value of nature lies not only in the estimated 5 to 15 million species that exist, most of which are still unknown to science (May, 1999), but in the evolutionary and ecological processes that allow these species to coexist. As Ehrenfeld (1976) argues, the study of intact functioning ecosystems provides blueprints for habitat reconstruction, design principles for new ecosystems, environmental baselines for monitoring threatened systems, and a wonderful teaching laboratory for the ecologists of the future. Finally, as regards *psycho-spiritual* value, nature offers rich meaning to human existence, for some an emotional touchstone, for others a sense of spiritual or intellectual purpose. This derives not only from the exquisite excitement of scientific discovery, but also from the beauty and more diffuse sense of awe and mystery that can be found in nature. This complex value E. O. Wilson (1984) calls "biophilia," invoking the strong bonds humans can feel with nature. A clear example of how conservation can be rationalized on utilitarian grounds, combining nature's goods, services, information, and spiritual value, is found in the Biodiversity Support Program's (BSP) emphasis on the links between health and conservation (Box 1.1).

Utilitarian approaches to the value of biodiversity find their purest expression in

a

b

c

d

Figure 1.2A–D. Material Lifestyles around the World.
Thirty statistically average families from thirty countries shared their lives by opening their houses to photographers who lived with them for a week. At the end of that period, the collaboration produced unique portraits of our material world at the end of the twentieth century. Each family was photographed with all their possessions outside their homes. a. Mali, b. Poland, c. Tibet, and d. United States (From Menzel, 1994). Photos copyright of Peter Menzel and Sierra Club Books.

TABLE 1.1
Four Categories of the Instrumental or Utilitarian Value of Biodiversity[a]

Category	Examples
Goods	Food, fuel, fiber, medicine[b]
Services	Pollination, recycling, nitrogen fixation, homeostatic regulation, carbon storage
Information	Genetic engineering, applied biology, pure science
Psychospiritual	Aesthetic beauty, religious awe, scientific knowledge, recreation, tourism

[a] Unsurprisingly, there are many other ways of classifying and labeling the values of nature. Barbier's (1992) terminology, splitting nature into "use" and "nonuse" categories, is often encountered. Nonuse values consist exclusively of "existence values" (including biodiversity, cultural and spiritual heritage). Use values include direct-use values (harvesting, recreation and tourism, genetic use, and education), indirect-use values (ecological services), and option values (future uses of direct- and indirect-use values). Another continuum often stressed is that between consumptive and nonconsumptive uses, for example, between hunting and variously consumptive tourism, education, and science. There is, however, considerable terminological slippage among authors; for instance, occasionally one sees nonconsumptive use employed interchangeably with nonuse value (IIED, 1994).

[b] Some goods in this category can be calculated as option values. For example, the value of plants used for medicinal purposes by local communities can be calculated on the basis of their possible future value on the global market (see chapter 8 for bioprospecting).

Source: Reprinted by permission from Sinauer Associates, Inc. From Meffe and Carroll: *Principles of Conservation Biology.*

ecological economics, a field that addresses the relationship between ecological and economic systems, focusing on environmental policy and sustainable development (Costanza, 1989; see chapter 9). Though global environmental-social modeling had been pioneered in the Club of Rome's doomsday forecasts of the late 1960s (Meadows et al., 1972; see also Hardin, 1968), ecological economics took the approach much further, essentially arguing that nature's monetary value is the key to its conservation. Ecological economists calculate the true costs of extinction, pollution, and environmental degradation by putting a dollar value on biodiversity, its goods, services, information, and aesthetic values (for an early provocative example, see Box 1.2). Particularly challenging has been the monetary evaluation of nature's diffuse roles in providing climate control, organic waste disposal, soil formation, nitrogen fixation, biological pest control, plant pollination, pharmaceuticals, and recreation. Considering the job of bees, Meadows poignantly asked over a decade ago, "How would you like the job of polli-nating trillions of apple blossoms some sunny afternoon in May?" (cited in Meffe and Carroll, 1997, 37). In recent years, ecological economists, in conjunction with environmental economists (who deal with nature as a commons) and resource economists (who focus on the harvest of renewable resources), have made important progress in quantifying the kinds of benefits that biological diversity provides (e.g., Barbier, 1992; Edwards and Abivardi, 1998; Fromm, 2000). For example, the answer to Meadow's question is US$1.6–5.7 billion annually, a range determined by estimating the gains to consumers through lower prices for crops pollinated by honey bees (Southwick and Southwick, 1992). More broadly, by quantifying the environmental benefits of such services worldwide, ecological economists have estimated the annual value of biodiversity at between $2.9 trillion (Pimentel et al., 1997) and $33 trillion (Costanza et al., 1997). Sometimes economists must devise imaginative methods for evaluating nature's goods and services, especially when these are not traded on conventional mar-

Box 1.1. The Links between Health and Conservation

Awareness of the links between the global environment and human health is not new. We tend, however, to think of the more negative associations, such as between industrial air pollution and respiratory disease, between dirty water and gastrointestinal illness, and between holes in the ozone layer and skin cancer. The Biodiversity Support Program (2001) encourages linking human well-being to the conservation of biodiversity, pointing out that there are many known benefits to biodiversity on a global scale.

Health: Biodiversity provides the raw material for the production of pharmaceuticals. Based on figures from 1993, 57 percent of the 150 most frequently sold prescription drugs in the United States had at least one major active compound derived from (or patterned after) plants and animals (Grifo et al., 1997), and sales of such drugs amounted to $15 billion worth of annual business in 1990, up from $4.5 billion in 1980 (Reid, 1997). For Europe, Japan, Australia, Canada, and the United States combined, the market value for both prescription and over-the-counter drugs based on plants in 1985 was estimated to be $43 billion. Though there are many problems with the calculation of such estimates (Principe, 1996), the magnitude of pharmaceutical benefits from plants is potentially large but hard to capture (see chapter 8).

Nutrition: All of the world's major food crops, including wheat, corn, and sorghum, depend on new genetic material from the wild to remain productive. Seventy-five percent of the world's staple crops, and even 15 percent of U.S. food crops, rely on wild animal species for pollination. Biodiversity in the world's oceans also has potential as a key food source; in 1995, alone 97 million tons of fish were commercially harvested for food and other products (UNEP, 1999).

Climate regulation: The Earth's oceans and standing forests serve as carbon sinks, fixing atmospheric carbon that would otherwise contribute to global climate change.

Moving from the global to the local scale, the links between health and biodiversity conservation are even more apparent. Many of the world's richest biological areas are inhabited by populations who rely on the conservation of biodiversity to ensure their health and survival in several ways:

Traditional medicines: Local cures, based on plant and animal materials, are the basis of health care for about 80 percent of people living in developing countries (Farnsworth et al., 1985; Grifo and Rosenthal, 1997; Shanley and Luz, 2003). For example, more than 5,000 species of plants and animals are used in China.

Food: Much of the world's rural populations rely on hunting and fishing for food, particularly for protein sources (WRI, 1992).

Ecosystem services: Intact ecological systems that protect biodiversity provide crucial services upon which local populations depend, such as clean drinking water, soil formation, and pollinators.

Box 1.2. Much More Than Stocks of Wood

In a pioneering study, Peters, Gentry, and Mendelsohn (1989) compared the economics of timber and nontimber forest products (NTFPs). Working on the Rio Nanay near Iquitos, Peru, and surveying only trees greater than >10.0 cm DBH (diameter at breast height) within a 1 ha plot, they found 842 trees representing 275 species. Of these, 72 (26 percent) species and 350 (42 percent) individuals yielded products with market value in Iquitos, including timber, fiber, latex, and fruit. Using estimates or observation of density, productivity of each species, and records of market prices over a year, they calculated the total commercial value (in US$) of a hectare of forest. Then, subtracting labor costs in collection and the cost of transport of products to appropriate markets, they calculated the net dollar value of 1 ha forest. Given that both fruit and latex can be collected every year, the total financial value of these resources was considerably greater than the current market value of one year's harvest. To calculate the net present value (NPV) of $6,330, they used a fifty-year time horizon, assumed that 25 percent of the crop was left in the forest for regeneration each year, and put in a 5 percent discount rate.

TABLE 1.2 BOX
Value of One Hectare of Forest on the Rio Nanay, Peru (in US$)

Product	Total Value	Net Value	Net Present Value
Fruits	650	400	6,330 (fruit and latex together)
Latex	50	22	
Timber (clear-cut)	—	1,000	1,000
Timber	—	310	490

The same kind of calculation was done for timber. If liquidated in one cutting, saw timber would generate a revenue of $1,000 on delivery to the mill. Since the trees are all gone there is no additional future value, and incidentally no value from fruit and latex trees that are destroyed in clear cutting. So NPV stays at $1,000. Under gentler logging practices (periodic selective cutting) the NPV of $490 is still far short of that resulting from fruit and latex collection. The authors concluded that fruit and latex ($6,330) represented 90 percent of the total forest value today (a combined NPV = $6,820). Note that this does not even include medicinal plants, small palms, and lianas.

Economic evaluations like this are of course shot through with potentially problematic assumptions and methods (see chapter 9). This study and others like it have been criticized for the assumption that plants are not damaged through harvest and for using high discount rates. More recent studies have found much lower values for Central American forests (Godoy et al., 2000), and a review of twenty-four such tropical forest studies worldwide established a median yearly NTFP of only $50 (range $1–$166 per hectare; Godoy et al., 1993). The high value for the Rio Nanay site no doubt reflects its unusually high fruit tree density and its proximity to market.

Peters et al.'s study nevertheless pioneered the idea of using not just ecological services and scientific value but market benefits to justify conservation (see also Myers, 1988), and as such lays the basis for extractive reserves (see chapter 10). Despite its flaws this and other early work were extremely important in stimulating ecological economics, a field that encompasses environmental valuation studies that put a dollar value on everything from a primitive relative of modern maize found in Mexico to a lion in Kenya (see chapter 9).

kets, as discussed further in chapter 9. Finally, in some instances economists can design markets for ecological services, for example, in carbon shares or debt-for-nature swaps (Box 9.4).

These economic approaches are useful. Not only do they offer powerful tools in advocacy, but they expose the pressures that work against conservation (Edwards and Abivardi, 1998). They are also rather controversial. Here we look only at the philosophical issues entailed, with the realpolitik getting closer scrutiny in chapter 9. In an early but extremely forward-looking attack on the idea of turning nature into a commodity, Ehrenfeld (1976) put his finger on possible dangers with utilitarianism. Most species and natural processes are probably not very useful. As such, option values are small and economic arguments for conservation weak. He also questioned how a utilitarian ethic can conserve species that are economic liabilities, species that are dangerous, or species whose medicinal products can now be fabricated synthetically. Furthermore, what if it really does make economic sense to drain a marshland for development, rather than leave it to provide ecosystem services or tourism? It is tempting but ultimately dangerous for a utilitarianist faced with such a dilemma to exaggerate a threatened species' (or resource's) economic significance, and try sneaking conservation in through a fabricated utilitarian rationale (Bell, 1987). Finally, there is the issue of the morality of pricing (Ehrenfeld, 1976; Sagoff, 1988). We generally try to protect an object's dignity by removing it from the market. Thus slavery was banned in respect of human life, and prostitution was outlawed to protect women. So, too, should nature be removed from the market, the argument goes. All in all, if we are to enlist the market in the cause of conservation this must be done very carefully, and on a case-by-case basis (Meffe and Carroll, 1997); utilitarian arguments are probably never sufficient, and science alone cannot protect biodiversity (Wilson, 1984).

1.4 Intrinsic Values

Intrinsic value is a much more subjective matter. While most people take the intrinsic value of humans for granted, the view that "Nature" (often personalized in this sense) has inherent rights and is as such subject to the same moral, ethical, and legal protection afforded humans is more controversial (Nash, 1989). Trying to put an anthropocentric spin on their case, advocates of intrinsic worth compare the immorality of the destruction of nature to the tossing of rocks at the windows of the Louvre, to the quarrying of stone from the pyramids of Egypt, to the burning down of a great library, or, in extreme manifestations, to genocide. But to what level of biodiversity does intrinsic value adhere—every single living being, only sentient beings, species, biotic communities, or evolutionary processes? And how can we judge this? Looking just at two writers whose contributions to the conservation movement are so prominent, Leopold attributes intrinsic value to all things "to include the soils, waters, plants, and animals or collectively: the land" (Leopold, 1970, 239), whereas Soulé (1985) claims that it is diversity itself that has intrinsic value, not its entities. The variety of modern natural rights (or eco-) philosophies that blossomed in the 1960s and 1970s predicated on intrinsic value differ on this dimension (Box 1.3). They also differ in the extent to which they view modern environmental problems as emanating directly from Western science and civilization, and to which they romanticize our preindustrial past, as incidentally do various strains of ecofeminism (Box 7.1).

Long prior to the emergence of these recent philosophically based movements, ancient traditions and scriptures guided recognition of moral and ideological (though not legal) rights of nature. As Callicott (1994) shows, most of the world's religions (including major faiths such as Islam, Hinduism, Jainism, Buddhism, Taoism, and Confucianism) furnish powerful worldviews that sanction spiritual interconnectedness among

Box 1.3. Ecophilosophies

Deep ecology. Inspired by Hindu metaphysics, Arne Næss launched deep ecology in 1973, calling for the liberation of land and nonhuman life from ownership and abuse. The central tenet of deep ecology is that all beings are manifestations of one Supreme Being (Næss and Rothenberg, 1989). Accordingly, the suffering of one creature is the suffering of all, guaranteeing a fundamentally biocentric worldview in which all organisms and entities in the ecosphere are equal in intrinsic worth (Sessions and Devall, 1990, 312). Deep ecologists define themselves in opposition to mainstream conservation policies or other "shallow, anthropocentric" approaches to environmentalism, that in their eyes amount to shuffling deckchairs on the *Titanic.* Indeed, deep ecologists argue that the only way to achieve meaningful and long-lasting conservation is through a fundamental change in human consciousness. Deep ecologists are deeply antagonistic to human society, a sentiment of intense alienation that fires some of their activist groups, such as Earth First! (Taylor, 1991). Note that "ecology" in this label derives not from the discipline introduced in chapter 3, but from religious and philosophical notions, and as such is more akin to environmentalism (Nash, 1989). Also, insofar as deep ecologists attribute moral worth to nonbiotic entities, such as rivers and mountains, their ecophilosophy goes beyond the postulates of Leopold (Meine, 1988; see also Soulé, 1985) and most contemporary conservation biologists (Meffe and Carroll, 1997). Deep ecology has also provoked intense criticism from non-Western radicals (Guha, 1989b).

Bioregionalism. Like deep ecology, bioregionalism calls for a substantial reorganization of human consciousness. Bioregionalists, however, see as the fundamental problem the separation between people and the land from which they both derive their subsistence and dispose of their wastes. Because most people in the developed world buy food produced far from where they live, use power produced elsewhere, and send their trash to still another area, they are insulated from their environment and oblivious to environmental impacts of their lifestyle. Bioregionalists advocate "living in place" (McGinnis, 1999). In this way people will become more attuned to what their local environment can produce, and adopt a lifestyle better suited to their surroundings.

Biophilia. Biophilia is a term coined by E. O. Wilson (1984; see also Kellert and Wilson, 1993) in his book of the same name. It is based on two related ideas: first, that humans have an innate interest in and attraction to all things living, and second, that understanding other organisms and the natural world fosters greater appreciation of their value. Through exploration and understanding, emotional bonds with the natural world are strengthened and a more meaningful conservation ethic is possible. Like the other ecophilosophies sketched above, biophilia calls for deep social change; in contrast, however, biophilia relies heavily on science rather than moral transformation to achieve these ends. Furthermore, it is concerned not with the intrinsic value of each and every biological specimen but rather with that of genes, species, ecosystems, and evolutionary processes.

Gaia. Gaia, the Greek word for Earth, was used by biologist James Lovelock (1979) to capture the idea that Earth is not a large rock with disparate biological and geological processes, but a single organismic entity with a unitary system of

(continued)

(Box 1.3 continued)

feedbacks and interrelationships, of which life is the most important. Gaia advocates view the Earth as a giant organism that has withstood the abuses of the last 4.6 billion years and may well be resistant to whatever humans can do to it. It is this deterministic element in the Gaia hypothesis that disturbs some environmentalists (Bennett, 1993). Gaia has also inspired a revival of utopian writing (e.g., Callenbach, 1977).

Spiritual ecology. Perhaps best thought of as a component of all of the above, spiritual ecology is defined as a mystical attitude toward nature, emphasizing a deep unity with land and animals (Nollman, 1990). It is often fueled by romantic readings of anthropology and religion that promote the attractive idea that while industrial society has lost touch with nature, tribal and indigenous peoples have not (see chapter 4). This spiritual component of all ecophilosophy spurs the formation of new cults as well as feeding the fanaticism of radical environmental groups.

---

all living beings and obligations of stewardship (see also Holm, 1994). For example, deep empathy with nature derived in part from Hinduism inspired an extraordinarily successful people's movement (Chipko) to halt the commercial exploitation of India's Himalayan forests (Guha, 1989a; see Box 8.1). Even the Judeo-Christian tradition, so commonly blamed for the current environmental crisis (e.g., White, 1967), can be read to embody a conservation ethic, with environmentally concerned Christians and Jews now challenging the conventional reading of Genesis as legitimating humankind's dominion over all living creatures. They champion biblical passages showing that God intended his creation to be teeming with life (as in the story of Noah's Ark, Figure 1.3), and the notion that humans were put in the world to act as stewards of nature (as in the story of the Garden of Eden). From these arguments has developed a Judeo-Christian Stewardship Conservation Ethic (Callicott, 1994). All in all, even though intrinsic belief in the value of nature is difficult to quantify, it is widespread, resilient, and quite easily disinterred from texts that may at first look as if they bear no conservation ethic.

Along these distinct rationales for conserving biodiversity based on instrumental and intrinsic value, conservationists often split into two mutually suspicious factions, anthropocentricists and biocentricists. The latter dismiss the former as shallow materialists who carve their case according to the demands of political expediency, and whose view of man as the measure of all things betrays the "arrogance of humanism," the very title of Ehrenfeld's (1978) book. The former eye the latter as unrealistic, unscientific, and sometimes too radical in their methods. Such a divide can be politically counterproductive, as when delegates to the 1992 Earth Summit in Rio de Janeiro became distracted with arguments over whether to emphasize nature's utilitarian or intrinsic value (Norton, 2000). Steps toward a more inclusive ethic could be taken. First, most conservation biologists would probably agree that neither utilitarian nor intrinsic arguments are sufficient alone, since different people and interest groups will respond to distinct kinds of reasoning (Caro et al., 2003). Second, there is an inescapable logic to the argument that intrinsic approaches are essential, not because they automatically confer protection (they do not), but rather because they place the burden of proof on the interest group planning to destroy biodiversity (Meffe and Carroll,

Figure 1.3. Noah's Zoo.
Ancient traditions and scriptures have upheld a moral recognition of nature throughout the ages. Even Judeo-Christianity, sometimes characterized as epitomizing man's domination over nature, has its tales of stewardship, as in the story Noah's Ark. Recently, a unique multifaith conservation initiative has emerged among Baha'is, Buddhists, Christians, Hindus, Jains, Jews, Muslims, Shintos, Sikhs, Taoists, and Zoroastrians. The Alliance of Religions and Conservation (ARC), funded in part by the WorldWide Fund for Nature (WWF), sprang from a 1986 gathering at Assisi, the Italian birthplace of the Roman Catholics' patron saint of ecology (Benthall, 1995), and is actively involved in investing up to US$30 billion religious assets in its "Sacred Gifts for a Living Planet" program, designed to use faith to combat forest and marine destruction and other environmental threats. Reproduced from Edward Hicks's *Noah's Ark* (1846). Philadelphia Museum of Art: Bequest of Lisa Norris Elkins.

1997). Third, an enormous step forward could be made if people recognized that whatever philosophical motivation (instrumental or intrinsic) lies behind their own private valuation of nature, it should not constrain the full range of strategies for conserving biodiversity that they are willing to support (e.g., Barrett and Grizzle, 1999).

At the margins this could be intellectually challenging; for example, a person who values each and every individual in nature may find a conservation project that advocates game ranching difficult to countenance, but for the most part we can be more open-minded. That the goals and means of conservation should be kept separate is one of the conclusions of this book (see below, and chapter 11).

1.5 The Changing Practice of Conservation: First, Protection

Given the very different ethical values people bring to conservation, it is not surprising that conservation, preservation, and management, words that have become household terms over the last few decades, mean distinct things to different people. In this and the next two sections we examine historically how these concepts have changed, accreting different nuances over time. As we will see later in the book, subtle differences in the meaning and interpretation of these terms can lead to very different expectations and conservation outcomes in the real world. We will trace the development of these potentially slippery terms, and in so doing provide a framework for revisiting these issues later in other more topically focused discussions. Specifically, we look at the development of the human dimensions of conservation, exploring how this history is interwoven with changing social and environmental philosophies.

For a number of reasons, not least simplicity, this survey is structured around the origin and range of strategies applied to conservation problems in the United States over the last 140 years. With its more recent European occupation and rapid environmental change, the United States' story provides an encapsulated view of the development of conservation thinking. Furthermore, the simultaneous existence of "frontier" areas and densely settled human populations forced the American conservation movement to deal with a variety of issues, many of which affect current conservation thinking around the world to this day. Not surprisingly, American conservation was strongly influenced by European perspectives, both in the metropolitan centers and in the colonies, and we allude to some of these similarities and differences in the boxes.

Many contemporary authors cite the establishment of "American-style" national parks as the beginning of the current conservation era. In 1832, after traveling through the American frontier, artist George Catlin wrote that the United States' natural heritage should be saved "by some great protecting policy of government . . . in a magnificent park . . . A nation's park, containing man and beast, in all the wild and freshness of their nature's beauty!" (Catlin, 1990, 35). And so the American national park paradigm, undeniably the most influential institution to come from the American conservation movement, was born. It is fitting that a nonprofessional conservationist is credited with the national park concept because in its early stages American conservation was heavily influenced, if not driven by, influential (and usually very wealthy) amateur conservationists. With help from these "hobby conservationists" and writers like John Muir who treasured the wilderness as an oasis in which to shelter from the evils of modern civilization, Catlin's vision slowly became a reality. In 1872 Yellowstone National Park was gazetted. Subsequently, and still with the help of wealthy, nonspecialist benefactors (Fox, 1985), Acadia, Yosemite, and Grand Teton National Parks and Muir Woods would all be established.

What made national parks different from earlier types of conserved areas such as royal forests (see chapter 2) was that they were protected simply for their intrinsic value, for the very fact they existed. They were established on the premise that all the people of a nation would collectively benefit more from their preservation than would a few from their exploitation. Today, national parks form the cornerstones of many countries' conservation programs and are still primarily predicated on intrinsic values. Many of the conservation successes and failures discussed in the remainder of the book stem directly from applying the American national park concept (even though national parks within the United States have evolved quite significantly away from some of their original caricatures; Schelhas, 2001).

There was, however, also something very unusual about Catlin's vision. From this same passage, in this magnificent park Catlin hopes to see "for ages to come, the native Indian in his classic attire, galloping his wild horse, with sinewy bow, and shield and lance, amid the fleeting herds of elks and buffaloes" (Catlin, 1990, 35). Here we have a man who, like Muir and Henry Thoreau, was a dissenter from the deeply rooted Western ideology that holds humankind in stark separation from (and yet in command of) nature. Catlin rejected the conventional view that celebrates humankind as created in the image of God and enjoying dominion over all the creatures, a view derived from at least one interpretation of the Bible's Book of Genesis. For Catlin, humankind was part of nature, and national parks should capture this harmony.

Perhaps because of its idiosyncrasy in this respect Catlin's dream was not durable. Human occupation quickly fell from the national park paradigm. Native Americans were evicted from Yellowstone only a few years after its creation. Only with the herding of native communities into reserves beyond the borders of the park could the image of wilderness lands untouched by human presence persist. The popular perception of national parks emerged as empty, pristine areas free from human influence. The conservationists' goal was to lock up these parks legally before commercial interests could spoil them, and humankind was to retain at least for a while their biblical dominion over nature (Box 1.4). The "man and beast" part of Catlin's paradigm had been forgotten, and the approach to protected area management that entailed excluding people from protected areas came to be known as the "Yellowstone model" (Stevens, 1997a).

This protectionist tradition that sought to eliminate all human influences in national parks contrasted with the more homely conservation ideas that were developing among common farmers and fishers in the eastern states (Judd, 1997). It contrasted also with European conservation traditions at the time. Because Europe was so densely settled, it had undergone substantial anthropogenic change and pristine areas free from human influence were rare. In addition, many parts of Europe had a long history of recorded land rights. As a result, European approaches to conservation emphasized the footprints of human presence, be these in hedges, hollows, stone walls, or copses (Harmon, 1991; Evans, 1997), and British conservationists explicitly developed a vision of nature with villages as an integral component of the rural landscape. As such, national parks in Britain not only fully recognize existing land rights but also seek to maintain and protect established farming systems. Interestingly, though, once Europeans moved to the colonies, they freely adopted the preservation and protectionism of the Yellowstone model, probably because these areas were perceived (often inaccurately; see chapter 2) as empty frontiers reminiscent of the American west (Neumann, 1998).

A final significant feature of early American national parks was that they were mostly established to protect spectacular scenery, not biological resources. As Stephen Mather, the National Park Service's first director (recruited to head the National Park Service from his role as a wealthy amateur conservationist), sought to expand the parks system, his instructions were to look for "scenery of supreme and distinctive quality or some natural feature so extraordinary or unique as to be of national interest" (Ise, 1961, 195). This focus on scenery and its effect on the design of parks and protected areas would also come to influence contemporary conservation strategies and is discussed in chapters 2 and 3.

1.6 Then Resource Management

Rapid industrialization and economic expansion of the late nineteenth and early

Box 1.4. Man versus Nature: From Hunters to Penitent Butchers

Man's dominion over nature, and the beginnings of his subsequent fall, is colorfully illustrated in the culture of hunting that was deeply embedded in the consciousness of European settlers and resource professionals, especially those with military backgrounds, all across the British Empire. In Africa the hunting of wild animals became a particularly important symbol of European dominance on the continent. Ritualized as "The Hunt" (MacKenzie, 1987), it served to distinguish social class within the settler society as well as to differentiate European hunters from other, obviously African hunters (MacKenzie, 1988; Neumann, 1996). In extreme cases Africans themselves became game, occasioning terminological slippage between "animal bags" and "human bags" (MacKenzie, 1988, 301).

Governor Sir Hesketh Bell with Hunting Trophies in Uganda, 1890.
Reproduced courtesy of the Royal Commonwealth Society.

"The Hunt" was hedged with rules and prescriptions: only choose males as quarry, deliver a coup de grace to injured animals, and do not "shoot from railway carriages, river steamers (except at crocodiles . . . always fair game), motor vehicles or aeroplanes" (MacKenzie, 1988, 299). Critical here was the notion of "sportsmanship." The idea of hunting "just for meat" was anathema to the ideology. African hunting practices were consequently deemed "unsportsmanlike," and their methods (traps, pits, spears, poison, etc.) inhumane, offending the sensibilities of conservationists in England, most notably the members of the Society for the Preservation of the Fauna (Neumann, 1998, 107). Through The Hunt, then, the subjugation of nature became a symbol of white dominance and a marker of manliness (MacKenzie, 1987), and through ritualized slaughter, the gruesome realities for the defeated lower orders (animal, social, and racial) were obscured.

(continued)

(Box 1.4 continued)

But this was not to last. In one of the more intriguing and tortuous twists in conservation history, from the erstwhile big-game hunters emerged ardent preservationists (Beinart, 1990), or in the terms of a Faunal Preservation Society history "penitent butchers." In some senses hunters have always liked to see themselves as conservationists and complained about the practices of others. In India and Africa a new generation of hunters decried the excessive destruction of animals that their forebears had engaged in, the scramble for museum specimens, and the whole "system of penetrating the country by feeding the natives" (W.D.M. Bell, cited by MacKenzie, 1988, 298). Later, at least in Africa, the blame for declining game resources was shifted to the natives, and legislative action was taken to destroy native hunting (chapter 2).

Subsequent shifts of public feeling in at least some Western countries have in recent years eroded the status and acceptability of hunting. This demise is illustrated by the changing relationship of the British monarchy to the chase (MacKenzie, 1988). While in 1924 the ideal honeymoon for a royal couple was a shooting safari in East Africa, the dispatching of a tiger on a trip to India and Nepal by the Duke of Edinburgh in 1961 caused worldwide outrage. And in 1986, when the Queen and Duke of Edinburgh returned to Nepal, the nearest they came to big-game shooting was to witness the tranquillizing of a rhino called "Philip" so it could be fitted with a radio device for conservation monitoring.

twentieth centuries that inspired the creation of national parks also brought concomitant increases in the consumption of natural resources. Already in other parts of the world, as global capitalism was taking hold, administrators were beginning to realize the finite limits to the productivity of the land and the need for central governments to take control of forests and zones with high economic potential. This sparked a new discipline—forestry. After depleting their own oak forests (and losing America's natural wealth after 1776), European colonialists needed to find timber—for ships and railways—to manage their empire and accordingly set up the East India Company (Guha, 1989a; Peluso, 1992). With clever language the plain exploitation of timber was hidden behind a starkly neutral label—"scientific forestry" (Gadgil and Guha, 1992). Scientific forestry was part of the newly emerging science of resource management, and developed initially in Germany. Though

it aimed at the sustained production of timber, its ecological consequences were largely negative. For instance, all across the Indian Himalayas mixed forests of conifer and broad-leaved species were replaced with single-species stands of commercially valuable conifers, destruction of traditional forest management practices occurred and large areas were converted into tea, coffee, or rubber plantations (Guha, 1989a).

In the United States at this time we see the beginnings of the same processes. A continent that had once seemed to offer inexhaustible forests and herds was beginning to show signs of decimation at the hands of increasingly hungry commercial interests. Accordingly the U.S. Congress allowed its conservation focus to wander beyond national parks, and passed the Forest Reserve Act of 1891. Within two years President Benjamin Harrison set aside 13 million acres (around 52,000 km^2) of forests; his successor, Grover Cleveland, added another

5 million (Meine, 1988). But no American president had a greater impact on conservation than Theodore Roosevelt. During his first year in office, Roosevelt established 13 new forest reserves. He followed with 53 wildlife reserves, 16 national monuments, 5 national parks, and 32 other reserves, totaling over 75 million acres (Fox, 1985); by the end of his tenure he had expanded the national forest system to 195 million acres (or about 780,000 km^2; Meine, 1988). Legend has it that when a Forest Service applicant was asked "Who created the first National Forest?" he thought for a moment and responded, "The first National Forest was created by God, but it was expanded by Theodore Roosevelt" (Meine, 1988, 77). Like the earliest conservation strategies, early national forests were perhaps best characterized as territorial claims seized to prevent them from being consumed by commercial harvesting. Initially, their management simply entailed prohibition of unauthorized uses and deciding what the authorized uses should be, in other words, a weak form of protectionism. Science had little or no role in the process, and conservation meant little more than showing restraint in using what nature provided in order to make it last longer. But during the first decade of the twentieth century, that began to change.

The name most commonly associated with early scientific management of forests is Gifford Pinchot. Strongly influenced by European "scientific" practices in the colonies, Pinchot advocated manipulating forests (i.e., management) to enhance the quantity and quality of their production, rather than just making them last longer. Science played a pivotal role in this process by helping foresters identify optimal growing conditions as well as planting and cutting strategies. In this way, science was combined with simple protection to shape the way conservation was carried out. The important implication was that the practice of conservation was no longer just about preserving the forest; it now entailed enhancing or improving the forest itself. The guiding principle behind this powerful new paradigm came to be known as "highest use: the greatest good for the greatest number in the long run." By applying European-style forestry to the large and relatively pristine forests of the western United States, the U.S. Forest Service became a key player in the development of conservation thinking.

Not surprisingly, the paradigm of highest use was as controversial as it was influential. It led to one of the earliest divisions within the conservation movement, one that would define a spectrum of conservation goals that still exists (the intrinsic/instrumental debate, discussed above). Under Pinchot's principle of highest use, the "greatest good" was based on the utilitarian value of nature, or the products that could come out of a forest. A successfully conserved forest would produce the maximum amount of timber for a rapidly growing and developing society. Challenging Pinchot's "utilitarians" was a growing segment of the conservation movement with John Muir as their figurehead, who championed the intrinsic value of nature. Suspicious of Pinchot's brand of conservation and scientific management, Muir and others argued that national parks and many of the country's national forests provided much more than just timber; they were worthy of strict preservation in their natural, unmanaged state. And so a philosophical divide between those who valued nature for intrinsic or instrumental values was born.

Nowhere did these opposing strategies attain more national prominence than in California, where the Hetch Hetchy controversy became an enduring symbol of the divide between preservationist and utilitarian rationales for conservation. At the turn of the twentieth century, the city of San Francisco had around 400,000 inhabitants and a grossly inadequate water supply. Even though the Hetch Hetchy valley was already part of Yosemite National Park, the city proposed that the valley be dammed to provide a more reliable and cleaner water source. What ensued was a thirteen-year battle polarizing preservationist and utilitarian perspectives. Leading the preservationists was John Muir, who had been instrumental in getting the valley included as part of Yosemite. Muir and his acolytes accused the utilitarians of purely economic shortsightedness by "trying to make everything dollarable" (Fox, 1985, 141). Leading the utilitarian infantry was the original proponent of highest use, Gifford Pinchot. In their eyes, the greatest good for the greatest number meant clean water for 400,000 San Franciscans, rather than an intact Hetch Hetchy valley for the whole nation. Muir, lamenting the precedent set by Hetch Hetchy and concerned that all national parks would meet similar fates, died a year after the final decision to build the dam. Under catchy titles ("Dam the Rivers, Damn the People"; e.g., Cummings, 1990) such battles continue to this day, often with recognition of a third stakeholder, the local community displaced or otherwise affected by the hydrological developments (e.g., Johnston, 2001; and see Box 4.6).

1.7 Leading to Game Management, Multiple Use, and Broader Conservation Goals

Until the 1920s approaches springing from utilitarian and intrinsic views remained separate, each settling for dominance in its own sphere: preservation for intrinsic value in national parks and utilization in the national forests, with notable exceptions like Hetch Hetchy, where the land of a national park was put to economic use. But with a more affluent society, greater numbers of people turned to outdoor recreation to escape the cities, and hunting became increasingly important in shaping conservation thinking. Reflecting the growing importance of sporting interests, a representative supporting a bill in the U.S. Congress to protect sport fisheries argued, "It is a

notable fact that of the twelve apostles selected by Christ, four were fisherman" (Fox, 1985, 168).

At the heart of this movement linking sporting and conservation interests was a forester named Aldo Leopold. Despite being trained in the Pinchot tradition of forestry, Leopold advocated a much broader approach to the goals of utilitarian management than simply the utilization of the trees therein. He recognized the national forests as important refuges for wild game and argued that their management should be based on more than just lumber production. In contrast to the European situation, where hunters paid large fees to hunt hand-reared game on private estates, American national forests provided hunting and recreational opportunities more equitably. Leopold considered the European system undemocratic, and wanted to steer national forests away from a similarly elitist system. His vision of highest use was not limited to timber or even economic benefits; it also included game species. Notably, the importance of game came more from its aesthetic and cultural value for hunters, rather than its market value. The dangers of seeing nature as a resource with a price on its head were as alive to Leopold as they are to Ehrenfeld (1976), and it is with Leopold that additional nonconsumptive objectives of preservationism began to creep into utilitarian conservation. The recognition of cultural value and the incorporation of nongame species into management ushered in the paradigm of "multiple use," still dominant in the U.S. Forest Service today. But at this stage in his career Leopold was far from a preservationist; he supported predator eradication programs in hopes of boosting game numbers. Ironically, however, these same predator control programs would push conservation thinking forward by inspiring Leopold to develop a broader ecological community-level approach to conservation.

The program in question was on the Kaibab Plateau, in the southwestern United

States. With predator control, coyotes, mountain lions, bobcats, and wolves had all been eradicated or nearly so. In 1905 the area was incorporated into the Grand Canyon Game Preserve, so hunting and livestock grazing were prohibited and the deer population was left to grow. What ensued was "one of history's most celebrated cases of game *mis*management" (Meine, 1988, 240). After twenty years without active management or predators the deer had severely overgrazed the plateau. Over the winter of 1925–26 over 60 percent of the herd died through mass starvation or was shot after emergency reinstatement of hunting. The Kaibab experience was a watershed for Leopold, demonstrating the importance of wildlife-habitat relationships.

Until that point, management had been based on simple cause-and-effect relationships: fewer predators meant more game, less hunting meant even more game. But through the Kaibab lesson Leopold realized that these simple relationships drawn from his experiences in forestry did not necessarily hold. His goal switched to providing a scientific foundation for game management like Pinchot had done for forestry twenty years earlier, recognizing that deer and other game species were bound in a broader web of interactions. He also realized that "habitat" meant more than just the vegetation and weather where game species lived. Predators, prey, competitors, and resources would all have to be considered when managing a single population like the deer of the Kaibab. Leopold's new science, called "game management," grew out of traditional forestry, but its view of natural systems was much broader than just the species being managed. By demonstrating their ecological importance, Leopold provided a convincing argument for conserving nongame and even predatory species, even though they were not directly useful to humans (Leopold, 1997). Here we see the beginnings of Leopold's enduring "land ethic," the notion that all species are part of a biotic community that includes soil, water, plants, and animals. While the science underlying Leopold's land ethic has been largely revised by modern ecology (chapter 3), Leopold himself remains a source of inspiration to many modern biologists and conservationists.

Game management significantly expanded the scope of conservation efforts in nonscientific ways as well. First, Leopold was a strong advocate for the protection of wilderness—large, relatively untouched, and inaccessible roadless areas—and the experiences to be had there. Second, realizing that wildlife relied on private as well as public land, Leopold argued that "conservation is not merely a thing to be enshrined in outdoor museums, but a way of living on land" (quoted in Meine, 1988, 310). In this context Leopold first ventured the notion of a "conservation ethic," as an inducement for private landowners to manage their property in ways that served not only their own interests but those of the public and future generations (Carpenter and Turner, 2000). This shift marked the beginning of a second phase of importance for amateurs in American conservation. This time, however, the amateurs were not wealthy or influential benefactors. Leopold argued that it was hunters and farmers who held the key to game conservation, as successful game management could be achieved with "*creative use* of the same tools which have heretofore destroyed it—axe, plow, cow, fire and gun" (Leopold, 1933, vii). These early steps, to include game species and then nongame in the management of national forests, to recognize the importance of "wilderness" and finally private lands, form the beginning of a larger trend toward expanding the goals of conservation. And, indeed, in other parts of the world a similar broadening of horizons was underway, as, for example, with the transformation of the colonial sport hunters into conservationists (as we saw in Box 1.4).

In the years following the development

of game management, conservation found a home in mainstream culture, both in the United States and in Europe (Evans, 1997; Fox, 1985). As this happened the movement swung back toward more preservationist ends. Amazingly, even during World War II, when natural resources were needed for the war effort, preservation for intrinsic value held strong and was subsequently bolstered by the unprecedented improvements in economic well-being during the postwar era. During the 1950s and 1960s the goals of conservation expanded enormously as concern for game and charismatic species captured the popular imagination and the environmental movement arose. With the birth of environmentalism, air, water aesthetics, and human health were added to conservation awareness, drawing an even wider circle of popular concern (e.g., Carson, 1962). Also in the 1960s more nonspecialists than ever became concerned with conservation issues. With the publication of Ehrlich's (1968) *The Population Bomb* and the Club of Rome's gloomy predictions about how much longer human appetites could be supported by the planet (Meadows et al., 1972), human population and resource consumption became global issues and consequently the concern of all (see chapter 2). Just as the conservation movement sixty years earlier was confronted with the finite nature of North American resources, the 1960s generation was alarmed by the suddenly tangible limits of the planet's resources, captured in pictures of Earth as a small, fragile entity entirely alone and vulnerable in space. Yet while many think of the 1960s and 1970s as the beginning of environmentalism, there is an interesting twist to the story, illustrative of the swings and pendulums that have characterized conservation thinking since its conception. Almost 200 years previously in far-flung colonial outposts environmental health had played a key role in stimulating conservation thinking and policy (Box 1.5).

Emphasis on the global environment combined with sensitivity to human interests set the stage for sweeping changes in conservation policy throughout the 1970s and early 1980s, developments that sparked the community-based conservation movement and our present conservation predicament(see chapter 2). Notably, however, it was not until the mid 1980s that the growing list of entities deemed worthy of conservation, and the ecological systems in which they had evolved, would stimulate the emergence of a discipline of conservation biology (chapter 3).

1.8 Conclusion

This history has been one of jostling values and interests, with conservation for either intrinsic or instrumental reasons taking precedence in North America at different periods or in different circles (Table 1.2). The highly preservationist ideals behind the formation of the world's first national park gave way to a period, lasting until the middle of the twentieth century, when artists, foresters, government officials, and laypersons alike wrangled over utilitarian and preservationist goals, as we saw in the history of Catlin, Pinchot, and Leopold. Then after World War II prosperity, international travel, population growth, and the media all conspired to ignite environmentalism, thereby vastly broadening the conservation agenda, lending more authority, at least in popular circles, to a preservationist mindset. By the end of the 1960s, the highly preservationist goal of comprehensive biodiversity protection (i.e., all species, and subsequently their genetic diversity, ecological interactions, and evolutionary processes) began to characterize the modern conservation era. In chapter 2 we will see how since that time yet another pendulum swing has brought the preservationist view under serious challenge, and how the old intrinsic/instrumental divide has become dressed up in development politics, ideology, and yet further ethical dilemmas.

Box 1.5. Early Environmentalists in the Colonies

Anxieties about soil erosion and deforestation have arisen at many periods in human history. As early as 450 B.C. Artaxerxes had attempted to restrict the cutting of the cedars of Lebanon in order to combat soil loss; and, as we see in chapter 4, indigenous strategies for environmental management have existed in many parts of the world since time immemorial. However, only in the mid eighteenth century did a coherent awareness of the ecological impact of the demands of emergent capitalism and colonial rule arise, as the colonial enterprise clashed with Romantic idealism in tropical lands from the Caribbean Sea to the Far East (Grove, 1995).

One such island paradise was Mauritius, evoking for Europeans the exotic south sea setting of Dante's *Divine Comedy*. By the mid seventeenth century, the Dutch, British, and French East India Companies were destroying the island's tropical ecology with agriculture, logging, mining, and hunting (Grove, 1992). Interestingly, it was scientists, often working as surgeons or curators of early botanical gardens for these trading companies, who spearheaded the burgeoning environmental concern, with the earliest developments in Mauritius. Initially Portuguese, then Dutch, it fell to French rule in 1721. Men such as Philibert Commerson, Pierre Poivre, and Jacques Henri Bernardin de St. Pierre, though charmed by the dreamy tropical allure of the island, were also sensitive to the rigorous empiricism of the French Enlightenment. In particular, they noticed the relationship between deforestation and local climate change. This led to a 1769 ordinance that limited the clearing of forests. The En-

Ebony Cutting in Progress on the East Coast of Mauritius, 1677.
One of the earliest portrayals of colonial deforestation in the tropics, with the only known illustration of the Dodo (on a tree stump) in its natural lowland ebony-forest habitat. Reproduced courtesy of the Nationaal Archief, The Hague.

(continued)

One thing that becomes very clear from this story is that the question "Why conserve?" insofar as it elicits different values for conservation has historically been associated with a range of distinct conservation goals; thus for Catlin it was a dream of wilderness, and for Pinchot an orderly and productive forest. Here we should emphasize a critical but often overlooked distinction in conservation: ends and means. Means are the tools used to reach conservation goals, whereas ends are the goals themselves. Though this chapter has focused primarily on goals, history shows (see again Table 1.2) that specific means have traditionally been associated with certain ends. Later in this book (see chapter 11) we will argue how important it is to keep these concepts distinct.

The two types of means revealed most clearly in this brief history are utilization and protectionism. Utilization entails some use of the area designated for conservation; utilization is generally consumptive in some way, like hunting, timber felling, or fruit collection, and is exemplified in the policies of Pinchot. Protectionism, by contrast, seeks to exclude human consumptive uses, most classically by keeping all people (other than tourists) out. The distinction is not absolute. For example, some consider tourism to be a wholly nonconsumptive endeavor, while others point to its negative impacts and label it consumptive (see chapter 10). For this reason, protectionism and utilization are best thought of as relative attributes rather than absolutely defined categories. Protectionism versus utilization is not, however, the only point of contention among conservationists, even though so much of the literature addresses this debate. There are other extremely important axes along which conservation, both ends and means, can vary, such as the importance attributed to anthropogenic landscapes as opposed to seemingly "untouched wildernesses," the role of humans in contributing to and protecting biodiversity, the focus on single species versus ecological processes, and the identity of managers—whether they are an elite band of hobby conservationists, sportsmen, farmers making a living off the land, or professional scientists. Each of these dynamics has been alluded to in historical context in this chapter, but is developed more analytically in later chapters. Another critical axis, the degree to which conservation strategies are centralized, is a principal focus of chapter 2, where we look at the post 1970s evolution of conservation policy.

Given the quagmire of ideological debates and practical obstacles surrounding conservation practice, how will we use the very word *conservation* in this book? Though it is impossible to strip any word of all its ethical and methodological implications, we aim to use the term in as neutral a manner as possible. For us conservation refers to an objective (goal) that is usually but not always based on the idea of nature having inherent/intrinsic value, and entails saving global representations of all unique populations, species, communities, and ecosystems within their natural context from uses that are deemed "not appropriate." It is good to be clear about this neutral definition when entering the highly interdisciplinary fray of players engaged with the modern ecological crisis, particularly given how ready different camps are to mischaracterize each other's position. Labels can become a millstone. Indeed, for some, conservation is a dirty word, evoking a stubborn will to preserve nature in some untouched and unchanging state at all costs. Modern conservation biologists have accordingly tried hard to disassociate the term from the implications of straight preservationism (see chapter

3). The latter notion was a serious casualty in the paradigm shift toward utilitarianism (chapter 2), on account of its apparent links with protectionism. Preservationism is also linked in some critics' minds with the idea of evolutionary stasis (an outmoded form of ecology; see chapter 3), and with wilderness fixation (a problematic approach given the importance of anthropogenic biodiversity discussed in chapter 7). As we use the term here conservation is neutral with respect to the means of achieving desired ends, open to the importance of anthropogenic as opposed to nonanthropogenic landscapes, and encompassing of change and dynamism. Conservation used in this sense implies neither the rationale for conserving (intrinsic/utilitarian), nor the means of so doing (protectionism, utilization, etc.).

Perhaps the clearest message of all to emerge from this brief history is that conservation is about choices, and choices are based on ethical values. The answer to the question of what kinds of biodiversity should be prioritized for conservation depends ultimately on why one thinks conservation is

TABLE 1.2
Changing Goals, Means, and Actors in U.S. Conservation Strategy, 1870–1969

Time Period	Goals	Means	People Involved
Pre-1900 (Claiming access to resources)	Utilitarian	Protectionism	Wealthy elites
1870s–Present (National parks)	Preservationist	Protectionism	Elites and professional managers
1890s–1910s (U.S. Forest Service) (Scientific management: production of a single resource)	Utilitarian	Utilization	Professional land managers
1910s–1920s (Game management: single resources and game species)	Utilitarian	Utilization	Professional land managers
1930s–1940s (Multiple use: single resources and game and nongame species)	Utilitarian and preservationist	Utilization	Landowners, land managers
1950s–1960s (All species: genetic variability)	Preservationist	Protectionism	International; nongovernmental organizations; communities

CHAPTER 2

The Evolution of Policy

2.1 Introduction

The last thirty years have brought all the conflicts outlined in the previous chapter to a head. Not only has the fate of the planet's biodiversity become a major source of concern for many different constituencies—science, health, economics, politics, recreation, and spirituality—but the world's human population has reached unprecedented levels, and the distinctions between wealthy and impoverished nations are greater than ever. To address these challenges an immense weight of institutional bodies, governmental and other, has descended on the environment, bringing to the fray a new suite of professional environmentalists and conservationists who venture to the ends of the earth in search of solutions. With all these new players there has been fracturing of conservation goals, essentially along the same preservationist/utilitarian split introduced in chapter 1, but deepened by intense threats to biodiversity and ecosystem health, expanding sources of money, and constantly shifting interest groups and alliances.

To catch a flavor of these changes, this chapter offers a historical overview of the evolution of conservation policy over the last thirty years, focusing on the use of protected areas. As such we shift focus from the philosophical rationales for conservation to the practical means employed by modern conservationists. We focus primarily on the establishment of protected areas because they hold frontline positions in the battle to conserve biodiversity. Protected area conservation has a long pedigree, from royal forests and hunting preserves, through game reserves and wildlife sanctuaries, to national parks. Since the very beginning of the twentieth century there have been international treaties that support the setting aside of areas of natural beauty or richness for conservation ends. In recent years, the establishment of protected areas has become a global business, with the proliferation of many governmental and nongovernmental organizations (NGOs) dedicated to their management.

We start out with a brief examination of the history of protected areas and their enormous expansion in the late twentieth century (2.2), before turning to the inherent limits and legacies of protected area management (2.3). Against this background we look at the alliance between conservation and development that emerged in policy circles in the 1980s and that became a serious challenge to protectionism (2.4), an associated move toward community-based conservation (2.5), and its backlash (2.6), before concluding with an overview of the challenge facing conservationists in the future with respect to the constant jostling between protectionism and utilization as a means of conserving biodiversity (2.7). As such, we set the scene for the subsequent chapters that examine the tools with which to tackle these challenges.

2.2 Global Conservation and Protected Areas

The earliest forms of conservation were barely disguised territoriality. Chiefs, kings, and other powerful families could lay claim to an area, and ensure that only a small coterie of privileged persons enjoy its resources (Box 2.1). This type of conservation, if it can be called that, was concerned mostly with rights of access rather than effective protection. It may nevertheless have been quite effective in some instances since, as we see in chapters 5 and 6, when rights of use are limited to small numbers of people and the powers of exclusion are paramount, protection from overuse is achieved relatively easily. And so, unlike in North America, where protectionism adhered to a romantic vision of nature and populist appeal (chapter 1), in the Old World it was deeply and unequivocally anchored in political inequality. Indeed until the mid nineteenth century large-scale conservation was a very private affair, a reemerging theme in the new millennium with the movement toward privately owned reserves (chapter 9).

By the end of the nineteenth century the first game reserves were set aside, in 1895 in South Africa and 1896 in Kenya. As noted in chapter 1, these were for the most part designed according to the Yellowstone model, in other words, without consideration of local land use, without local consent, and primarily for the enjoyment of outsiders. In 1900, the first international conservation treaty, the Convention for the Preservation of Animals, Birds and Fish in Africa, was signed in London, and this became the basis of colonial wildlife legislation in Anglophone Africa and subsequently worldwide. According to this treaty land could be demarcated for parks and reserves with the principal goal of "saving from indiscriminate slaughter . . . the various forms of animal life existing in a wild state which are either useful to man or are harmless" (cited in Bonner, 1993, 40). With this and subsequent legislation in place, the number of protected areas grew steadily throughout the colonial and postcolonial periods.

Protectionism escalated in the 1970s and early 1980s, decades that saw a rapid and somewhat bumpy transition into the modern conservation era, as conservation transmuted into a global endeavor. As early as 1962, the modern parks and protected areas movement was initiated with the first World Parks Congress; at the time there were only 1 million km^2 of protected land, mainly in North America and colonial Africa. Since that time land protection has increased thirteen-fold (Green and Paine, 1997). More protected areas were gazetted after 1970 than had previously existed (WCMC, 1992), and the trend continues (Figure 2.1). Currently terrestrial and marine reserves cover 7.9 percent and 0.5 percent of the Earth's land and sea area, respectively (Balmford et al., 2002). Many of these protected areas are in tropical zones that comprise the world's most species-rich countries, often where biologically rich ecosystems are still largely intact (see chapter 4). The longstanding objectives of the protected area movement are enshrined in the Convention on Biological Diversity (CBD, article 8, *in situ* conservation). This 1992 agreement, reached at the United Nations Conference on Environment and Development (UNCED) in Rio de Janeiro, requires that signatory parties establish a system of parks and protected areas and promote appropriate development policies in and around these areas that will contribute to the conservation of biological diversity.

Though many appellations such as *national park, nature reserve,* or *wildlife sanctuary* are still used locally to designate conservation units, in 1978 the (then) International Union for the Conservation of Nature introduced *protected area* as a cover term, and developed a typology of internationally recognized categories that are ranked with respect to the number of restrictions on human activities. These categories were simplified and revised in 1994 (Table 2.1), primarily to acknowledge that owners

Box 2.1. Ancient Royal Forests

Reserves for hunting and riding first appear in recorded history in Assyria in 700 B.C. (Dixon and Sherman, 1990). As early as 500 B.C. reserves in the Indian subcontinent were kept not just for royal hunts but to protect elephants, the essential war machine of the expanding Mauryan state (Gadgil and Guha, 1992). Over subsequent centuries the Moghuls reinforced this tradition across Central Asia. In Britain, after the Norman barons' eleventh-century invasion, the setting aside of royal

Cheetah and Stag with Two Indians. George Stubbs, 1765.
Copyright Manchester Art Gallery.

forests for hunting and other forms of recreation gained such momentum that within 100 years nearly 25 percent of England's countryside was dedicated Crown Land; forests (or parks) were the supreme symbol of royalty, and as such were avidly imitated by Princes of the Church and Earls (Rackham, 1986). Bitter resentment emerged among the local populace (Westoby, 1987), captured in the stories of Robin Hood, which tell of the resistance of Saxon yeomen to the ban on hunting the "King's deer." Perhaps because of this fierce resistance, the royal forests did not totally extinguish deeply vested traditional rights in land (Rackham, 1986), some of which are based on written documents dating back to Viking days, specifically King Canute's Conservation Laws, laid down in Winchester in A.D. 1016 (Manwood, 1615).

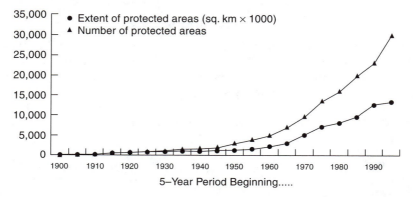

Figure 2.1. The Growth of Protected Areas.
Cumulative growth in the number and extent of protected areas (1900–1999), with year 2000 data from http://wcpa.iucn.org/wcpainfo/protectedareas.html. The slight decline in growth of numbers, in conjunction with the continued growth in extent of land protected, shows how emphasis is being placed on establishing larger protected areas. In fact, 87 percent (1995 figures) of the global network is made up of the six largest protected areas: Greenland National Park (Greenland, 972,000 km^2), Ar-Rub'al-kjali Wildlife Management Area (Saudi Arabia, 640,000 km^2), Great Barrier Reef Marine Park (Australia, 344,800 km^2), Qiang Tang Nature Reserve (China, 247,120 km^2), Cape Churchill Wildlife Management Area (Canada, 137,072 km^2), Northern Wildlife Management Zone (Saudi Arabia, 100,875 km^2). Redrawn from Green and Paine, State of the World's Protected Areas at the End of the Twentieth Century. Copyright 1997 World Conservation Monitoring Centre.

and managers of such areas (particularly Categories V and VI) need not be governmental agencies but NGOs, commercial companies, individuals, local communities, and indigenous peoples. As can be seen in these new categories, management for species and genetic diversity is given more emphasis in some categories than others. Direct human modification of the environment is allowed in IV–VI, restricted in I–III, and banned only in Category Ia; furthermore, the revised guidelines recommend recognition of indigenous peoples within all six categories (with the exception again of Category Ia), so long as this does not undermine basic management goals (Stevens, 1997a, 15). As such, there is a graduated scale with respect to the extent that goals such as education, recreation, and resource extraction are combined with biodiversity conservation.

As protected areas proliferated the Old and New World legacies became thoroughly intertwined. From the United States persisted the idealized image of nature untouched by humans, offering a sanctuary from the evils of civilization. From the royal forests of Europe and Asia derived the implicit legitimacy of elites protecting and enjoying nature, as if this were a private good, often against the wishes of others. Intriguingly there was a third vision of national parks that had emerged primarily in the British Isles over many centuries, that never spread far beyond the metropolitan culture. There conservationists embraced a vision of nature that incorporated long-established customs of land tenure and cultivation (Harmon, 1991), and put humans at the heart of the rural landscape (Figure 2.2). As the national parks movement began to filter across the world, protected area managers showed little cognizance of such traditional rights and uses. Furthermore, they were often unaware of the fact that local groups and communities had, sometimes for thousands of years, protected their areas from resource depletion, often through cultural, religious, spiritual, and customary practices

TABLE 2.1
Protected Area Management Categories

	Title[a,b]	Stated Management Objective	Management Objective[c,d]								
			SR	WP	SD	ES	NF	TR	ED	SU	CA
Ia	Strict nature reserve	Science	***	**	***	**	—	—	—	—	—
Ib	Wilderness area	Wilderness protection	*	***	**	***	—	**	—	*	—
II	National park	Ecosystem protection and recreation	**	**	***	***	**	***	**	*	—
III	National monument	Conservation of specific natural features	**	*	***	—	***	***	**	—	—
IV	Habitat/species management area	Conservation through management intervention	—	*	***	***	*	*	**	**	—
V	Protected landscape/seascape	Landscape/seascape conservation and recreation	**	—	**	**	***	***	**	**	***
VI	Managed resource protected area	Sustainable use of natural ecosystems	*	**	***	***	*	*	*	***	**

[a] IUCN uses six categories for classification of protected areas, according to the management objectives of the sites (IUCN, 1994). These 1995 data are taken from Green and Paine (1997), who also defined and scored the management objectives.

[b] Using data from Green and Paine, land use is apportioned in the following order to II (national parks, 4,000,605 km^2), VI (managed resource protected areas, 3,601,440 km^2), IV (habitat/species management areas, 2,459,703 km^2), and V (protected landscape/seascape, 1,057,448 km^2), with Ia, Ib, and III all having less than a million km^2.

[c] SR, scientific research; WP, wilderness protection; SD, species/genetic diversity; ES, environmental services; NF, natural/cultural features; TR, tourism and recreation; ED, education; SU, sustainable use; CA, cultural attributes.

[d] *** primary objective; ** secondary objective; * potentially not applicable; —not applicable.

(chapters 4 and 5). These oversights, failures to show sensitivities that were respected in the metropolitan culture, were undoubtedly responsible for some of the severe challenges protected area managers were to face in the postcolonial period.

2.3 The Limits and Legacies of Protectionism

The spate of new national parks and reserves set up in the last thirty years have in places been very successful in safeguarding some highly visible megafauna, such as many of the mammals of southern and eastern Africa. In one sense, then, protected areas are the jewels in the crown of international conservation policy; furthermore, in addition to preserving species diversity they maintain essential ecological processes; safeguard historic, cultural, and economic features important to local populations; secure landscapes and wildlife that enrich human experience; provide opportunities for community development, research, education,

Figure 2.2. Protecting English Countryside.
Agriculture plays a huge part in shaping the landscape of the Lake District, now a National Park. Hay or silage may be made from the grassy lowlands, whereas the upland fells are used for sheep farming. To preserve areas such as these, John Dower, the architect of the national park system in England and Wales, devised a unique definition in his pioneering 1945 report: "A national park is an extensive area of beautiful and relatively wild country in which, for the nation's benefit and by appropriate national decision and action, (a) the characteristic landscape beauty is strictly preserved, (b) access and facilities for public open air enjoyment are amply provided, (c) wildlife and buildings and places of architectural and historical interest are suitably protected, while (d) established farming use is effectively maintained" (cited from J. Dower in Harmon, 1991, 34). As such national parks in Britain are classified as Category V areas (Protected Landscapes). Photo courtesy of David Norton/naturepl.com.

and recreation; and serve as sources of national pride and human inspiration (McNeely, 1995). In another sense, however, they are mine fields. In McNeely's words, "protected areas are on the front lines in the battle to conserve biodiversity, and since these areas are often selected and managed specifically to protect species and ecosystems of outstanding value from human degradation, they are also sites of highly contentious debates between local and national interests" (cited in Brandon et al., 1998, 10). There are many reasons why this is so, which pertain to biology, economics, politics, and ethics.

First, protected areas do not always target sites of high biological diversity. Ice sheets and deserts are relatively easy places to designate national parks because of their spectacular scenery and lack of development pressure, but they are not sites of great species diversity. Indeed, most of the increase in protected area coverage noted above is due to a few huge areas in lightly populated desert and high mountain areas, such as the empty quarter of Saudi Arabia and the Qiang Tang Reserve in western China (see legend to Figure 2.1). As a result, new moves have been made afoot to map the world according to its "ecoregions," geographic areas defined by the unique biodiversity they contain, in order to guide decisions as to where to locate future protected areas. Conservationists are also mapping the distribution of ecological communities against existing protected area networks, looking for communities not represented in existing reserves (see chapter 3). In this way protected areas can more efficiently conserve biological diversity.

Second, new scientific information about the complexity and true extent of effective ecosystems (see chapter 3) has challenged the utility of strict protectionism, because entire ecosystems can never be protected. Wrestling with the question "Why are there so many kinds of animals?" ecologists turned in the 1960s and 1970s to the study of floral and faunal communities on islands (island biogeography; chapter 3). They found that both the size of the island and its isolation were key determinants of the number of species it supported. The implication for a protected area, now recast as a kind of island, was that without a constant flow of plant and animal immigrants from outside its borders it risked incrementally losing species and eventually becoming an insular, biologically depauperate refuge in a sea of degraded land. While zero protection might well precipitate this outcome even more quickly, the point is that biological difficulties inherent in managing reserves were becoming apparent. At the same time, the conservation bar was being raised even higher. In addition to all species and their genetic diversity, scientists argued for protection of ecological processes like migrations and predator-prey cycles so that evolution could continue unfettered. Exacerbating this situation even further was the problem mentioned above, that many protected areas were established around scenic rather than biological resources; only parts of ecosystems are protected, leaving migrations or seasonal ranges unprotected. In short, some ecologists were themselves losing faith in protectionism, at least as a unique strategy for conserving nature.

Third, turning to the fiscal and political issues, as the budgets of postcolonial governments dwindled so did their ability to enforce protectionist policies. Often managers are expected to use military-style strate-

gies to ensure protectionism, and these are expensive. Anti-poaching guards need to be paid, and if their salaries are not forthcoming it is tempting for them to collude with villagers whose illegal activities they are supposed to police (e.g., Neumann, 1998). The effectiveness of protectionism is often directly tied to expenditure: in Zambia the annual decline in black rhino and elephant numbers closely matches the patrol efforts of anti-poaching squads in different parts of the Luangwa Valley, and across different African nations such declines mirror per-kilometer government spending within conservation areas (Leader-Williams and Albon, 1988). Funds to run reserves and national parks usually come from the national budget, but most government agencies now find it difficult to justify increased expenditures on protected area management, particularly because of the social and political costs this incurs at the local and regional levels. Indeed, average budgets for protected areas in developing countries are only 30 percent of the minimum amount required for conserving those areas (Spergel, 2002), and in different calculations the estimated global costs of protecting critical percentages of global terrestrial and aquatic ecosystems (US$45 billion/year) fall even farther short of current annual spending of US$6.5 billion on protection (Balmford et al., 2002). In this context of spending shortfalls emerged the notion of "paper parks," protection that exists only in the lines drawn on a map. A comparison of protected area spending between African countries and industrialized states of the North (Table 2.2) indicates how little African nations pay per square kilometer of protected area (US$373/km^2; US$117/km^2 if Kenya and South Africa are excluded) compared to industrialized nations (US$2,768/km^2). On the other hand, the percentage of land in protected area status is roughly comparable (9 percent versus 12 percent), and rather remarkably African nations contribute a higher percentage (0.19 percent) of their national budget to protected areas than do industrialized Northern nations (0.08 percent) (Wilkie et al., 2001).

TABLE 2.2
Protected Area (PA) Spending in a Sample of Nations around the World[a]

Country	Total Area (km²)	PA (km²)	Percent of Total Area Protected	PA Spending (US $1,000)[b]	Percent of Budget	Unit Area Spending US$/km²[b]
Germany	356,910	58,579	16	45,968	0.01	785
Netherlands	37,330	3,500	9	19,635	0.01	5,610
United Kingdom	244,820	46,271	19	161,073	0.03	3,481
Canada	9,976,140	496,812	5	308,470	0.25	621
United States	9,372,610	982,192	10	1,864,565	0.12	1,898
Angola	1,246,700	81,812	7	30	0.00	0
Botswana	600,370	100,250	17	5,654	0.27	56
Burkina Faso	274,200	31,937	12	261	0.02	8
Cameroon	475,440	25,948	5	771	0.03	30
Central Africa	622,980	46,949	8	505	0.02	11
Republic Cote d'Ivoire	322,460	19,929	6	2,524	0.05	127
DRC	2,345,410	100,262	4	768	0.17	8
Ethiopia	1,221,900	32,403	3	4,806	0.12	148
Gabon	267,670	18,170	7	178	0.02	10
Ghana	238,540	13,049	5	3,011	0.07	231
Kenya	582,650	32,726	6	69,685	0.50	2,129
Malawi	118,480	10,585	9	2,069	0.10	195
Namibia	824,290	112,159	14	14,170	0.66	126
Niger	1,267,000	84,163	7	143	0.02	2
Nigeria	923,770	34,218	4	12,310	0.02	360
South Africa	1,221,040	57,638	5	157,065	0.33	2,725
Tanzania	945,090	258,997	27	52,074	0.78	201
Zimbabwe	390,580	50,736	13	18,090	0.23	357

[a] From Wilkie et al. (2001), who use data from published and unpublished sources.
[b] Currencies are presented in US$, controlled for purchasing power.
Source: Wilkie et al. (2001), *Biodiversity and Conservation* 10:691–709. Copyright Kluwer Academic Publishers.

These fiscal difficulties can in some cases belie deeper problems, such as lack of national commitment to conservation, our fourth reason why the success of protected areas has become controversial. The problem is that protected areas' benefits are seldom obvious to the general public, especially when these areas appear to be no more than exotic vacation spots for foreigners. Political commitment is particularly weak when the economic returns from protected areas are meager compared to alternative human-settled land uses, as in contemporary Kenya (Norton-Griffiths and Southey, 1995; see Box 9.1). Lack of a strong national and local constituency can translate into insufficient human and financial resources being devoted to protected area management, as noted above. In such situations parks can become open-access areas (see chapter 6), with degradation occurring in response to an expanding agricultural frontier, illegal hunting, fuel collection, and uncontrolled burning. Especially acute are situations where protected areas harbor natural resources whose harvest can bring huge economic windfalls to the government. The will to exploit the resource often outweighs the will to conserve, as seen in the sad tale of Peninsular Malaysia's protected areas (Box 2.2). Finally, political apathy can stimulate counterproductive outcomes, as in the case

Box 2.2. Reserves: Their Comings and Goings in Peninsular Malaysia

A typical tale of weak governmental will as regards protectionism comes from the Peninsular Malaysia's protected areas coverage between 1903 and 1992 (Aiken, 1994). Under British rule, Malaysia's habitats for wildlife were already in trouble: rubber cultivation (adopted in 1905) was causing extensive deforestation, and the older activity of tin mining continued to cause localized but extensive land degradation. With economic development wildlife came under increasing pressure because of forest fragmentation, a burgeoning use of firearms, and commercialized hunting, trapping, and fishing. As in many other parts of the world at the time the colonial government, with encouragement from London, legislated to protect overexploited animals, initiated a forest reservation program, and established protected areas. In the early 1930s, recognizing that many species of fish, birds, and mammals were still severely depleted (including the one-horned Javan rhinoceros [*Rhinoceros sondaicus*] and two-horned Sumatran rhinoceros [*Dicerorhinus sumatrensis*]) much larger areas of land were brought under protection. By the end of the colonial period, about 5.3 percent of the peninsula's land mass was under protection.

Since the late 1950s, one-third of the peninsula has been deforested, and the percentage devoted to protection has increased by only one percent. Not only have very few protected areas been established since 1957 (the beginning of the national period) but the already-existing area under protection dwindled as a result of excisions (e.g., great tracts removed for rubber plantations) or intrusions (land clearance for agriculture, logging, mining, poaching, and settlement). Aiken attributes this poor environmental record to three factors. First, commercial interests have outweighed conservation will. The Malaysian constitution provides for state jurisdiction over lands and forests, and since revenue derived from timber sales finances recurrent and development expenditures, state governments are understandably reluctant to sacrifice productive lands and forests to conservation. Second, protected areas in Peninsular Malaysia have insecure legal status, a situation dating back to the colonial period when in 1937 Comyn-Platt observed that areas were "allotted to wild life just so long as they are not required for anything else . . . So long as the cultivator and prospector have not need of them, so long will they remain intact" (cited in Aiken, 1994, 53). Third, the peninsula's aboriginal peoples have no legal title to remain in the reserves, and therefore little commitment to protecting their lands (see chapter 4).

The grim conclusion is that Peninsular Malaysia's protected area coverage is probably less now than it was in 1940 (when, unlike today, most of the peninsula was forested). Furthermore, it is not representative of the different habitats of the area, and many of the reserves are small and isolated, jeopardizing their effectiveness in protecting biodiversity (see chapter 3). Degazetting of parks and protected areas is commonplace; while Malaysia is hardly alone in this regard, the case points to the importance of political will.

of the current international conservation target of protecting 10 to 12 percent of each nation's land; for example, Soulé and Sanjayan (1998) describe how the Canadian province of British Columbia, to avoid political difficulties, simply converted to the parks system's unproductive and already well-represented lands to meet the target.

Fifth, and finally, increased sensitivity to human rights makes the old Yellowstone practice of relocation, repeated with abandon across the globe throughout the twentieth century, politically very unpopular. While not all protectionism entails population displacement, the ethical issues raised by forcible eviction (the "fences and fines" approach) must be addressed. The literature is brimming with reports, books, and articles deploring the old practices of eviction, offering a massive cataloguing of past, recent, and ongoing abuses, in the form of highly influential early policy documents (Colchester, 1997; IIED, 1994; Kiss and World Bank, 1990), closely argued scholarly works (DiSilvestro, 1993; Dixon and Sherman, 1990; Gadgil and Guha, 1992), committed anthropological investigations (Alcorn, 1994; Brockington and Homewood, 2001; Brosius, 1999), and passionately written indigenous pleas (Kemf, 1993; Lynge, 1992). Very often eviction had been justified under the premise that the protected area was previously "empty" (or at least "practically empty"), but this designation often includes pastures or forests that are infrequently or seasonally used yet critical for a community's survival. Eviction elicits all kinds of resistance (see chapter 8), as demonstrated by the Tanzanian Meru peoples' subtle maneuverings on the hillsides of their mountain (Figure 2.3). The ethical issues associated with eviction have captured the popular imagination with remarks, such as that of Bonner's (1993), that the predatory objectives of conservation are sometimes quite difficult to disentangle from those of colonialism, capitalism, and, in some places, even Christianity. Furthermore, they are as relevant today as ever. First, the

pattern of forcibly removing resident populations repeats itself again and again, for example, in northern Thailand, where Karen and Hmong hill tribe villagers still live under the threat of eviction from forests that were classified first as reserve forests, then national parks, and now in some cases world heritage sites (Nepal, 2002). Second, with the spread of democracy even long-established protected areas now face vigorous challenges from erstwhile-displaced local people (Lindsay, 1987; Neumann, 1998). Third, in the new ethical climate governments that formerly bought into the colonial ideology of eviction and preservation can now no longer afford to jeopardize their local support with a fences-and-fines approach to conservation; as such, they are hamstrung by international commitments to expand areas under protection, particularly as land becomes ever more scarce and uninhabited areas increasingly difficult to find. Continuing attempts to displace resident populations, now through enticements rather than threats, will need very careful monitoring (Box 2.3).

All of these factors—global conservation awareness, the recognition that parks are not islands and that land outside protected areas also needs conservation, and land scarcity—conspired to shift conservation interest to the developing world and to human-dominated landscapes. At the same time, the pragmatic, political, and ethical critique of protectionism sharpened, in part because of the role it was to play in justifying newer conservation strategies (see below). Unwittingly, local people living in biologically rich areas were enlisted as key players, ushering in a third stage of amateur involvement in conservation reminiscent of Leopold's engagement of farmers and hunters. But this new conservation initiative faced an ethical, fiscal, and political climate that was very different from anything that had preceded it, raising particularly complicated questions: if national governments cannot afford (either fiscally, politically, or ethically) to relocate people living in biologi-

Figure 2.3. Tricks on Mount Meru.
To be successful the Meru peoples' incursions into protected forest on Mount Meru had to go undetected. Neumann recounts a particularly innovative style of encroachment that approaches an art form. In 1955 a government surveyor described how local residents were carefully repositioning concrete beacons 30 to 45 meters parallel to the legal forest boundary. They dug guide trenches, erected guideposts, and built cairns with a beacon in the center as an "exact replica of the original. . . . To fully comprehend this master-piece of encroachment is difficult," the bewildered surveyor wrote. Mount Meru seen from modern Arusha town. Photograph by Thomas Spear as reproduced in his *Mountain Farmers* (Oxford, 1997).

cally rich areas, how could these ecosystems be saved? And how could rural people in developing countries, arguably the poorest people in the world, be expected to forgo consumption of natural resources to protect shrinking wild areas?

2.4 Conservation "with a Human Face"

In an early response to these problems, the United Nations Educational, Scientific, and Cultural Organization (UNESCO) launched the Man and the Biosphere (MAB) Program in the late 1960s. With this program UNESCO hoped to encourage interdisciplinary research to form the foundations for sustainable resource use worldwide. Toward that end, they proposed a worldwide network of protected areas, or biosphere reserves, where both pristine and anthropogenically modified areas could be studied and conserved (Box 2.4). The inclusion of human-modified areas within these reserves

was an important departure for conservation thinking. For the first time since the national park movement had dropped Catlin's phrase "man and beast," human settlement and land use were explicitly included in a protected area framework. An equally important feature of the biosphere model was MAB's contention that "conservation of environmental resources could and should be achieved alongside of their utilization for human benefit" (Batisse, 1982, 101). That consumption of a resource could actually lead to its conservation was a novel and contentious idea, and one that remains so until this day. The final defining feature of the biosphere concept was that it engaged, even empowered, local communities, and so firmly departed from the "neocolonial" legacy of protectionist conservation. Conservationists, eager to avoid the large-scale and expensive failures of top-down development projects, embraced the idea of conservation from the bottom up.

The biosphere concept caught on rapidly.

Box 2.3. Coercive Conservation: Tigers, Lions, Carrots, and Sticks

In India, perhaps more than anywhere else, large-scale development projects have brought misery to rural populations. As a result of having to leave their homelands to make way for dams, plantations, afforestation programs, and national parks, more than 600,000 tribal peoples were resettled between the mid 1980s and early 1990s (Ghimire and Pimbert, 1997). With the new emphasis on enlarging the amount of land under protected areas status, this trend is only increasing. We look at two cases.

The Gir Maldharis, who had lived in the Gir Forest for generations, were in the 1970s resettled to make way for Gir National Park, the last sanctuary of the Asiatic lion (*Panthera leo persica*) (Choudhary, 2000). Maldharis had lived as herders of cattle and buffalo, basing their economy on the sale of dairy products; these domestic animals grazed in the forest by day. The people, while vegetarian in diet, also depended heavily on the forest for materials for fuel and house construction. In 1972 the Gir Lion Sanctuary Project began to move all Maldhari families to resettlement areas. There are different stories about the reasons for their removal. One stresses fodder shortages in 1972, catastrophic contagious livestock disease, predation by lions both on people and on dwindling livestock (livestock constitute 40 percent of the lions' diet), and a need to resettle communities whose life expectancy was a mere twenty-four years. Another, from the management plan for the Gir Sanctuary and National Park produced by Gujerat's Wildlife Division, implicates the negative impacts of the Maldharis' consumptive use of forest fodder and fuel products, sharp declines in lion numbers, and cultivation creeping up the forest's fertile valleys. Resettlement, however, was not a success, voluntary as it was. The financial needs of the program were grossly underestimated. Twenty-eight years later, many families still remain in the forest, and 45 percent of the resettled families have left their new sites, with some returning to the forest. Those that remain in the forest are increasingly tied to the market economy, keep larger herds, and are more mobile than formerly (necessitating greater use of housing materials). Since 1996 a World Bank–sponsored project has become involved, but there is still great hostility among the Maldhari toward the conservation initiative and much talk of forced relocation.

From Karnataka State we hear a very different story (Karanth et al., 2001; see also Karanth, 2002). Nagarahole National Park (644 km^2) offers prime habitat for threatened tigers, but in the 1990s over 6,000 Kurubas had homes inside the park. Living in inaccessible villages, these families lacked basic amenities like health, water, and education; made insufficient money from the sale of forest products; suffered severe conflict with tigers and other wildlife; and were forced into working cheaply as laborers at nearby coffee plantations. In 1997 a group of tribal leaders visited the state capital, along with fifty tribal families, demanding a participatory, land-based rehabilitation program. Given how such resettlement would help consolidate the highly fragmented tiger habitat across the Western Ghats (a biodiversity hotspot), conservation authorities jumped at setting up a voluntary resettlement scheme, offering as incentives to each family farmland, housing, drinking water, agricultural extension, a fuel and fodder lot, and electricity. Opposition from local NGOs focusing on potential violations of local culture and human rights was swept

(continued)

(Box 2.3 continued)

aside, and in 1998 the first tribals moved into new homes in an area on the edge of the park ecologically similar to their original village. By 2000 over 200 families had moved out of Nagarahole, another 435 families had accepted resettlement from a neighboring reserve, and most remaining families had expressed willingness to move. The project is lauded for having both delivered social justice and consolidated prime tiger habitat, and is endorsed by Rajendra Singh, one of India's leading human rights advocates, although plenty of criticism can be found elsewhere (Guha, 1997).

What are we to make of these cases? Clearly, critical questions to pursue in all such cases are: Who did the idea of resettlement really come from? What are the relative strengths of the carrots of incentives versus the sticks of enforcement? What is the legitimacy of NGOs that either facilitate or protest protectionism?

In 1976 MAB designated 57 biosphere reserves, 61 more were added in 1977, and by 1986 there were 261 in 70 countries (Batisse, 1982, 101). Today there are 356 biosphere reserves in 90 countries (Lasserre, 1999). This remarkable growth no doubt reflects how closely the program's concern with the human communities living in areas of special biological importance, its focus on utilization as a means of conservation, and its grassroots orientation fitted with the emerging social and scientific climate of the 1970s and 1980s.

This was also a time of enormous ideological and political shifts with respect to the relationship of conservation to development. We can see this first in the changing concept of national parks (Hales, 1989). In contrast to all previous World National Parks Congresses, where attention had been trained inward from the park boundaries, the 1982 meeting convened in Bali sported banners emblazoned with "Parks for Sustainable Development," and in Caracas ten years later with "Parks and People." Community development was now seen as key to protecting the integrity of national parks and protected areas (McNeely, 1995; Western and Wright, 1994). This ideology reached far beyond the domain of parks. With the publication of the Brundtland Re-

port (WCED, 1987), a new buzz word, *sustainability*, gained prominence (Box 2.5), with sustainable development defined as "development that meets the needs of the present without compromising the ability of future generations to meet their own needs" (WCED, 1987, 43). By the end of the 1980s, the idea that conservation and development could be linked through sustainability was becoming mainstream, almost approaching dominance in international conservation circles. Furthermore, the Brundtland Report strongly advocated a utilitarian rationale for conserving biodiversity.

Brundtland's message dovetailed with many of the themes in the World Conservation Strategy (IUCN/UNEP/WWF, 1980), a conservation manifesto that had been crafted by the world's leading conservation organizations a few years previously. Though the stated aims were to maintain essential ecological processes and to preserve genetic diversity, this document argued that conservation is not the opposite of development insofar as human welfare depends on nature. The logic was simple and far-reaching: "to be sustainable, development needs conservation of natural resources, and conservation, to succeed in low-income countries, requires development" (Abbot et al., 2001, 115). But it was not until 1991 when the

Box 2.4. The Evolution of Biosphere Reserves

The original biosphere reserves and UNESCO's Man and the Biosphere (MAB) project grew out of the 1968 Biosphere Conference, the first intergovernmental worldwide meeting to adopt recommendations concerning the environment (Batisse, 1993). The goals of MAB were to promote nature conservation by developing new approaches to human resource use that would not undermine the integrity of natural ecosystems. To do so, the MAB project needed large "natural laboratories" representing the diversity of ecosystems and land uses for each country (Batisse, 1982). These areas became biosphere reserves, designed under the tripartite strategy of conservation, sustainable utilization, and a strong research, education, and monitoring component. As such, utilization was stressed twenty-four years prior to the Rio CBD convention.

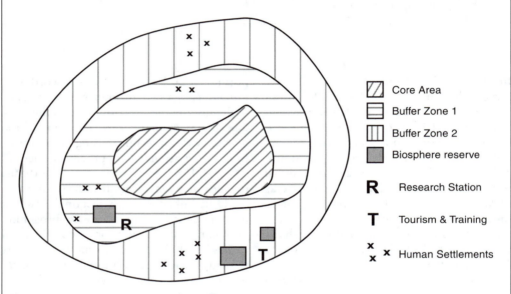

Core Area

Buffer Zone 1

Buffer Zone 2

Biosphere reserve

R   Research Station

T   Tourism & Training

x x x   Human Settlements

Typical Design of a Biosphere Reserve.
Reprinted with permissiona of Cambridge University Press from *Environmental Conservation* Volume 9, No. 2 (1982): 101–111.

Key to the original concept was a protected area core surrounded by a series of buffer zones where increasing amounts of human use were allowed (see figure above). Initially, most biosphere reserves were formed around existing national parks or other protected areas, and followed a standard model. As this was applied in new areas, many of the assumptions underlying the model were relaxed, and new kinds of reserves were established. The most common of these was the "cluster reserve," where a number of core areas were surrounded by transition and buffer zones embedded in a larger matrix of human uses and areas under lower levels of protection (Batisse, 1982; 1986). Biosphere reserves are currently found from downtown San Francisco (United States) to Lake Ichkeul (Tunisia), and from Fontainebleau (France) to the Ciénaga de Zapata in Cuba's Bay of Pigs. In such diverse

(continued)

(Box 2.4 continued)

settings the biosphere reserve model has been pushed into new roles and is continuously changing. Perhaps unsurprisingly they have been controversial. Some argue they give an inflated sense of protection; for example, the Fontainebleau biosphere reserve covers over 69,000 ha but of that area only 3,900 ha are uninhabited. Others fear that the management challenge of requiring local residents to tune their resource use according to new and often quite complex zoning laws is unrealistic. Note the complexities of using a biosphere reserve for commercial extractive purposes in Box 10.6.

same organizations updated their manifesto in "Caring for the Earth" (IUCN/UNEP/ WWF, 1991) that the magnitude of this change in thinking became obvious (Holdgate and Munro, 1993; Robinson, 1993). This new manifesto revealed how dramatically the conservation tide had turned away from preservationism, and how strongly some conservationists now embraced the utilitarian approach advocated by the Brundtland Report; undoubtedly, it also reflected the philosophy of its funders, with the World Bank now allocating a full one percent of its budget to conservation. The principal aim of "Caring for the Earth" was "to help improve the condition of the world's people" (3), and conservation was defined as "the management of human use of organisms and ecosystems to ensure such use is sustainable" (210); the preservationist goals of conservation had almost become something of an afterthought: "besides sustainable use, conservation includes protection, maintenance, rehabilitation, restoration and enhancement of populations and ecosystems" (210). It was this new meaning of conservation that was formalized in the CBD (see chapter 9), the agreement signed at the UNCED in Rio de Janeiro.

Another legacy of the Brundtland Report was its strong advocacy for conservation strategies relying on utilization, captured in the quip "use it or lose it." To ensure community welfare the economic value of nature was to be harnessed to economic devel-

opment. The enduring debate that arose from the elision of conservation and development, and more specifically from striving for preservationist ends through utilization, is captured in the highly publicized debates over ivory, where harvest of one species (elephant) can be promoted to fund protectionism for others (Box 2.6). This still unresolved controversy illustrates how persistent long-standing conservation debates are often at the heart of contemporary conservation challenges.

Widespread acceptance of a potential link between biological conservation and economic development opened the door, as we have seen, for a suite of new players in the global conservation arena. International development agencies, eager to move beyond environmentally checkered pasts, speedily signed up for this new brand of conservation. In many ways, their new role made perfect sense. Their focus remained on improving people's lives in the underdeveloped world, but now through sustainability this goal carried the added benefit of contributing to biological conservation. One consequence of this arrival of jostling new players on the conservation scene was that the term *biodiversity* increased in political weight and everyone began tacking their agendas (noble or not so) onto it (Table 2.3). Another consequence, somewhat ironic, was that these highly centralized international organizations working through national governments were going to have to become the

Box 2.5. Sustainability—Mere Hopes about the Future

With the Brundtland Report sustainable development became the sine qua non of any project entailing economic activities based on exploiting renewable resources. Since the critical definition of sustainability entails survival and persistence of a system into the future, what pass as definitions of sustainability are ultimately merely hopes that actions taken today will lead to predicted outcomes in the future. Three facets of sustainability are usually identified:

Ecological sustainability implies practices that do not irreversibly deplete resources or degrade the habitat. Keeping harvest rates of a resource below the rate of natural renewal should lead to ecological sustainability—but this is a prediction based on maximum sustainable yield theory, not a description (see Box 3.5). This is because ecosystems are difficult to monitor and claims about the sustainability of offtake are based on models rather than empirical observations (Folke and Costanza, 1996). Additional problems in ensuring ecological sustainability reflect the fact that ecosystems are dynamic (it is difficult to predict their future course, even without the exploitation by humans), and that definitions of depletion or degradation are often open to numerous interpretations (Box 3.3).

Social or institutional sustainability is often used by rural development agencies to suggest that a project will not fall apart after the explicit development activities have ended. In this sense it implies that a set of practices are ideally all of the following—socially (or culturally) acceptable, technologically appropriate, and institutionally stable. Again, this is more often a hope than a specific state, but socially sustainable systems can be promoted by good consultation, grassroots inputs into planning, and other collaborative procedures.

Economic sustainability refers to a set of practices that will make money over a certain period, and not cause any major collapse or instability within the local economy. Components of economic and institutional sustainability can be combined to produce yet other facets of meaning found in the literature: efficiency (the pareto-optimum whereby no user can take more of the resource without other users suffering; chapter 6), equity in harvests among users, livelihood security, and empowerment of the users to manage the resource in question (Berkes, 1996).

It is worth emphasizing, first, that ecological sustainability addresses much longer time scales than do social and particularly economic sustainability. Given the short time periods over which businesses calculate their profits, the *ecologically* unsustainable harvest of slow-growing species such as mahogany may well be rated as *economically* sustainable. Second, claims of sustainability can conceal crucial gaps in logic. Thus, a few years after the publication of the Brundtland Report, social scientists were protesting that sustainable development was an impossibility because of the logical incompatibility between growth (voracious for inputs) and sustainability (e.g., Goodland et al., 1993); furthermore, they objected to its political conservatism, observing that it was a mere anesthetizing label under which international business could continue as usual to exploit people and nature in the developing world (Heang, 1997; Lele, 1991). Third, sustainability is not of course an inherent characteristic of any practice, but depends on the fit between the practice and its particular ecological, economic, social, and political context. Thus swidden (once derogatively termed "slash and burn" farming) can be an efficient, productive, and sustainable form of production in tropical forest, but only with mobile and low-density populations, adequate fallow periods, and large areas of forest in reserve. To claim that a practice is or is not sustainable requires specification of its application. Use the term at your peril!

Box 2.6. The Ivory Wars: Debates over Utilization

Few if any species have received the attention, or generated the controversy, that the African elephant (*Loxodonta africana*) has. Hammered by a seemingly insatiable global demand for ivory, African elephant populations collapsed during the 1970s and 1980s. In Kenya, for example, 300 elephants were poached in a single year (1989). Responding to this decline, the Convention on International Trade in Endangered Species of Wild Fauna and Flora (CITES) sought to outlaw international trade in elephant tusks, moving the African elephant from Appendix II (which allows regulated trade in elephant ivory and skin) to Appendix I (total ban). But not all African states with elephants benefited from this policy. In the southern African countries of Zimbabwe, Zambia, Botswana, Namibia, and South Africa, declines in elephant numbers were less precipitous, and state-sanctioned harvesting of elephants was still taking place. Wildlife managers in these countries argued that controlled harvests of elephants and sale of their ivory were necessary because they provided funds for the management and protection of other species. Further, they argued, there simply was not enough habitat available to allow elephant numbers to grow unchecked. For them, an international ban on the sale of ivory meant the end of their current conservation and management funding. East African governments and conservationists held to their campaign to ban all trade, because in their eyes any international trade in elephant products provided opportunities for illegal offtake. They feared that a reopening of legal trade would send signals to illegal traders and poachers that the entire market in ivory, including Asian ivory, was now legitimate. The ban held, and in the following year (1990) only forty-six elephants were poached in Kenya, but the price of ivory rose from US$2–$3 per kg to US$40–$50 per kg (Barbier, 1998) as a result of these changes.

Ivory Tower: Kenya's Burning of Tusks as an International Display against Utilization.
Photo courtesy of Iain and Oria Douglas-Hamilton of Save the Elephants.

The situation is still not resolved. The annual CITES conferences are regularly distracted, even derailed, by the debate (Micklebugh, 2000; Swanson, 1999a). In 1997 a cartel of three southern African states (Namibia, Botswana, and Zimbabwe) gained permission from CITES to sell their ivory to one buyer (Japan). In 1999 southern African countries had their elephants temporarily downgraded to Appendix II status, and were allowed a one-time sale of ivory. At the 2002 CITES conference they requested (successfully) another one-off sale and an

(continued)

(Box 2.6 continued)

annual sale quota, while countries like Kenya and India continue to seek a reclassification of all elephant populations back to Appendix I.

The ivory wars represent a present-day example of the preservation/utilitarian debate (see chapter 2), but with a new twist. In the past, utilization was a tool for mainly utilitarian ends, and preservation had always been pursued exclusively through protectionism. In the last fifteen to twenty years, however, preservationists have begun to enlist utilization as a strategy. Thus, southern African countries are arguing that utilization of one species (elephants) can fund large areas of land being taken out of agricultural or domestic livestock production, thereby providing habitats for other more endangered species (see chapter 9). In other words, utilization can further preservationist ends.

primary advocates of a decentralized, utilitarian approach to conservation. The link between conservation, rural development, and decentralization forms the crux of a new conservation paradigm to emerge in the 1980s: community-based conservation.

2.5 The Rise of Community-Based Conservation

With the innovative experiments of biosphere reserves, the changing concepts of national parks, and the new relationship be-

TABLE 2.3
Biodiversity: Contested Meanings

Actors	Focal Feature of Biodiversity	Rationale
Ecologists and evolutionary biologists	Species richness	To ensure continued existence of as many species as possible (chapter 1)
Plant breeders	Crop germplasm	To ensure agricultural food sources in the future (chapters 1 and 8)
Ethnobotanists	Local knowledge of useful species	To retain traditional ecological knowledge and protect cultural identity of local populations (chapter 8)
Pharmaceutical companies	Potentially medicinally useful species	To retain option values of medicinal products (chapter 8)
Anthropologists	Cultural and linguistic diversity	To emphasize the links between cultural and biological diversity (chapter 4)
Indigenous peoples	Cultural and linguistic diversity	To find representation at international conservation fora by noting the role of indigenous peoples as stewards of nature (chapters 4 and 8)
Human rights activists	Powerless and under-represented constituencies	To defend local environmental activists (chapter 8)
Western corporations	Supplies of forest and other natural products	To protect biodiversity by finding sustainable commercial applications, such as nuts, shade-grown coffee, etc. (chapter 9)
Rock stars, media figures, etc.	Natural (or wild) images	To stimulate activism, or to promote image/market (chapter 8)

tween conservation and development, the scene was set for the emergence of community-based conservation (CBC; or community conservation, CC) and related developments, community-based natural resource management (CBNRM), co-management, and integrated conservation and development (ICD) projects. The benchmark publication for CBC was Western and Wright's (1994; see also West and Brechin, 1991) collection of case studies and two influential technical reports (IIED, 1994; Kiss and World Bank, 1990). Initially defined very loosely as "natural resources or biodiversity protection by, for and with the local community" (Western and Wright, 1994, 7), a more precise definition comes from Barrow and Murphree (2001, 31): "the sustainable management of natural resources through the devolution of control over these resources to the community as its chief objective." This investment of control with the community was based on two assumptions: that local communities have more knowledge about local resource dynamics than anyone else, because of their intimate relationship with the environment, and that they have greater incentives to manage their resources sustainably, because they depend on these for their survival (Ostrom, 1990).

Early much publicized examples of CBC were Zimbabwe's Community Area Management Programme for Indigenous Resources (CAMPFIRE) and Zambia's Administrative Design for Game Management Areas (ADMADE). By 1999 proponents were hailing CBC, when utilized in concert with sustainable development and market orientation, as the "New Conservation" (Hulme and Murphree, 1999). Here we simply outline CBC's three dominant themes (e.g., Getz et al., 1999; Hulme and Murphree, 1999) and explore their impact on ongoing conservation policy developments, such as ICD projects.

First is the devolution of authority to local communities to manage their natural resources. The aim is to shift conservation projects from their traditional role as exoge-

nous structures imposed upon local people to endogenous ones that sustain themselves from within (geared to *social* sustainability, as defined in Box 2.5). Today, the distinction between top down and bottom up (IIED, 1994) has been replaced by the more refined concept of centralization, which refers to the degree to which the administration and decision making of projects or protected areas are concentrated under a single agency. The emphasis of CBC is on the community as the locus of action. In reality, the extent to which community-based projects are truly participatory is highly variable: in some cases community members are true managers, in others a few lucky villagers may find temporary employment as guards or tourist guides, and in yet others community "involvement" is limited to the occasional handout of meat or to services generated by the revenues of conservation, a strategy more accurately referred to as outreach (see chapter 10).

The second theme to epitomize CBC is utilization which, as we have just seen in the Brundtland Report (WCED, 1987), was already a prominent emerging concept in international conservation. The idea is that people cannot undertake conservation, undeniably a *long-term* strategy, when their *short-term* needs are not met. Based on this premise, conservation problems are perceived as human rather than biological issues. If conservation is to take hold in the developing world, it will have to become economically as well as ecologically sustainable (see Box 2.5 again). Rural people simply can not afford the luxury of leaving pristine areas alone.

Third, there is the reliance on neoliberal economic thinking. The idea here, and we go into it in more detail in connection with ecological economics (chapter 9), is that if a species or habitat is to be conserved it should be exposed to, not protected from, the market. Indeed, Hulme and Murphree (1999) define the "New Conservation" as CBC conjoined to the market. The presumption here is that only in the market will biodiversity draw high value, thereby enhancing its chances of conservation; this is seen as preferable to placing biodiversity in the hands of a government liable to bribery and inefficiency.

There are several reasons, then, why the CBC initiative became so popular (Adams and Hulme, 2001): first, it triggered all the homely and politically correct ideas that were challenging the top-down and state-based development discourse that held sway in the 1970s; second, it captured a huge upwelling of policy commitment spurred by the concept of sustainable development; third, its link to the market offered a direct means of delivering development; and finally, as we discuss in chapter 3, biologists were realizing that small populations with limited genetic diversity were exposed to extinction risk, necessitating the conservation of remaining viable populations in landscapes in which human communities lived.

From within the intellectual and political climate of CBC ICD projects emerged. Seeking to formalize a particular kind of relationship between conservation and development, these projects were based on the premise that protected area management must reach beyond traditional protectionist conservation activities within protected areas to address the needs of local communities outside (Wells et al., 1992). The schematic design of the ICD project concept is shown in Figure 2.4; there is always a "core" protected area in which uses are restricted. There are three major strategies whereby ICD projects attempt to meet their goals (Brandon, 1997). First, there is emphasis on the design or enhancement of management plans for the core area (and buffer zones where these exist). Second, plans are developed to mitigate the costs to local communities of biological conservation within protected areas; these might include providing local people with substitutes for the materials previously taken from the protected area, and sharing direct benefits from the protected area (meat, entrance

fees in the core area, and other forms of tourist revenue) within the local community. Third, appropriate forms of economic and social development in local communities are encouraged, such as technical agricultural improvements, income-generating projects, and provision of social services, like schools and hospitals, again with the rationale of providing alternative and improved livelihoods to the affected population. A fairly typical example of an ICD project is the Annapurna Conservation Area Project (ACAP), where many of these ideas have been developed in response to environmental threats that have arisen from using tourism as a general strategy for protection (Box 2.7). Some of the more ambitious ICD projects work in conjunction with NGO-led attempts to change the legal status of wildlife and/or protected areas within the national constitution, often a critical step to the sustainability of the initiative (see chapter 10). The common thread to ICD projects, diverse as they are, is the interdependence between economic development and biological conservation (Abbot et al., 2001). In conceptual terms ICD projects can be thought of as a subset of CBC; all are implemented (or supposed to be implemented; see chapter 7) at the community level, but not all CBC projects involve the scale of economic development entailed in ICD projects.

Following development agencies into the now sprawling CBC fray were campaigners for social justice, who sought to secure indigenous peoples' land tenure on the grounds that they were (and still are) successful stewards of biodiversity. For them, the shift toward decentralization presented an ideal opportunity to piggyback human rights issues onto environmental conservation (chapter 8). Since land tenure is often compromised under pressure for large-scale projects like commercial logging, the benefits for biodiversity conservation of securing community land rights are obvious. After all, local and indigenous communities cannot be expected to act as responsible guardians

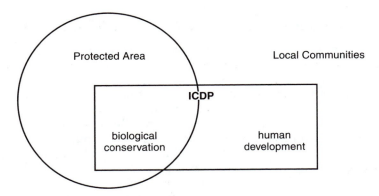

Figure 2.4. Schematic Design of Integrated Conservation and Development Projects.
The Protected Area constitutes a core, the Local Communities represent the affected population, and the ICD project attempts to link biodiversity conservation and human development by promoting planning and zoning, designing compensation schemes, and instigating appropriate development. Redrawn from Alpert (1996), *Biodiversity* 46:845–855. Copyright Tropical Conservancy.

of their land if they have no long-term secure access to these areas (see chapter 6), and they often have sophisticated knowledge of local ecological and social conditions that can be effectively used to manage natural resources (see chapter 4). Consequently, many reserves have been designed primarily to protect indigenous and local communities and their traditions (examples can be found in the boxes of later chapters). In Ecuador's Yasuni National Park and Biosphere Reserve, for example, access is denied to everyone other than the 1,500 Huaorani families who still reside in the area (see Box 4.2), as well as to employees of a company extracting oil in the area and to researchers; indeed, an even more restricted zone has been established to protect an area where several uncontacted indigenous communities are believed to live (Koester, 2001; Lu, 1999). Such indigenous or "anthropological" reserves mark an end point in the continuum of protected areas (falling into International Union for the Conservation of Nature [IUCN] Categories V and VI), and are closely related to CBC; in a sense they are like biosphere reserves dominated by the settled transition zone, with no core. Generally, emphasis is on maintaining tradi-

tional livelihoods rather than biodiversity per se (Brandon, 1997).

There are all kinds of twists associated with CBC that should be distinguished. First, community-based national resource management (CBNRM) (Getz et al., 1999) refers to rural programs concerned more with utilization than protected area management, though often the emphasis on natural resource management is linked to efforts to mitigate previous conservation evictions (or preempt future evictions); as such, the "targets" of CBNRM projects are sustainable natural resource systems, rather than wildlife or biodiversity writ large, and CBNRM is often perceived as a more general form of CBC. Second, programs developed for the sustainable management of New World tropical forests are known as extractive reserves (see chapter 10); they share many attributes with CBC, and often overlap both functionally and geographically with indigenous reserves. Finally, there is co-management. Conservation practitioners implementing CBC projects increasingly realize that it is not sufficient to focus only on the affected community as a locus of action; they use the term *co-management* to highlight the critical importance of coordi-

Box 2.7. Integrated Conservation and Development in Action: Annapurna

ICD projects can be linked to biosphere reserves, buffer zone projects, large-scale regional planning units, and multiple-use areas. An example of the latter is Nepal's Annapurna Conservation Area Project (ACAP). Established in 1986, it consists of 4,600 km^2 of a unique mix of ecosystems ranging from dry alpine deserts with blue sheep and snow leopards to lush subtropical lowlands full of diverse orchid flora, bamboo jungles, and extensive rhododendron forest. In addition, it contains one of the world's deepest gorges, the Kali Gandaki, as well as some of its highest mountains, including Annapurna and Dhauligiri. The 40,000 inhabitants living in 300 villages belong to many different ethnic backgrounds, with distinct languages, customs, and subsistence. The infrastructure is very limited, with few airfields and the regional center, Pokhara, a three-day walk from the conservation area boundary. Until the flood of tourists began arriving in the 1970s, the principal contact with the outside world was trade with Tibet. The people relied heavily on forest resources to meet an estimated 97 percent of daily needs, in cooking, heating, wood for construction and fencing, fodder for domestic animals, wild fruits and vegetables, medicines, fibers, and cloth (Bunting et al., 1991; Gurung, 1995; Hough and Sherpa, 1989; Wells, 1994).

Nepal was opened to foreigners in the 1950s, and tourism soon became a top foreign exchange earner, with people drawn to the country's cultural heritage and religious sites, the Himalayas, wildlife viewing, and trekking. Annapurna accounted for 36,000 of the 105,000 visits to Nepal's protected areas in 1990. In its wake tourism brought massive environmental problems, particularly accelerated deforestation because of the trekkers' needs for food and lodging. Other problems were soil erosion, garbage accumulation, inappropriate disposal of human waste, and contamination of water supplies. Additionally, the rise in the local resident population promoted wildlife hunting (tourists may not hunt, but locals do), expanded agriculture, and overgrazing of the cattle and yak herds.

On King Birendra's initiative in 1985, and with subsequent NGO support, a multiple-use management plan was devised in close consultation with communities, breaking from the government's previous history of evicting local people from fragile areas of high tourist appeal. The area was zoned into (1) special management zones, to protect highly degraded areas along trekking routes; (2) wilderness zones, upper-elevation seasonal grazing to be fully protected from human exploitation; (3) protected forests and seasonal grazing zones, to meet the needs of local people for building, hunting, and firewood collection; (4) intensive agricultural zones around villages; and (5) biotic/anthropological reserves where communities and ecosystems are still largely unaffected by tourism, and only scientists focusing on mountain ecosystems have access.

The biggest victory of the project was the decision by a newly formed lodge managers' committee to clean up the trails, improve sanitation, educate visitors on the natural and cultural impact of their visit, and, most important, to use kerosene fuel for all cooking and heating purposes. A fuel depot was set up in a central village (Chhomrong). In addition, farmers are being helped to plant trees for fuelwood and fodder, students are being trained both nationally and internationally in conservation area management and leadership, and entrance fees are being directed toward projects that provide seed money for activities such as chicken farming and vegetable patches. The approach appears to be viewed well by local people (Mehta and Heinen, 2001), though its overall success in ecological restoration does not seem to have been evaluated (Nepal, 2002) and recent reports suggest kerosene supplies are rarely available.

nating decentralized initiatives with relevant regional and national administrative bodies (see chapters 6 and 11).

The enthusiasm for these various kinds of community-based projects promoting utilization was in some instances heavily predicated on arguments for the ineffectiveness of parks. As such, there are CBC proponents who have called for something close to the complete abolition of strictly protected areas (Ghimire and Pimbert, 1997; Janzen, 1994; Wood, 1995). The stage was set for counterattack.

2.6 Imperiled Parks

In the 1990s a feisty backlash (e.g., Brandon et al., 1998; Inamdar et al., 1999; Kramer et al., 1997) arose against the fashionable tendency to downplay the role of protected areas and to favor utilization over protectionism, with Oates (1999) and Terborgh (1999) giving more personal autobiographic perspectives. While these conservationists acknowledge the unfortunate legacies and practical limits of protected area management, they argue strongly that protectionism should not be sacrificed on the altar of political expediency or political correctness. As such they offer a critical eye on policy directions that link conservation to development, scrutinize some of CBC's hottest slogans, and question its fundamental assumptions. There are several strands to their argument.

First, they point out how patchy protectionism really is. Despite the goal of protecting 10 to 12 percent of each nation's land (already too low a target; see Soulé and Sanjayan, 1998), countries like India and Mexico (havens of remaining biodiversity) have less than one percent of land in protection (Soulé, 2000). Furthermore, despite the initially impressive protected areas coverage internationally (see Figure 2.1), only Categories I to IV include biodiversity conservation as a management goal. For example looking at Latin America, Sayre et al. (1998)

found that of the 20 percent of the total land area under protection, almost two-thirds is in resource and anthropological reserves (older IUCN terms similar to Categories V and VI that give little priority to biodiversity). Additionally, many of the protected areas falling into I to IV are mere "paper parks," seriously underfinanced as discussed above. As such, what appeared to be a recent trend toward greater protectionism is neither as encouraging nor as distressing (depending on one's viewpoint) as it might at first appear.

Second, these conservationists scrutinize the logic of the claim that preservation and utilization are mutually interdependent, a notion anchored to the Brundtland Report, "Caring for the Earth," and most CBC initiatives. Specifically they question "win-win" scenarios, development strategies that encourage utilization as a way of preserving natural resources. Ever since the Marshall Plan was launched in 1948, the key ingredient of development has been economic growth (Redclift, 1987), and growth is hardly compatible with resource preservation. While over large timescales it is indeed true that the welfare of nature and humans is intricately interdependent, with conservation not affordable without economic development and economic development unsustainable without conservation, short-term conservation dilemmas inevitably entail trade-offs. A locally threatened species cannot prosper if harvesting continues at unsustainable levels. To claim otherwise may be politically expedient insofar as it implies that conflicts can easily be solved, but it is not well grounded in biological and ecological knowledge (Robinson, 1993; Milner-Gulland and Mace, 1998). At minimum a deep understanding of ecosystem dynamics at specific sites is necessary before sustainable utilization strategies can be recommended (chapter 3). Additionally, not all parties can win. Program designers (local or other) must determine who gets to harvest, consume, distribute, or sell the resource (see chapter 7), just as protected area managers may have to favor some species over others (see Box 3.6). Furthermore, not all things can be preserved through use (as philosophers point out; see chapter 1). In short, those leading the parks in peril initiative believe the strategy often dubbed as "use it or lose it" is not merely risky, but fundamentally flawed. So much emphasis is placed on the means of conservation (utilization) that the original goal of biodiversity conservation is lost.

Third, with the tectonic shifts in policy and ideology that have occurred in the last twenty years, protected areas are increasingly obliged to act as vehicles that balance local biodiversity protection with social and economic welfare. As Brandon (1998) observes, with each World Parks Congress the agenda becomes more crowded. In 1982 the Bali conference called for increased support for protected areas through education and revenue sharing in the local community. In 1989 parks were asked to move beyond local development and become engines of regional growth, and by 1993 expectations had escalated to national and international levels. For managers faced with the practicalities of day-to-day conservation the bar has been raised too high, crippling parks under a Sisyphean task (Redford et al., 1998). Parks cannot be held responsible for alleviating every structural problem—from corruption to poverty, or from market failure to injustice.

In these three senses, then, the dilution of real protection afforded in so-called protected areas, the salience of the belief that protection comes uniquely through utilization, and the constantly elevated expectations of how parks must serve humanity, the very notion of parks and reserves has become imperiled. There is an almost existential sense to this threat: if conservation is only possible through utilization, parks are at risk precisely because their resources are not being harvested!

Much of the exchange over these issues has become emotional, angry, and deeply ideological. At its worst, biological and social scientists line up against one another and engage in predictable and deeply engraved interdisciplinary skirmishes. With regards to ends, social scientists deplore the neocolonial tendencies of conservation biologists and their imperialist yearnings to preserve the world in its "natural state"; regarding means, they question the right of outsiders, touting themselves as experts in wildlife, forestry, or conservation, to dictate how environmental objectives can be achieved (Guha, 1997). On the other side biologists accuse social scientists of arrogance, romanticism, empirical vacuity, and flagrant misrepresentation (e.g., Spinage, 1998). The debate is fundamentally agenda-driven, with little effort apparently exerted in trying to understand the other side.

In the meantime comparative studies at both local and national levels show that protected areas can do a good job protecting large mammals (Caro et al., 2000). Indeed, a worldwide analysis suggests that even "paper parks" are important in protecting biodiversity (Bruner et al., 2001). Questionnaires conducted with park agencies and NGOs of ninety-three parks in twenty-three countries show that parks, though often underfunded, understaffed, and challenged by immense pressure of human populations, are generally successful at mitigating the impact of land clearing, and reasonably successful at countering logging, hunting, grazing, and fire damage (Figure 2.5). Though ecological health as reported by protected area personnel may be somewhat biased, Bruner et al.'s study emphasizes the importance that parks still play in conservation strategy. Nevertheless, and this is a big limitation of the Bruner study, the key question of how parks compare with CBC systems and other IUCN categories under different kinds of conditions remains unaddressed (see chapters 10 and 11).

2.7 Conclusion

In this chapter we have looked at some major pendulum swings in the history of pro-

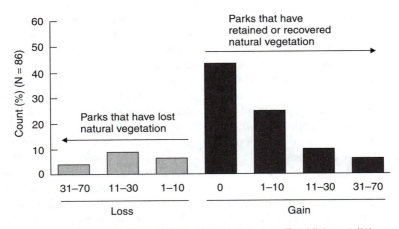

Figure 2.5. Vegetation Changes since Establishment of Eighty-six Tropical Parks.
The majority of parks have either retained or recovered natural vegetative cover. The effectiveness of parks at preventing land clearing was assessed by comparing the current extent of clearing with clearing at the time of park establishment. All parks analyzed are over 5000 ha, and mean park age is twenty-three years. Reprinted with permission from Bruner et al., Volume 291:125–128. Copyright 2001 American Association for the Advancement of Science.

tectionism as a means to conservation over the last thirty years, focusing largely on protected areas. The drama has been told as a history, but in fact there were always players who spoke out of time, such that undercurrents to particular debates often long preceded their notoriety. For example, it would be unfair to imply, contrary to many current claims, that conservationists in the colonial period were to a man (and only men made such decisions in those days) deaf to the needs and rights of the local people. As Neumann (1998, 106) discovered from his research into controversies surrounding Tanganyika's 1940 Game Ordinance, the colony's solicitor general lamented "there can be few such subjects which are capable of rousing such violent passions in the breasts of otherwise quite reasonable people than that of game preservation." The colonial administration was actually quite willing to concede limited hunting rights to villagers (besides, allowing customary rights to game and forest products had the added benefit of keeping African wages depressed); interestingly, it was not the Africans they

were fighting for the most part but a set of serried interests within European society, both in the colonies and at home, that were anxious to see the hunting rights of the "unsporting" (see Box 1.4) extinguished once and for all. Moving to more recent times, we saw that the MAB project too was far ahead of its time in emphasizing the traditional management of resources within zoned protected area authorities, and when we look at new conservation directions (chapter 11) we find that in fact many of the ideals of CBC are being incorporated into a new style of protectionism. In short, the pendulum shifts have in reality never been as sharp or regular as a historical account like this would imply.

There is nevertheless a clear trend to developments over the last thirty years—a marked shift away from protectionism toward utilization. This has sparked a splintering of conservation goals and means, such that the interests of the "conservation world" are neither homogeneous nor unified. With such diversity among so-called conservationists it is not surprising that debates can be found between preservationists and utilitarianists that are just as sharp, if not sharper, than over Hetch Hetchy. This heterogeneity has been exacerbated by the fact that an immense institutional apparatus has descended on the environment, much as it once did on development (Escobar, 1995), spawning new ranks of conservation players, very different kinds of conservation practice, and heaps of money.

And so, at the beginning of a new century, there is still no consensus over how to manage protected areas, and probably as much rhetoric and intolerance as that which sixty years ago troubled the Tanganyikan colonial officer quoted above. The ardor of commitment on both sides of the fence, together with the complexity of issues entailed, pose serious challenges for conservation in the future. There is, however, a brighter side. There is an awful lot of knowledge to draw on from the social and biological sciences that, if we can keep our heads cool, can guide a search for compromise. We now turn to this knowledge. In the following chapter we begin assembling a theoretical toolkit by examining how changing conceptualizations of natural ecosystems affect conservation strategies, specifically looking at the changes wrought in the discipline of ecology as conservationists begin to deal with human-dominated landscapes.

CHAPTER 3

The Natural Science behind it All

3.1 Introduction

Conservation biology grew out of the tradition of scientific wildlife management (described in chapter 1) and its marriage to the discipline of ecology in the 1970s climate of growing environmental awareness (see chapter 2). It has since become the scientific cornerstone of current conservation initiatives. This chapter offers a high-speed "overflight" of ecology and conservation biology, focusing first on the role of ecology in shaping the field of conservation biology, and then exploring some of the key tools and models that natural scientists bring to contemporary conservation problems. As such, the reader gains a practical grasp of the natural science landscape and how it will affect the journey ahead. Our goal is four-fold. First, we dispel the notion of conservation biologists as obsessed with preservationism, a notion still strongly held in some social science literature. In fact, we demonstrate how shifting views in ecology and conservation biology have laid intellectual foundations for community-based approaches to conservation, even beyond the ethical, pragmatic, political, and fiscal considerations explored in chapter 1. Second, echoing a theme from previous chapters, conservation is about choices. Conservation biologists' preferences have evolved over time, and even the mathematical approaches to prioritization build on fundamental value judgments in their algorithm building. Third, we

demonstrate how, as conservation issues are increasingly addressed at broad regional scales, social science becomes part and parcel of conservation biology. Conservationists must draw on the insights of experts in policy, law, culture, and economics; in this respect the stage for the rest of the book is set. Finally, we show that one's perspective of an ecological system affects the way one chooses to conserve it.

The chapter starts by going back more than 100 years to trace the development of the science of ecology through its Golden Age and its major legacies for conservation biology, most notably island biogeography theory (3.2). We then look at how ecologists began to recognize the role of disturbance and disequilibrium as key factors shaping ecological systems, and the implication of this shift in thinking for policy and management (3.3). Having identified these core concepts within ecology, we turn to the emergence of conservation biology as a new and applied discipline designed to avert the erosion of biodiversity loss (3.4), before describing in more detail developments that are most relevant to community-based conservation initiatives (3.5) and conservation planning (3.6). Our conclusion (3.7) can be summarized in a single sentence: how ecological communities and ecosystems work, or how we think they work, affects the way we go about conserving them. More particularly, radical shifts in how ecologists and conservation biologists think about the mechanisms maintaining species diversity, and the scale at which these occur, have direct implications for management decisions. All in all, questions about human populations and their significance for conservation will undoubtedly turn on local ecological characteristics and our perception of how they fit together.

3.2 From Natural History Comes Ecology and its Golden Age

The term *ecology* is often equated with environmentalism. This is an understandable mistake—after all, many ecologists are motivated by a desire to save the world, or at least a moral obligation to protect the study systems that provide their livelihood. But whereas environmentalism is a philosophical and normative ideology rooted in any combination of science, ethics, religion, or even guilt (see Box 1.3), ecology, the study of natural systems, is a scientific field.

As with most sciences, ecology has no distinct starting point. The word *ecology* comes from the Greek word *oikos*, which means house. While the early Greeks, and indeed a variety of ancient non-Western cultures, all had insights and worldviews that could be characterized as ecological insofar as they appreciate the interrelatedness among living things (and humankind's responsibilities within this web; see chapter 4), these views do not constitute the establishment of a science. The precursor to modern ecology lies in the study of natural history, which flourished in several western countries in the nineteenth century. The growing middle class to emerge from the industrial revolution helped provide an abundant supply of hobby and professional naturalists, eager to identify and catalogue the flora and fauna of Europe. As European colonialism established itself around the world, the naturalists' universe mushroomed. So rapid was this expansion that biblical scholars were forced to reconsider and re-

calculate the length of the cubit, the unit of measurement that described Noah's Ark, to make the accommodation of all these new creatures more plausible (Quammen, 1996).

As the number and variety of species recognized by natural historians rose, scientists began wondering what factors could create or maintain the patterns that were starting to emerge. For example, why do some species range throughout entire hemispheres while others have small, isolated distributions? This shift was an important one because it marked a transition from the descriptive focus of natural history, cataloguing and comparing species, to an analytical one concerned with why species are distributed the way they are. The German developmental biologist Ernst Haeckel modified the word *oikos* to form *Oecologie,* which later became *ecology* to describe this growing area of inquiry (Haeckel cited in Bowler, 1993). Though early studies were restricted to the physiological limits of species' ranges (indeed, in its infancy ecology was considered a subdiscipline of physiology), deeper and broader aspects of how species interact with their environments soon gripped the curiosity of early ecologists. For example, after determining that arctic foxes inhabited the arctic because more temperate latitudes were too warm and red foxes did not reach so far north because of the cold, ecologists began puzzling over the much bigger question of why there are species at all, and why they differ as they do. This demanded a far more comprehensive explanation than physiology could provide.

Enter Alfred Russell Wallace, Charles Darwin, and the theory of evolution by natural selection. Their novel insights were that variation among individuals within a population is the raw material of evolution. Selection pressure—for example, a tendency for lions to catch the slowest or least agile gazelles—then "acts" on this variation to remove unfavored (slower) individuals, leaving more "fit" (faster) individuals to survive and reproduce in greater numbers than the less fortunate members of their species. To the extent that traits that affect fitness are heritable (i.e., passed from one generation to the next), natural selection shapes and changes the full suite of traits (physiological, morphological, and behavioral) of any species over time. This simple recipe of variation plus selection creates the extraordinarily powerful force of evolution. Key to the theme of this chapter, the theory of evolution by natural selection dramatically broadened ecologists' notion of a species' "environment." Now it had to include competitors, diseases, predators, prey, and symbionts. The potential set of interactions considered within ecology now expanded well beyond the physiological requirements of individuals, and though it would take decades for ecologists to focus on all of these factors, the conceptual roots of modern ecology were in place.

By the last decade of the nineteenth century, ecology was in full stride. Plant, animal, and aquatic ecologists were exploring countless ecosystems. With a focus on ecological communities (groups of species living close enough to affect each other's abundance, distribution, and behavior), the early ecologists identified many of the fundamental patterns over which modern ecologists still ponder. The triumph of these early ecologists was their recognition that even with tremendous complexity and myriad ecological interactions, natural communities were dynamic systems, organized by regular and understandable forces. This emphasis on the prime movers in natural systems formed the backbone of ecology's "Golden Age" in the second half of the twentieth century.

Robert MacArthur, one of the most influential of all ecologists, exemplifies this period. By describing in meticulous detail how a suite of warblers used separate parts of conifer trees he explained an apparent paradox—how could a set of very similar species, all using the same trees, manage to coexist? The explanation for the persistence of each species lay in the partitioning of resources within individual trees (Figure 3.1). Soon a suite of other inter- and intraspecific relationships came under scrutiny: competition between different species, the dynamics whereby predators and prey interact, and the factors that keep a population at a given size at any particular time, many of which could be reduced to elegantly simple models. Numerous quantitative and experimental studies began to make sense of seemingly overwhelming diversity in ecological systems, and the science of ecology gained momentum with each pattern explained. Fundamental to these models was the notion that ecological systems exist in some sort of stable equilibrium, and that if they are perturbed they behave in predictable ways, either returning to their previous equilibrium or finding a new state (Box 3.1). Armed with these powerful new models and quantitative methods, ecologists shifted from description of the natural world to explanation, analysis, and even prediction of ecological phenomena.

The Golden Age left many contributions to modern conservation biology; the most significant of these was the equilibrium theory of island biogeography (MacArthur and Wilson, 1967). The idea on which island biogeography theory (IBT) rests is that ecological communities on oceanic islands are shaped by two fundamental processes: immigration of new individuals (and therefore species) from mainland sources, and the local extinction of populations on islands (Figure 3.2). The number of new species living on an island is, not surprisingly, a function of both its distance from the mainland and its size; more isolated islands receive fewer immigrants than less isolated ones, and smaller islands have higher rates of extinction than larger islands. These two, now famous, curves intersect at a dynamic equilibrium (similar to supply and demand in economics), and the equilibrium point dictates the number of species on an island. The elegance and simplicity of IBT undoubtedly helped spread its message: small and/or isolated islands are

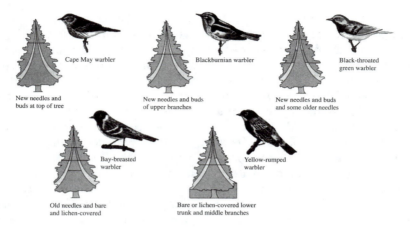

Figure 3.1. Warblers in Their Niches.
MacArthur's (1958) analysis of warblers' feeding positions revealed an organized and neatly partitioned world within each conifer tree. Such a stylized vision of ecological systems dominated ecology until the 1980s and persists in nonscientific perceptions of the field today. Reprinted with permission from MacArthur, RH, *Ecology* 39(4): 599–619. Blackwell Science Ltd., Publishers.

doomed to harbor fewer species, while larger and/or less isolated ones will be more species-rich. The implications of these ideas for national parks and reserves, islands of remnant habitat in terrestrial ecosystems often surrounded by a sea of humanity, would come to have great significance for conservation biologists (see below).

Two other contributions from this Golden Age are worthy of mention here. First, foraging theory (sometimes referred to as "optimal foraging theory") posited that evolution should favor individuals that collect food in the most efficient manner. This would afford them a variety of benefits like avoidance of predators by spending less time foraging or the ability to invest more energy on reproduction. In its early days foraging theory focused primarily on two questions: (1) which items should an optimal forager collect or ignore? (2) how long should a forager spend in different parts of a patchy landscape? These two questions and their application to resource-procuring strategies of humans in the context of natural resource conservation are explored in chapter 5. A second contribution, that came more directly from the fusion of ecology

with resource economics, was a model for determining maximum sustainable yield (MSY; Box 3.2). The MSY model identifies the largest harvest that can be taken from a population indefinitely, without driving the population to extinction. Though a heavy reliance of this model on equilibrial dynamics curtails its usefulness in many systems, it nevertheless continues to play an important role in guiding both the objectives and the pitfalls of sustainability (Milner-Gulland and Mace, 1998; and see below).

During the Golden Age the dominant view of natural systems was one of stylized, elegant, and finely tuned systems, working in efficient, regular ways, a worldview labeled "apollonian" ecology by R. P. McIntosh (1985). This view of ecology has persisted. Notions about the "economy of nature," "balance of nature," and "interconnectedness of all life" still pervade the popular view of ecology and natural systems, and are intimately embraced by many environmentalists. In many respects these worldviews yielded powerful new perspectives on ecological communities and natural ecosystems, but the perspective from within ecology was about to change.

Box 3.1. Stability and Equilibrium

Holling et al.'s (2002) schema illustrates different kinds of ecosystem dynamics. The ball, which symbolizes the state of the system, could represent the rate of population growth, a species' abundance, or some ecosystem process like productivity. The first scenario (A) illustrates a system that has no stable equilibrium, so perturbations cause the system to fluctuate in unpredictable ways. The second (B) has a clear equilibrium point to which the system reliably returns when perturbed. The third (C) illustrates an unstable equilibrium that, when perturbed, experiences positive feedback and spirals away from its initial state in a hyperbolic process of collapse. The fourth (D) represents a system with multiple states, only some of which are stable; such systems are characterized by a wide range of internal dynamics, and will exhibit periods of growth, stasis, and collapse. Human-caused disturbances are analogous to forces acting on the balls in the different scenarios. Obviously, human exploitation will have drastically different effects, depending on the situation.

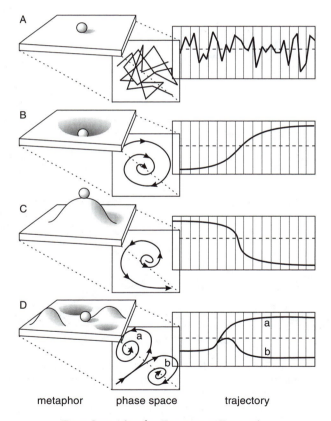

metaphor   phase space   trajectory

Four Scenarios for Ecosystem Dynamics.
Redrawn from Holling et al. (2002). From *Panarchy: Understanding Transformations in Human and Natural Systems*. Copyright Island Press.

(continued)

(Box 3.1 continued)

With the emphasis on disturbance and stability came a broader language with which to describe stability (Rahel, 1990). At the most basic level, persistence stability, all species within a community remain in the system through a specified time period. Rank stability implies that all the species in the system not only persist, but their ranks remain constant; that is, the most abundant remains more numerous than the second, and so on, even though their numbers may fluctuate. Numerical stability implies that all species persist and their ranks and numbers remain constant through time. To this nested hierarchy, one could add an even higher level: ecological stability, where rates of interactions between species also remain constant (Redford and Feinsinger, 2001). When we discuss the impact of conservation strategies in later chapters, it should be remembered that stability may mean different things to different players.

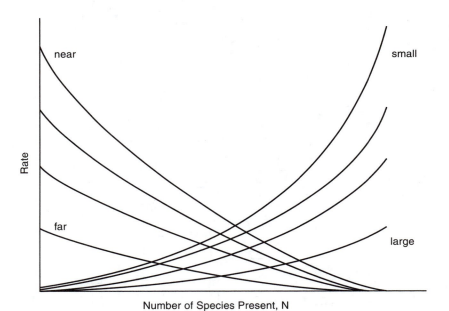

Figure 3.2. Dynamic Equilibrium of Island Biogeography.
The equilibrial species number is reached at the intersection point between the curve of rate of immigration of new species (not already on the island) and the curve of extinction of species from the island (families of falling and rising curves, respectively). An increase in distance (near to far) lowers the immigration curve, whereas an increase in island area (small to large) lowers the extinction curve. Accordingly, the curves for large and/or near islands cross at a higher N (number of species present) than do the curves for small and/or distant islands. MacArthur, R. H. and Wilson, E. O.; *The Theory of Island Biogeography*. Copyright 1967 by Princeton University Press. Reprinted by permission of Princeton University Press.

Box 3.2. Maximum Sustainable Yield

Biological populations can be thought of as natural capital, but living off the interest is not simple. This is because, unlike economic systems, the interest they yield is not proportionate to the size of the capital but is density-dependent, as captured in the logistic growth curve (Figure 3.3). The concept MSY, fundamental to the notion of sustainable harvesting (see Box 2.5), is that a population is held at a constant level by harvesting the individuals that would normally be added to the population, allowing it to continue to be productive. The crux of the model is that intermediate-sized populations with a high potential for growth produce the highest yields.

When populations are small, resource limitation does not constrain individuals' reproductive rates, but because there are few individuals, the overall yield is small. At intermediate population densities, also represented as half the carrying capacity (or K/2), individuals are still able to breed at their maximum rate, but because more individuals are breeding, the population produces more individuals to be harvested. Above K/2, density-dependent factors increasingly limit breeding until at K, there are no surplus individuals to be harvested and yield drops to zero.

An important feature of the MSY model is how harvested populations respond to environmental fluctuations or illegal offtake. Consider a population at N_b harvested at a constant harvest level H_1. If the population falls (due to poor rainfall or illegal harvest) this will ease density-dependent population regulation and increase

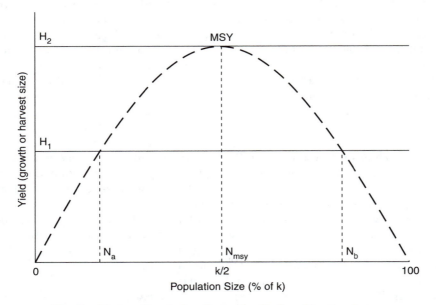

The Equilibrium Population Size under Various Hunting Rates.
The solid line represents the number of individuals removed from the population (H^1 and H^2). The dashed line is the population growth rate (r) at a given value of N, based on the logistic growth curve. Equilibrium population sizes occur when the number of individuals removed by harvesting equals the population growth rate (where the growth curve and harvest rate intersect). Harvesting at K/2 represents the level known as MSY.

(continued)

(Box 3.2 continued)

yield, moving the population back to N_b, a stable equilibrium (negative feedback creates inherent stability, as in scenario B, Box 3.1). The lower equilibrium (N_a) is not stable however; a population crash or illegal harvesting will decrease population yield farther below the current harvest level, creating a positive feedback (as in scenario C, Box 3.1). Harvesting at N_{msy} is also potentially unstable: a small decrease in the population can lead to positive feedbacks and extinction if the harvesting regime (H_2) is not reduced; thus, harvesting at N_{msy}, poised on a knife edge, is dangerous.

The MSY model, predicated as it is on equilibrium, became a casualty of the shift in ecology toward disequilibrium theory because of its sensitivity to disturbances (Ludwig et al., 1993); additionally, harvesting MSY is unsafe on economic grounds (Milner-Gulland and Mace, 1998). The parameters that go into estimating MSY are nevertheless important, and are still fruitfully used in providing guidelines for assessing the sustainability of particular harvesting practices in situations where harvesting intensity and resource abundance are assumed to be coupled. For example, when biologists propose a sustainable harvest of tropical forest game species, they need to know the maximum rate of population growth (r_{max}; for difficult-to-observe animals this must be estimated from body size, Robinson and Redford, 1991), and (ideally) how this is affected by population density (Bennett and Robinson, 2000a,b). Such methods can usefully be employed in field studies to determine which species are likely to persist under extraction and which are likely to be overharvested (see the Amazonian study in Table 4.2; also FitzGibbon et al., 1995; Freese, 1998; Robinson and Bennett, 2000). Additionally, the MSY model itself can be modified to harvest a certain percentage of the population rather than an actual number, thereby avoiding some of its instabilities. Generally, resource managers appreciate that the farther below MSY harvests are set, the safer the population (Milner-Gulland and Mace, 1998).

3.3 Things Get Messy: Disturbance and Disequilibrium

As in virtually all sciences, the number of ecologists has increased exponentially over the last five decades. With so many new practitioners, the theoretical predictions from the Golden Age could (and would) be tested across a broad array of diverse biomes and circumstances. With this trend came a similar increase in ecologists' ability to collect, store, and analyze quantitative data, effectively elevating the standards to which ecological theories were held. The result was that ecologists learned an awful lot about ecosystems, particularly how messy they really are. Soon the stylized models of finely tuned systems were overwhelmed by complex interactions, confounding variables, and environmental variation. Ecologists responded by focusing on the "noise" or disturbance within the model rather than on the "signal," and questioned the notion of equilibrium, both of which demanded more attention to scale.

Historically, ecologists had worked at small spatial scales and tried to minimize the amount of disturbance and heterogeneity that affected sampling designs (Diamond, 1986). But the relationships that appeared clean and strong within 5 m^2 plots became fuzzier and often broke down when considered in 100 m^2 plots, or when the same question was considered over a longer timescale. In many cases, the problem was that disturbances of all sorts crept into ecological observations: drought, fire, mechanical disturbance from animals, and countless anthropogenic effects. For example, in 1992 Hurricane Iniki eliminated five bird species and subspecies on the small Hawaiian island of Kauai (Simberloff, 1998a). With such cases ecologists began to realize that the disturbance (or "noise") itself may be a key factor or process structuring natural communities, not the mere inconvenience it had once seemed (Connell, 1978; Lubchenco and Menge, 1978; Sousa, 1979). From this intellectual ferment emerged the "intermediate disturbance hypothesis," which held that ecological communities, in the absence of disturbance, would eventually be dominated by the competitively superior species. Intermediate levels of disturbance can actually increase species diversity by allowing other, less competitively dominant species to persist. Illustrating the strength of the paradigm shift, Reice (1994) claimed that biological communities are always recovering from the last disturbance. Disturbance and heterogeneity, not equilibrium, generate biodiversity.

The emphasis on disturbance revived an old and unresolved debate within ecology: how populations are regulated. On one side were ecologists who argued that the primary factors limiting populations were density-dependent, such as food shortages or an overabundance of predators; in other words, they argued for an equilibrium model. On the other side were those who argued that disturbances and density-independent factors (i.e., things that happen regardless of how abundant a species is, like the weather) were the strongest factors limiting the numbers of animals. Density-dependent population regulation is quite straightforward, and can be captured by the famous logistic growth curve (Figure 3.3). By definition, density dependence produces negative feedbacks; when populations get large, food becomes scarce and the rate of population growth (generally labeled r) declines from

its maximum (r_{max}) until it eventually levels off at an equilibrium level where r is zero; the implications of these dynamics were explored in Box 3.2. The density-independent model of population regulation arose from empirical studies showing populations to be limited by infrequent, but severe events (Andrewartha and Birch, 1954; Davidson and Andrewartha, 1948). Under this "nonequilibrial" scenario populations would be constantly increasing from periodic reductions, as claimed by Reice. As with most debates in ecology, data were marshaled in support of both positions, but the balance began to tip in favor of nonequilibrium dynamics. And so continued the march away from the clean and simple relationships of the Golden Age into the chaotic and noisy ecological dynamics of the contemporary era.

These interlinked debates regarding population regulation, the extent to which systems are equilibrial or disequilibrial, and the importance of disturbance had, and continue to have, significant implications for conservation, particularly with respect to human impacts. First, noise and disequilibrium make the diagnosis of human impacts difficult. If ecological systems are inherently noisy, then an observed decline in a species' abundance may simply reflect environmental variation or some other noise, rather than anthropogenic effect or overharvesting, and be nothing to merit management intervention. The debate over global amphibian declines exemplifies the difficulties in detecting temporal trends in populations, trends that are notoriously variable over time and rarely reduce to a single cause (Alford and Richards, 1999).

A second, and related, implication draws the very nature of human impacts into question. An equilibrial, density-dependent system is likely to have tightly coupled interactions, where one species' abundance is closely linked to that of its competitors, predators, or prey, as envisioned by the Golden Age ecologists. Under this scenario increases in human numbers, or changes in their behavior, would likely have significant

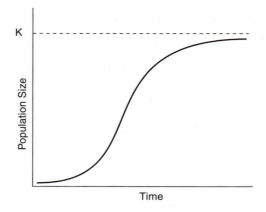

Figure 3.3. The Logistic Growth Curve.
A sigmoid or logistic growth curve is the most widely used population model in ecology. The model assumes that at low population densities, individuals can breed at their maximum rates, but because there are fewer individuals in total, overall population growth is slow. At intermediate population densities, population growth is at its maximum (the steepest part of the curve) because individuals are still not resource limited and there are many individuals breeding. At higher population density-dependent factors (like food shortage or a thriving predator population) cause the population's growth rate to decline until it levels off to zero at the population's carrying capacity (K).

impacts on the species being exploited and most likely on other species in the system; an example of this kind of thinking is presented in Figure 5.2. However, in systems that are disturbance-driven and disequilibrial, the impacts of people should be less significant. In the absence of tightly coupled interactions, the effects of a disturbance by one species (or a big change in its behavior) would be less likely to percolate through communities. The implications of these ambiguities for conservation and development policy are enormous. For example, the appropriate management of East African rangelands fundamentally depends on whether or not East African livestock populations are regulated by density-dependent mechanisms (Box 3.3).

A third major implication for conservation from these shifts in thinking can be seen in the rise of adaptive management. As ecologists came to understand that ecosystems are more complicated and unpredictable than imagined during the Golden Age, uncertainty took the front seat with respect to management. These intellectual shifts generated a new form of management designed to deal with, and indeed capitalize on, uncertainty: adaptive management. Here the goal was to emphasize flexibility and responsiveness, and to recognize that managers should design their interventions in such a way as to be able to learn from their mistakes (Box 3.4).

Fourth, the dynamics of ecosystems, and in particular how tightly coupled their interacting elements are, have implications for the design and effectiveness of reserves. Under equilibrial dynamics it is possible to conceive of a protected reserve (if large enough) that functions quite adequately in complete isolation and which, if by chance disturbed, will bounce back to its former equilibrium (as in diagram B, Box 3.1). However, a small, isolated reserve may no longer benefit from the large-scale ecological processes (e.g., migrations, grazing, or disturbance from animals) or biophysical processes (e.g., fire, flooding, or nutrient deposition) that originally shaped it. In these cases, the system may be in a state of disequilibrium, "wandering" irregularly (as in diagram A,

Box 3.3. Stability, Pastoralism, and Opportunism

The land use practices of pastoralists, or people who depend primarily on livestock for their subsistence, have generated intense debates about ecological systems, human behavior, and the economics of making a living in arid and variable environments. At issue is whether grazing by pastoralists' livestock leads to irreversible degradation of rangelands (often called "overgrazing"), or whether the dynamics of rangelands are largely independent of livestock numbers. On one side were countless biologists and ecologists. Their argument was that the rangelands of East Africa were so badly overgrazed (and lacking in biodiversity) that they might never recover (e.g., Lamprey and Yussuf, 1981; Lamprey, 1983), and graphic allusions were made to the Sahara Desert rolling inexorably southward. Appealing to an outmoded style of anthropological thinking, some have attributed this degradation and loss of productivity to pastoralists' irrational attachment to cattle as a marker of prestige (Brown, 1971). This viewpoint has led to the policy recommendation of forced destocking.

Anthropologists were shocked by both the ethical abuses with destocking policies and the inadequate understanding of human ecology. They argued that the large herds kept by pastoralist communities represent a rational, safe, and highly specialized way of making a living in arid and unpredictable environments (Homewood and Rodgers, 1987; review in Mace, 1991). Empirical data suggested that livestock mortality was more strongly affected by unpredictable droughts (particularly multiyear droughts) than by density dependence (Ellis and Swift, 1988), and Sandford's (1983) innovative simulation modeling showed that an opportunistic strategy (keeping large herds in order to maximize productivity when resources are available) was better suited to the highly unpredictable arid zones in which pastoralists struggle to survive than a conservative strategy (defined as maintaining herds at a relatively constant level, without overgrazing, through good and bad years alike). From this perspective, traditional pastoral production systems constitute a classic disturbance-driven, disequilibrial ecosystem (Ellis and Swift, 1988). Increasingly, anthropologists in the pastoral zones of Africa and elsewhere (Australia, Pickup et al., 1994) appeal to disequilibrial dynamics to account for the subsistence strategies of the communities they study, claiming that livestock numbers are uncoupled from long-term vegetation change.

Ecologists and anthropologists have different concerns (biodiversity conservation versus human welfare), hence their arguments and policy recommendations pass like ships in the night. But, and here is the rub, they both appeal to the science of ecology, anthropologists to disequilibrial models as justification of human behavior and ecologists (some at least) to equilibrial models as evidence of "overgrazing." The "new ecology" espoused by anthropologists (e.g., Scoones, 1999, and underlying Ellis and Swift's argument) is used to support the claim that any perceived ecological effects of grazing are spurious or simply insignificant, whereas biologists continue to lament irreversible vegetation changes and/or reduced productivity on savanna lands, which they attribute to overstocking (Prins, 1992; Ward et al., 1998). Accordingly, debates about pastoralist management persist, with biologists supporting destocking and social scientists proposing more extensive and secure land rights for pastoral populations, measures that will allow herders to respond to local conditions of range deterioration in their traditional manner (splitting their herds and grazing them in different areas). A balanced review of the evidence shows that plant community dynamics and productivity are not insensitive to grazing pressure; both density-dependent and density-independent factors are important, and their effects segregate by scale and season (Illius and O'Connor, 1999). More generally this debate illustrates the fundamental disconnect that can result from incompatible social and biological viewpoints, and more generally how strongly perceptions about the nature of ecological systems can affect conservation and land use policies.

Box 3.4. Adaptive Management

Complex and unpredictable systems can defy nearly every effort to manage them. Countless are the stories of well-intentioned natural resource managers who in attempting to solve one problem only create another, sometimes more serious problem, a classic early example being Leopold's dilemma over the Kaibab Plateau (see chapter 1). Historically, managers responded with more vigorous efforts to control the systems they were appointed to manage, but eventually most realized the futility of such an endeavor (Holling and Meffe, 1996). Rather than wasting resources on inviable management strategies, managers needed a way to learn from their mistakes, improve their understanding of ecosystems, and change course when necessary; and so "adaptive management" was born.

Adaptive management is built upon a few key principles:

1. *Uncertainty is inevitable.* Management decisions must be made arbitrarily or with incomplete information. Instead of relying on information-intensive strategies, managers should search for more robust principles to guide decision making. Adaptive management was widely adopted and refined in fisheries, where information about the size of populations, their age structures, and their movements is often limited.
2. *Management within bounds is more likely to succeed than managing for a narrowly defined state.* Conventional wildlife management sought to hold ecosystems in an optimal state (e.g., at the equilibrium point in scenario "B," Box 3.1), but this kind of management relies on strong negative feedback. A more robust and efficient approach may be to allow the system to fluctuate within broader boundaries akin to "natural" ranges of variation.
3. *Maintaining processes is more realistic than stasis.* Rather than manage for a narrowly defined stasis, adaptive management entails maintaining known key processes, like fire or flooding regimes affecting habitat quality. We see an example of this with the support of traditional burning practices in Australian national parks (Box 11.4). This focus on process is intended to keep the system regulating itself so that managers do not have to force the system in unrealistic directions.
4. *Use interventions as experiments.* Management actions should be structured in such as way as to reveal more information about the ecological system under management. This might mean applying management actions as replicated treatments instead of uniformly in a management area, or leaving some areas unmanaged to serve as controls against which to compare managed areas.
5. *Monitoring is critical.* Without tracking the effects of interventions on populations, communities, or ecosystems, it is impossible to know whether conservation efforts are having their intended effects. There is also no point monitoring if managers are not prepared to change course (i.e., adapt) when interventions are not working.

Two related topics are ecosystem management and the precautionary principle. Like the idea of managing for processes, ecosystem management is built on the

(continued)

(Box 3.4 continued)

notion that individual species or habitats cannot be managed in isolation from the context in which they are found. Ecosystem management seeks to account for these contextual issues by operating at larger spatial scales, incorporating a broader set of stakeholders, or considering entire systems in which conservation targets (like a particular species) are found. Related to the idea of adaptive management, but with its own roots, is the precautionary principle. Simply put, the precautionary principle reverses the burden of proof from conservationists to developers. Instead of conservationists demonstrating that an ecological modification will have a negative effect, their opponents must demonstrate the absence of such an effect.

Some biologists remain quite skeptical of adaptive management's current popularity. Accepting its value as a form of management that enhances flexibility and responsiveness, they note that an absence of fixed standards inhibits accountability of managerial actions. "With no general rules we risk the criticism that every project is claimed a success even if it results in environmental degradation or diversity losses" (Schwartz, 1999, 89).

Box 3.1) or rapidly shifting to another state (as in diagrams C and D, Box 3.1). With these dynamics in mind, reserve managers have increasingly turned their attention from within reserves to the activities over their borders, as we saw in chapter 2; although initially this was primarily to minimize or control the impact of human disturbance there is nowadays a sense in which managers of reserves (especially small reserves) must capitalize on processes that occur at scales that extend beyond reserve boundaries (see below).

Finally, the dynamics of an ecosystem make a tremendous difference in how people can make a sustainable living by harvesting natural resources. In equilibrial systems, people are able to closely track the availability of resources, as well as their impact on these resources. Under these conditions it should be quite easy for outsiders to monitor changes in the availability of resources (on the basis of the MSY model and its modifications; Box 3.2), or for local communities to evaluate their own impacts (and perhaps devise management institutions accordingly; chapter 6). But in systems subjected to periodic shocks and the vicissitudes

of climatic variability, it is vastly more difficult for both outsiders and the harvesters to track resources and monitor changes, especially since harvesting intensity and resource abundance may be uncoupled (see chapter 5). The shortcomings of the MSY model under the uncertainty associated with nonequilibrial conditions were pointed out by Ludwig et al. (1993). In a devastating critique of the model they showed how purportedly responsible exploitation systems based on MSY can be utterly unsustainable, using as an example the collapse of numerous fisheries worldwide (McGinn, 1998; Jackson et al., 2001). Increasingly, managers realize that they need some grasp of the real size of the population, the unpredictable shocks to which a population might be subjected, and, perhaps most important, the preferences and behavioral shifts of the human predators (Winterhalder and Lu, 1997) as will become clear when we consider extractive reserves (chapter 10) and the bushmeat trade (chapter 11). All in all, understanding ecosystem dynamics is critical when planning a reserve that allows utilization as a means to conservation, such as the anthropological or indigenous reserves

---

Box 3.5. Ecological Sustainability: Still a Slippery Term

Environmentalism has elevated the word *sustainable* into a household term, encouraging practices that are "environmentally friendly." As such, the word has lost nearly all its meaning. In Box 2.5 we discussed different facets of sustainability. Here we focus on its ecological sense, defining sustainability as the ability to maintain something undiminished over a specified time period. The devil lies in three particular details: the "something," the definition of "undiminished," and the time period.

For some, the "something" might be an open space with natural or semi-natural vegetation. The loss of a few species or the invasion of exotic species may be of little consequence for "sustaining" one person's vision of open space. At the other end of the continuum, "something" is an intact ecological web that includes not only all the species in the open space, but also their interactions. For adherents of this position, a significant decline in the density of one or two species might be "ecologically unsustainable" because the interactions have been compromised (Callicott and Mumford, 1997).

Equally treacherous is the definition of "undiminished" and the timescales implied. Recall the hierarchy of stability from Box 3.1, where species persistence is easier to sustain than rank, and rank easier than numerical stability. The question inevitably arises of whether it is the mere persistence of a species, its ranking, or its number of individuals that is to be undiminished. Finally, the most exploited loophole in the definition of sustainability is the question of timescale. Ecologists and environmentally minded economists are tempted to play the game of ever-receding timeframes to justify sustainable use over intensive exploitation. Take species like the African mahoganies (*Entandrophragma* spp.) that regenerate and grow so slowly. In theory a sustainable harvest could be achieved: only a tiny number of trees would be cut each year, but over hundreds of years the total value could, in theory, exceed the amount of a single harvest. While this reasoning is commonly used, it is badly flawed. People tend to devalue the importance of resources in the future (see chapter 5), and cannot be expected to defer harvests unless extremely lucrative alternatives are made available. In other words, even if ecological sustainability can be determined, social and economic realities must be addressed.

---

discussed in chapter 2. And it is precisely because of these issues, particularly the difficulty in monitoring human impacts on harvested resources over the relatively short lifespan of most projects, that sustainability (inherently a loose concept; see Box 2.5) retains its slipperiness even when considered from an exclusively ecological standpoint (Box 3.5).

The messy developments within ecology outlined in this section radically restructured the way ecologists thought about the world. This intellectual shift had major reverberations for policy and management style, changes that themselves had profound implications for the new community-based approaches toward conservation that began to emerge in the 1980s (see chapter 2). Many of these developments have been extremely positive—acknowledging the complexity of ecological systems, abandoning what were perhaps ecologically unsound policies such as destocking (see Box 3.3 again), and instigating flexibility (see again Box

3.4). But there are potential dangers. In fact, disequilibrium theory can be called on to justify all kinds of anthropogenic disturbance, on the pretext that everything is uncertain and feedbacks are weak (e.g., see Behnke, 2000); we saw, for example, in Box 3.3 the tendency of anthropologists to view the effects of livestock as spurious, insignificant, or even nonexistent, and in chapter 10 we will see the same issues arising over extractive reserves. Similar worries have been voiced by some regarding the slippery benchmarks and the lack of accountability inherent in the extremely popular move toward adaptive management (see again Box 3.4).

3.4 A Brave New Science: Conservation Biology

Up to this point, this chapter has focused on the ways that ecological thinking has affected the practice of conservation, and through the 1960s ecology was the only discipline to offer a scientific perspective on conservation. However, growing environmental awareness and the need for scientific information to slow the escalating loss of biodiversity led to the formalization of a new field. In the 1970s, the tradition of scientific wildlife management (described in chapter 1) married itself to the ecological tradition described earlier in this chapter, and conservation biology emerged as a new discipline in the early 1980s (Soulé, 1985).

The new field was distinctive in a number of ways. First, conservation biology, the child of 1970s environmentalism, is explicitly dedicated to halting the decline in biological diversity. Focusing on the conservation of *all* biological diversity, rather than on the effective protection and/or use of any *specific* resource, it is easily distinguishable from fields such as scientific forestry, fisheries biology, or resource economics. From the outset Soulé (1985) nominated four ethical postulates: (1) "diversity of organisms is good"; (2) "ecological complexity is good";

(3) "evolution is good"; (4) "biotic diversity has intrinsic value, irrespective of its instrumental or utilitarian value." In other words, conservation biology's original manifesto subscribes to the philosophy of inherent value, as outlined in chapter 1. In this sense, then, conservation biologists use science to inform and carry out a normative value structure. This disciplinary espousal of values was a novel departure, seen by some as "a brave new science" (Sarkar, 1998, 37). Indeed, in Sarkar's analysis the emergence of conservation biology closely mirrored the values of Westerners at a certain juncture in their history, with environmentalism, populist issues, and legislative concerns each playing a strong role in the development and direction of the new discipline.

A second novel feature of conservation biology is its synthetic nature. Because it is a mission-driven field, conservation biology has established a broad, catholic scope that includes investigations that traditionally fall within demography, ecology, genetics, public policy, and systematics, as anticipated in Soulé's initial picture (Figure 3.4). Furthermore, since Soulé's portrayal the social science slice has thickened, with anthropologists, sociologists, philosophers, political scientists, economists, legal scholars, educators, and others entering the conservation field.

The new discipline struggled to incorporate the messy developments within ecology (reviewed in the previous section), and profited enormously from theoretical and methodological developments in at least four subfields of contemporary ecology (Table 3.1). These fall under two dominant themes identified by Caughley (1994): the "small population" and "declining population" paradigms. The former focuses on the dynamics and persistence of small populations, while the latter examines the factors that reduce populations to small sizes in the first place. For instance, community and population ecology provide models for the design of individual protected areas and protected area networks, by providing in-

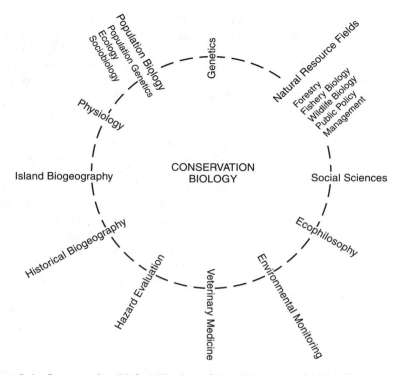

Figure 3.4. Conservation Biology Envisaged in 1985 as a Multidisciplinary Science.
Redrawn from Soulé (1985) What is Conservation Biology? *BioScience* 35: 727–734. Copyright
BioScience Magazine/AIBS.

sights into how ecological communities are structured and how differently sized populations prosper; landscape ecology examines the context and scales within which to analyze human disturbance and the design of individual reserves; and behavioral ecology yields the tools with which to examine decision-making processes and behavior of predators (including humans foragers and harvesters) and their prey (see chapter 5). Studies of genetics and systematics (not mentioned in the table) also play a role in determining the "species" status of a particular population—whether it constitutes a distinct species or incorporates unique genetic variants (a potentially important issue raised in chapter 1).

A final key feature of the new discipline, and one that grew directly out of Caughley's small population paradigm, was population viability analysis (PVA), which used quanti-

tative techniques to assess the extinction risk faced by small populations. Small populations were seen to be at risk not only because of their susceptibility to stochastic shocks, but because of their small gene pools. Attention accordingly turned to the importance of genetic factors in precipitating extinction. But things were to change as the debates became excessively academic. Though the role of genetics in conservation remains a vigorous area of interest, many researchers are now more concerned with Caughley's declining population paradigm, and are investigating the broader reasons for population declines rather than processes (such as inbreeding depression) that account for the final extinction (Caro and Laurenson, 1994; Harcourt, 1995).

And so the "fire brigade" period of conservation biology (Edwards and Abivardi, 1998), which had emphasized inventory, res-

TABLE 3.1
Ecology's Four Subfields and Their Contribution to Conservation Biology

	Subfield	Primary Analytical Focus	Areas of Relevance to Conservation Biology
Small Population Paradigm	Community ecology	Groups of species proximate enough in time and space to affect each other's distribution and abundance	• Island biogeography theory, and associated concepts of competition, predation, trophic webs, etc. • Indicators of species diversity
	Population ecology	Interbreeding groups of individuals	• Persistence and extinction of populations • Spread of genes and individuals between groups of subpopulations (metapopulations) • Population viability analyses • Dynamics of population growth, including determination of maximum sustainable yield[a] • Protected area design (population persistence
Declining Population Paradigm	Landscape ecology	Relationship between spatial patterns and ecological processes	• Protected area design (size, shape, connectivity, edge: interior ratios, etc.) • Emphasis on the broader-scale context in which conservation takes place
	Behavioral ecology	Individuals	• Individual foraging—where, how, how much (see chapter 5) • How foraging effort is related in fitness and status (see chapter 5)

[a] Span, both the "small" and "declining" population paradigms.

cue, and protection (e.g., Simberloff, 1998b), gave way to a broader focus on ecological interactions, often in human-dominated ecosystems (Redford and Padoch, 1992; Milner-Gulland and Mace, 1998; Prins et al., 2000), as might be expected given the emergence of new subfields within ecology (outlined above) and the developments in conservation policy (covered in chapter 2). In recent years the targets of conservation have extended from genes, populations, and species to communities, habitats, ecosystems, and ecological processes (Meffe and Carroll, 1997; Redford and Richter, 1999). Ac-

cordingly, some conservation organizations changed their names to reflect these shifts: the World Wildlife Fund became the World-Wide Fund for Nature (WWF), reflecting a focal shift from species (wildlife) to ecosystems (nature), and the IUCN added "and Natural Resources" to its name (before changing it altogether to the "World Conservation Union"). Further reflecting these shifts, conservation biologists increasingly turned to landscape-level conservation planning (Noss et al., 1997).

Despite this marked shift in the discipline's focus, there is still is a common perception (among social scientists and the general public) that conservation biology is closely allied with preservationism (as a goal) and protectionism (as a means; see chapter 2). As such, the discipline has become a target of the new wave policy critiques outlined in chapter 2. Guha (1997) admonishes Western conservation biologists for their obsession with the goal of wilderness preservation (see also Sarkar, 1999), emphasizing that as preservationists conservation biologists fail to recognize the conservation value of anthropogenic landscapes (see chapter 7) and consequently the potential role of local people in conserving their landscapes. Striving to distance the discipline from these critiques, Meffe and Carroll (1997, 179) stress the distinction between a "preservationist approach" and a "conservationist approach." In their view the conservationist approach allows change in a dynamic, evolving system (which might well include anthropogenic influences), whereas the preservationist approach entails maintaining the status quo and preempting evolutionary (genetic) change; in Meffe and Carroll's sense, the preservationist approach has stasis as an objective, and does not admit ordinary evolutionary processes. Clearly, the current paradigm in conservation biology is very far from the pejorative stereotype associated with preservationism, and in this respect the social science critiques are off base. Nevertheless, what constitutes an ordinary or legitimate evolutionary process is of course still at issue; is it "ordinary" for !Kung to hunt in the Kalahari Game Reserve with guns rather than bows (Box 8.3) or for European sportsmen to prowl the savannas of Africa (Box 1.4)?

Conservation biologists are also identified, again in the minds of critics, with protectionism (Guha, 1997). Here it becomes critically important to keep the ends and means of conservation distinct. While most conservation biologists might loosely refer to their goals as preservationist, not always making explicit the somewhat technical

distinction offered by Meffe and Carroll (above) or even explicitly preferring to exclude certain disturbances (particularly those that are of recent anthropogenic origin), many are becoming increasingly catholic with respect to their favored means of conservation, and eschew the reliance on protectionism as a unique conservation solution (see chapters 10 and 11).

3.5 The Fire-Brigade Discipline Comes of Age

As conservation biologists trained their gaze outside of reserves into human-dominated ecosystems, a number of new, intricately linked research topics emerged, many of which are directly pertinent to CBC initiatives. Here we trace landscape approaches, metapopulations, habitat fragmentation, corridors, source-sink systems, and buffer zones, each of which have implications for not only the planning and design of protected areas, but also broader landscape-level land use planning; we also consider the importance of scale.

Recognizing the importance of landscape configurations (how patches and ecosystems are arranged spatially), ecologists broadened their focus from *content* to *context*. In other words, they began to explore whether ecological processes that take place within a patch (or protected area) might reflect the properties of surrounding patches as strongly as those of its own patch (Lidicker, 1995). For example, flood control and conversion of land to cattle pastures in the headwaters of the Florida Everglades have reduced wetland coverage in this national park from 25 to 15 percent, directly affecting the diversity of aquatic invertebrate communities and thus the ecological integrity of the wetland (Steinman et al., 1999). This broadening in focus dovetailed with policy shifts in the late 1970s that switched attention from the inside of national parks to the activities outside their borders (see chapter 2). Conservation biology now needed more than ever before to incorporate the contributions of the social sciences, including economics, anthropology, sociology, and the political sciences into their solutions. Increasingly, the goal of conservation biology became understanding natural ecological systems well enough to maintain their diversity in the context of exploding human populations, changing livelihood patterns, global markets, and a shrinking base of natural resources and habitats.

With growing numbers of larger-scale analyses and their integration with human land uses, landscape ecologists and population biologists increasingly study the spatial structure of populations and how this structure affects population growth and persistence. Even undisturbed populations are unevenly distributed, sometimes forming discrete patches. This situation is exacerbated as habitats become fragmented. In this context ecologists revived the notion of metapopulations, a concept first formalized in the late 1960s (Levins, 1969). The critical feature of a metapopulation is that none of the population subunits are viable on their own (i.e., they become locally extinct), but because fluctuations in each subpopulation are out of phase with one another, extinction in one subpatch is balanced by colonization from other subpatches. The overall metapopulation persists so long as the subunits are linked by routes of immigration and emigration. Metapopulation theory has two very important implications. First, a temporarily unoccupied habitat patch may be important for the persistence of a metapopulation, particularly when it serves as a stepping-stone between other patches. Second, by affecting the flow of individuals between subpopulations, the "matrix" in which subpopulations are embedded will dictate the fate of a metapopulation.

The importance of matrix habitat and movement between reserves and subpopulations reaches well beyond metapopulation theory. Conservation biologists consider "connectivity," or the extent to which fragmented patches are linked, in a variety of

contexts. For example, global climate change may alter biophysical environments so drastically that in order to persist species and entire communities may have to shift their ranges to higher or lower elevations or latitudes. When this is the case, an insular "core" reserve embedded in a hostile matrix has little hope for survival. The giant panda provides a troubling example of such a situation (Figure 3.5). Pandas face numerous threats in the wild, and most are exacerbated by the fragmented nature of their distribution (Loucks et al., 2001). Because they live in montane habitats, a cooler or drier climate may push them to lower elevations, which are currently inhospitable. Pandas also face a lack of genetic diversity within each subunit, a problem that could be mitigated if greater connectivity allowed individuals to move between populations. Finally, panda's main food source—bamboo—flowers irregularly, only once every fifty years or so, depending on the species. When this happens all bamboo plants in a particular patch of forest flower and die, potentially leaving a panda subpopulation to starve. Without connectivity, subpopulations occupying patches smaller than the area of a typical bamboo flowering event will eventually die out themselves. For pandas to survive in such small patches, they will have to be able to reach younger patches as old ones die, and to do this they will have to move through human-dominated corridors. It is easy to see, then, how purely ecological considerations can force conservationists, even those with a purely biological focus, to deal with economic and social conditions outside of the technically protected areas (Peng et al., 2001).

In many cases the distinction between "core" and "matrix" may be less stark than in the panda example. Often, a strictly protected core and a heavily harvested or disturbed matrix appear as a single, unbroken block to a wild animal. Returning to the spatial structure of populations, it is easy to imagine a situation where a single patch of suitable habitat is large enough to harbor

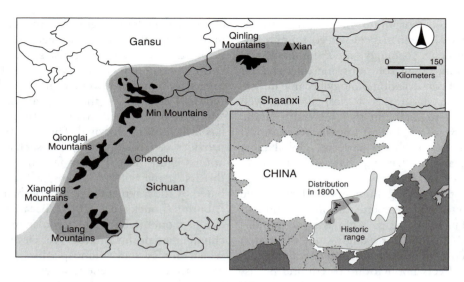

Figure 3.5. Historic and Current Distribution of Giant Panda.
China's estimated 1,100 wild giant panda (*Ailurpoda melanoleuca*) survive in only a fraction of their historic range (shown in inset), and are currently restricted to about twenty-four montane forest populations at the edge of the Tibetan Plateau, where most populations have fewer than fifty individuals, too few to be viable over the long term. Their future depends on increased protection and restoration of corridors among remaining forest fragments. Thus, the most urgent issue is conservation of remaining habitat outside of the reserve system. Recent policy changes in China banning logging in natural forests will provide strict protection to all the remaining forests throughout the panda's range. Reprinted from Loucks et al., Volume 294: 1465. 2001 American Association for the Advancement of Science. Copyright IUCN.

a viable population, and also to supplement subpopulations in surrounding patches. This situation is called a "source-sink" system when populations in the surrounding sub-patches depend on the large, high-quality "core" patch for their persistence. Source-sink systems can be important in cases of human exploitation. For example, a group of hunters may heavily harvest wildlife (in an apparently unsustainable way) within a few hours' walk of their settlement, but if this area is contiguous with an unhunted area, natural movements of animals from the source to the sink may render the heavy offtake sustainable when both the harvested and source areas are considered together (see Novaro et al., 2000). The implications of this kind of situation are discussed further in the analysis of traditional hunting in the Neotropics (Box 5.5). More broadly, these

scenarios highlight the importance of examining ecosystem dynamics at a broad landscape scale, as well as of determining where exactly human foragers go to harvest their resources (chapter 5) and what precise property rights they have to the territory that lies beyond the zone immediately adjacent to their settlements (chapters 4 and 6). We link these ideas to the initiatives of local communities to demarcate their own non-hunting areas in chapter 11 in an attempt to maintain the sustainability of their harvests.

Of course, hunting is not the only anthropogenic effect influencing remnant habitats and protected areas. Numerous studies show that the outer boundaries of protected areas are exposed to all kinds of disturbances not experienced in the interior, disturbances ranging from (in the case of a forest) drying from increased sunlight, treefalls from ex-

Figure 3.6. Edge Effects and Swidden.
Swidden cultivation, known locally as *tavy*, on the edge of Ranomafana National Park, Madagascar. Photo, Monique Borgerhoff Mulder.

posure to wind, invasions of exotic species, to selective cutting (e.g., Gascon and Lovejoy, 1998). All these processes are referred to as edge effects. While edge effects may sometimes increase species diversity (through an influx of new species), they more often threaten the integrity of protected areas, particularly when they consist of human activities like firewood collection or swidden agriculture (Figure 3.6). Buffer zones, introduced in the context of biosphere reserves (chapter 2), are the primary tool reserve designers use to mitigate edge effects. Theoretically, buffer zones insulate a protected area by prohibiting potentially destructive activities like wood-cutting or agriculture, and tolerating less invasive activities such as beekeeping or seasonal regulated hunting offtake. In practice, however, buffer zones are sometimes created from existing reserve boundaries inward, effectively diminishing the area under protection. Buffer zones can also be politically costly, since to excluded populations they can look indistinguishable from expanded protectionism.

All of these issues come to a single focus

when we consider the general principles governing reserve design: larger reserves are more efficient at conserving biodiversity than smaller ones; landscape heterogeneity in the reserve is a good thing; connected reserves are generally preferable to isolated ones; and reserves must be managed as an integral part of their surroundings (Figure 3.7). In the most recent edition of Primack's (2000) textbook, human presence is added as a new design feature (even though no explicit rationale is given), no doubt reflecting the developments in policy and practice traced in chapter 1 that began with biosphere reserves.

We end our discussion of conservation biology's coming of age by highlighting the importance of scale in all aspects of conservation biology, a feature that has been developed primarily within the subfield of landscape ecology. Scale refers to the "spatial or temporal dimension of an object or process" (Turner et al., 2001, 27). Ecological processes can operate at tiny scales, for example, within thimblefuls of soil, or at enormous scales encompassing continents

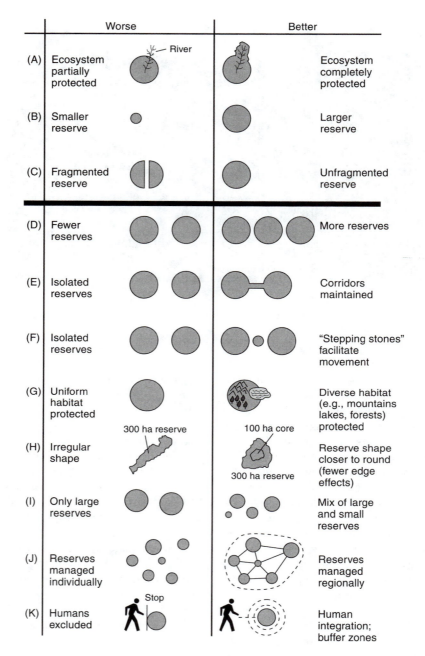

Figure 3.7. Reserve Design.
Reprinted by permission from Sinauer Associates, Inc. From *Primack: A Primer of Conservation Biology.*

and ocean basins. Scale has two components: grain and extent. Grain refers to the maximum resolution at which one looks at a phenomenon or question, and extent to the total area (or time) considered. For ecologists these two dimensions represent a critical trade-off. Fine-grained analyses have better (higher) resolution and more information, but require greater amounts of data and, because they tend to focus on smaller extents, may not be representative of the broader area. In contrast, coarse-grained analyses cover larger areas and are more representative, but by averaging over these areas key processes are potentially obscured. Large extents generally necessitate coarse grains and vice versa, but with more efficient data acquisition and more powerful tools for data analysis, storage, and manipulation, ecologists are less tightly constrained by grain/extent trade-offs than they used to be. In evaluating the results of ecological studies, attention must be given to these trade-offs. For instance, a species or population may be threatened locally because of overharvesting, as with some primates in the Amazon (see chapter 4), whereas at a broader scale its viability is compromised almost entirely by habitat conversion. When the grain is fine and extent is small, we need to ask: are the results generalizable, or do they simply reflect local conditions? Alternatively, when the grain is coarse and extent is wide, we need to question whether the study has uncovered mechanisms that are appropriate to smaller-scale questions.

Ecologists are committed to understanding how local interactions at one level of biological organization (small scale) influence phenomena or patterns at higher levels of organization (large scale), and to determining how different processes may be entailed at different scales. Accordingly, there is no "right" or "wrong" scale, only more and less appropriate ones for a given question (see chapter 11). Here we point out only that a relationship or process that seems clear at one scale may become less clear, disappear, or even reverse at larger

spatial or longer temporal scales. Questions of scale weave back and forth across the human/biological interface, and are returned to at several junctures in this book.

3.6 Conservation Planning

Recognizing the complexity of ecological systems and the need to work and plan at larger spatial scales, conservation biologists often adopt a "coarse filter" approach (Hunter, 1991). Rather than focus on individual populations or species occurrences, this approach seeks to represent habitat and ecosystem types and in doing so, protect their constituent species. In this context a new challenge arose: how to prioritize the relative conservation value of different ecosystems when value can be assessed by a whole range of independent criteria, including how endangered the habitat is, how diverse the species it harbors are, and how critical its ecosystem functions are (Schwartz, 1999). Inevitably, individuals and institutions hold different views on these issues, reflecting diverse values, motivations, and disciplinary specializations. Accordingly, much attention is now given toward developing objective technical methods for determining conservation priorities (Box 3.6). These methods do not eliminate subjectivity, since choices of methods and conservation targets ultimately reduce to value judgments (Robinson and Redford, 2004; see also Colchester, 1997); nevertheless, systematic methods to identify conservation priorities are a major step forward because they expose the logic underlying management decisions and increase transparency in deliberations, as the issue of hotspots illustrates.

The species-based hotspot (see Figure 1.1) approach has usefully, if somewhat coarsely, served to identify twenty-five global sites where there are high concentrations of endemism (based primarily though not exclusively on plant species) and high levels of threat (a hotspot must have lost 70 percent

Box 3.6. Prioritizing Conservation Efforts

With respect to developing objective methods for determining conservation priorities, several have been suggested—those concentrating on individual species or populations, habitats, or ecosystems, and those that approach the problem from a socioeconomic perspective.

Individual species approach: Focusing on assessing threats of extinction rather than prioritizing conservation efforts per se, Mace and Lande (1991) devised a widely accepted set of criteria for classifying species as critical, endangered, and vulnerable on the basis of the likelihood that if current circumstances prevail the species will go extinct within a given period of time. Their method depends strongly on population viability analyses, and does not include economic, political, and logistical factors that might affect the success of restorative action, nor the taxonomic distinctiveness of the species under review. Rather, it provides an objective assessment of extinction risk that can contribute to prioritization efforts. Species-based approaches to prioritization are key to saving certain highly endangered species (Ceballos et al., 1998) but cannot, by definition, include all species. Furthermore, when single focal species (or suites of focal species; e.g., Lambeck, 1997 who emphasizes focusing on species most vulnerable to disturbance) are used as surrogates for biodiversity, questions inevitably arise over what is the best, or most appropriate, surrogate species (van Jaarsveld et al., 1998). The answer to this question depends on what we know about cross-taxon congruence (in other words, how the diversity of, say, butterflies maps onto the diversity of, say, birds; Howard et al., 1998), and on how effective so-called surrogate species (such as indicator, umbrella, and flagship species) are as a shortcut to monitor or solve conservation problems (Caro and O'Doherty, 1999).

Habitat approach: Habitat conservation plans (Noss et al., 1997) are intended to provide comprehensive coverage of multiple species within the habitat. This approach focuses on habitats as harbors for species, rather than as providers of ecosystem services. Such a strategy is particularly appropriate when knowledge about individual species is lacking (Hunter, 1991). Planning may also seek to protect a minimum number of occurrences of each habitat type (called "representation"), or focus on particularly important, irreplaceable ones (sometimes called "special elements").

Ecosystems approach: This ideally combines species, populations, and habitat approaches with ecosystem services (Ayensu et al., 1999) but is loosely defined. In practice it is extremely difficult to rank on the same axis ecosystems according to widely differing criteria (e.g., carbon sequestration and biodiversity); as a result, conflicting or even contradictory conservation priorities emerge. Ecological economics provides a potential framework for resolving these discrepant priorities (though not all criteria can be identified and compared using economic terms; see chapter 9).

Using algorithms within habitat and ecosystem approaches: If biologists want to represent the maximum biodiversity in protected areas (Pressey et al., 1993), conservation planners must prioritize sites on a diverse suite of criteria—the number and rarity of species the site harbors, its vulnerability (or level of threat), the irreplaceability of the particular site, how well the ecosystem is represented else-

(continued)

(Box 3.6 continued)

where, and its complementarity with other protected sites (Margules and Pressey, 2000). None of these criteria are adequate on their own; furthermore, different criteria (reflecting different conservation goals) lead to distinct and often nonoverlapping areas to be designated for protection (Sarkar, 1999), only exacerbating the competition among conservation proponents for limited funds. Accordingly, conservation planners have proposed various algorithms to rank areas for protection at local, regional, and even global scales. Thus, for the United States Noss (2002) develops an algorithm to work out the different ways that various combinations of rare, irreplaceable, and representative species can be protected in a reserve system, using simulation to minimize the cost of such a reserve system—cost in this algorithm is the amount of land that must be bought, its price, and its connectedness with other protected areas. Critics of such approaches suggest that urgent conservation issues are more effectively dealt with by the pragmatism of managers, but in reality the expertise and judgment of managers is enhanced rather than replaced by algorithms (Pressey and Cowling, 2001).

Socioeconomic context approach: In prioritizing conservation effort it may also be necessary to balance representation and conservation efficiency against a suite of socioeconomic costs such as the economic and political consequences of converting land into a reserve (Margules and Pressey, 2000). For example, in Papua New Guinea areas were set aside for reserves on the basis of not only maximizing biodiversity but also minimizing lost opportunities for timber extraction, and avoiding areas of high population density and intensive community agriculture (Faith et al., cited in Margules and Pressey, 2000; see also Faith et al., 1996; Balmford et al., 2000).

or more of its primary vegetation to qualify; Myers et al., 2000). There are, however, limitations to this approach: only terrestrial sites are included, and since the criteria are species-based we have no certainty that the diversity of ecological processes is represented (Balmford et al., 1998). More worrying, however, are the haunting questions evoked by any such prioritization: in focusing all conservation effort on, for example, these twenty-five hotspots, what are we missing and what are we giving up? For example, Dobson et al. (1997) find little evidence in the United States that the hotspots for different species overlap; thus, hotspots for endangered plants are most efficient in protecting large numbers of species, whereas hotspots for endangered birds and reptiles better protect other endangered species. Given these kinds of observations it is not surprising that there are many other priority-setting efforts afoot, in addition to those of Myers and his collaborators at Conservation International (CI), including WWF-US, BirdLife International, IUCN, World Resources Institute (WRI), and The Nature Conservancy (TNC). These employ a range of different scales, use different criteria (or combinations of criteria, such as representation and biological importance rather than endemism and threat), and focus on different conservation targets (Redford et al., 2003; at times entailing unnecessary duplication of effort, Mace et al., 2000).

Priority-setting exercises can also take place at smaller spatial scales. Particularly interesting from the perspective of both CBC and landscape ecology are those rankings that take into consideration the opportunity costs to the people who lose their land. In the algorithm used by Noss (2002) (see Box 3.6 again) the only cost (to be minimized) was that of buying the land, which is a function simply of the amount of land needed for the reserve system and its price. How well land prices reflect the opportunity costs for those who lose their land will vary. In highly monetarized economies it may provide quite an accurate measure of what is being lost. But as we see in our discussion of environmental valuations (chapter 9), compensatory payments may be gratefully accepted in the developing world despite the fact they fail to compensate adequately for opportunity costs. These opportunity costs might entail productive use of the land that local communities had never heretofore considered, as yet unappreciated touristic potential or irreplaceable cultural sites—all very difficult things on which to put a monetary value.

The picture becomes even messier when the deeper socioeconomic time horizon is considered. Nagendra (2001) proposes incorporating into the prioritizing algorithm not only the cost of landscape transformation but also the probability of further land transformations over a specified time frame. Thus, in the Western Ghats of India, once a forest (high conservation value) is reduced to a savanna (reasonably high value) this savanna is very likely to be converted to rice paddy within a few years (very low value). Under these circumstances reducing forests to savanna in a rice-growing area is much more costly to biodiversity, in the sense of a greater net loss within the specified time frame, than in an adjacent area where land is not under pressure from rice. Incidentally, while Nagendra's method suggests that conservation priority should be given to the area under pressure from rice (to maximize the effectiveness of the conservation effort), a triage-style argument could also be extended to focusing on areas with less human pressure and more likely long-term chances of success.

Incorporating social and economic data may help conservation practitioners be more efficient by focusing on less expensive, or less intensely contested, areas, but these considerations raise an inevitable trade-off. To the extent conservation planners use such data to set conservation priorities, they may risk watering down more objective (though still poorly agreed on; see Box 3.6 again) criteria for biodiversity conservation to meet more politically expedient goals. This is one of many slippery slopes some conservation biologists are nervous about approaching. A very similar dilemma applies to the species-based prioritization approach too. For example, in chapter 4, we see anthropologists and ethnobotanists claim that traditional horticulturists' careful management of their multispecies forest plots is equivalent to biodiversity conservation. While it is true that such management practices do protect certain plants, undoubtedly these are species that are viewed as economically or culturally valuable. What about rare and apparently useless plants that are not human commensals? Are they to be given such protection? As economic, political, and cultural considerations are brought to the prioritization of conservation efforts, purely biological priorities tied more closely to species, populations, habitats, and ecosystems may be sacrificed.

All in all, whatever technical means are devised for prioritizing areas for special conservation efforts, a few general points emerge. First, as we noted at the outset, these methods do not eliminate subjectivity since choices of conservation targets and methods must be determined by the people charged to solve a particular problem; they are useful, however, in forcing an explicit articulation of conservation targets and the principles used to identify priority areas (Groves et al., 2002). Second, priorities tell conservationists where to do conservation, but do not indicate how to get it done. Third, there are no hard-and-fast rules for deciding between these strategies, so conservation practitioners must proceed with constant vigilance and interpret the results of priority-setting exercises with caution. Fourth, increasingly these exercises provide a backdrop for local and regional priority-setting workshops, offering a useful framework within which trade-offs can be posed and negotiated. Finally, despite the evident broadening of conservationists' concerns to embrace the full range of biodiversity (as we outlined above and in chapter 2), large charismatic mammals are still in practice major targets of conservation effort (e.g., Seidensticker et al., 1999). For some species this is justified because they range widely, require large blocks of undisturbed habitats, or may be severely persecuted (Woodroffe and Ginsberg, 1998). In other cases large beasts dominate conservation thinking to satisfy public expectations and to draw funds from the general public (Bonner, 1993). However, this strategy can sometimes backfire, when, for example, the charismatic species tolerates higher levels of threat than other species. Also, species like elephants or tigers may be considered charismatic on one side of the globe, but widely hated as dangerous crop-raiders or livestock thieves in their natural habitats. Focusing on these species may immediately alienate the human populations affected by the conservation initiative (see chapter 8).

3.7 Conclusion

There are four main conclusions we draw from this chapter. First, ecology matters! All management systems, from that of small-scale traditional societies to large-scale global policy, have inherent ecological models at their heart. How ecosystems (and ecological communities) work, or how we think they work, affects the way we go about conserving them. This is especially true with respect to utilization of natural resources and monitoring of impacts. Questions about local people and their significance for conservation will undoubtedly turn on local ecologi-

cal characteristics. Second, by expanding the scale at which conservation dilemmas are examined, social science has become part and parcel of conservation biology, offering distinct niches for lawyers, anthropologists, policy experts, and the like. Third, as a consequence of the importance of ecology and the number of unanswered questions, conservation biologists can never afford to lose sight of the basic science, despite the applied nature of the discipline. Finally, conservation biology offers no clear agenda about prioritizing conservation efforts. Whether decisions are in the hands of amateur conservationists, professional managers, policy advocates, or just ecologically aware members of local communities, it is quite clear that conservation priorities boil down to value judgments. This is consistent with the historical overview of the previous two chapters. Conservation biologists' preferences have evolved over time: first they wanted to protect valuable resources, then save threatened species, more recently intact ecosystems, and currently they debate how to rank these. Local communities' preferences are also dynamic but, as we will see in later chapters, they are often more concerned with preserving resources like crop diversity, the quality of watershed forests, and other broad ecological processes than with the fate of individual species. As conservation biology broadens its agenda there is potential for a congruence of interests to emerge, but equally as it expands into human-dominated ecosystems there is much scope for conflict.

It is easy to leave this chapter feeling overwhelmed. Ecosystems are messy, noisy, and larger than at first envisioned. Conservation biologists are increasingly forced to work at larger scales, address more threats, and reconcile more interests. Even more daunting is the fact that just describing the effects of human land uses is not enough. We must understand the dynamics of the human dimensions of ecosystems if we are to reduce threats to biodiversity successfully. But fear not! The chapters that follow

CHAPTER 4

Indigenous Peoples as Conservationists

4.1 Introduction

The future of indigenous peoples and the fate of the Earth's remaining areas of natural beauty are intricately entwined. This is largely because they inhabit vast zones of Africa, Asia, the Americas, and the Pacific into which larger-scale metropolitan societies, with their associated economies, have yet to penetrate fully. The claim is often made that these indigenous (a term that includes native, traditional, rural, and local) communities are natural guardians of biodiversity. This assertion provides both a practical and a moral foundation for new policy initiatives that treat local people as partners and allies in conservation efforts. It is of practical significance because, as we saw in chapter 2, with exclusionist policies losing favor policy makers are seeking the active cooperation of local communities in attempts to conserve biodiversity. It is of moral significance because increasingly the romantic notion of humans living in harmony with the natural world is used as a moral touchstone for calls to environmental action; indeed, this idea lies at the heart of many modern environmental philosophies (chapter 1).

In this chapter we scrutinize the role of traditional, local, and indigenous groups in protecting biodiversity within their territories, in light of recent claims that this is a naïve and romantic view of human nature. The chapter starts by noting that much of the remaining biodiversity is found in areas inhabited by small-scale traditional and indigenous cultural groups (4.2), and some of the reasons why this might be (4.3). We follow this with a summary review of past and present impacts of human foragers on the natural world that tells a somewhat different story (4.4). This in turn provokes a more analytical look at where the notion of "indigenous conservationist" comes from (4.5), and a reconsideration of the relationship between cultural and biological diversity (4.6). We conclude that although the so-called ecologically noble savage debate has usefully exposed problems with romanticizing traditional communities, it has generated acrimonious debate among scholars who otherwise share strong commitment to indigenous peoples' welfare and the health of their lands (4.7). However, wherever different conservation practitioners fall in this still ongoing debate, most appreciate that the ecological footprint left by traditional and relatively undeveloped rural communities may be a small price to pay for their increasingly aggressive role in protecting broader habitats. This renders them invaluable partners in national and international conservation initiatives, as discussed further in chapter 8.

4.2 Cultural and Biological Diversity

Of the 6,526 languages currently spoken in the world, 5 percent are classified as virtually extinct (i.e. they have less than 100 native speakers), and between 52 and 75 percent are spoken by communities of 10,000 or less, with many of these heading for virtual extinction. In fact, almost half the people in the world speak, as a mother tongue, one of the world's ten largest languages (Harmon, 1996), and recent analyses suggest that language diversity is more threatened than that of birds or mammals (Sutherland, 2003). Since language is intricately linked to culture (though not necessarily coterminous with cultural difference), this erosion of linguistic diversity signals a decay of cultural diversity. In the last decade we have witnessed a global trend of rollercoaster proportions toward cultural homogeneity, captured in Barber's (1995) allusion to a McWorld. Indeed, you can trek deep into the rainforest only to find indigenous inhabitants watching a soap opera on a car battery-powered television (Bennett, 1996).

On grounds of human rights alone, the disappearance of cultural diversity is a sad and dangerous state of affairs. Furthermore, whether or not this loss is viewed as a social problem, some argue that it may be contributing directly to the erosion of biodiversity. This is because the knowledge embedded in these disappearing communities contains a reservoir of skills, institutions, and values that have evolved over millennia, much of which concerns environmental practices. As we discuss later, cultural diversity codes a vast store of knowledge, just as biological diversity reflects a rich pool of biotic wealth and information.

Intriguingly, it turns out that cultural and biological diversity often go hand in hand. As we saw in chapters 1 and 3, species richness is highly patterned, concentrated in areas of high human (Cincotta et al., 2000; Balmford et al., 2001) and linguistic (Moore et al., 2002) diversity. Over a decade ago the National Geographic Society produced maps of Central America which show that virtually all the remaining indigenous groups live in two areas, the jagged volcanic high-

lands of Guatemala and the heavily forested Caribbean coast stretching from Belize through to the border of Panama with Colombia. These zones also harbor an abundant and diverse flora and fauna, constituting a vegetational corridor between the ecosystems of North and South America (Nabhan, 1997).

Are these maps representative of a more general pattern? With the goal of setting conservation priorities across the whole of South and Central America, Wilcox and Duin (1995) made a broad examination of whether biological and cultural diversity are related. They found that the pattern is quite robust across the whole region, not just Guatemala. Habitat types used by the largest number of distinct ethnic populations were also those ranked highest in terms of biodiversity and utility value (Box 4.1). Looking more specifically at endemicism (languages and species restricted to a single country), linguists find the same pattern (Harmon, 1996, 97). Sixteen of the twenty-five countries highest ranked in terms of endemic mammals, reptiles, and amphibians were also so ranked for endemic languages (Table 4.1). These findings are intriguing, and have been replicated at broader scales (Maffi et al., 2000; Moore et al., 2002), albeit with exceptions; for example, restricting analysis to contact period Native North American populations, Smith (2001) finds little evidence for correlated diversity. The overall pattern is nevertheless open to a wide range of quite loose interpretations.

4.3 Guardians of Biodiversity

An appealing explanation for the apparent coincidence of cultural and biological diversity is that biodiversity persists because of cultural diversity. The general argument for this position is that small-scale, traditional and indigenous communities protect the biological resources on which they depend for subsistence and survival by means of customs and practices that limit or at least dis-

Box 4.1. Cultural and Biological Diversity in Central and Southern America

Wilcox and Duin (1995) divide South and Central America into eleven major habitat types (MHTs), based on 182 terrestrial ecoregions developed in biodiversity priority-setting initiatives commissioned by the World Bank and the U.S. Agency for International Development (USAID) (see Olson and Dinerstein, 1998). They rank these MHTs in terms of cultural diversity (measured as the number of distinct languages and/or ethnic indigenous populations living in that MHT, and shown as dots in the figure), biological diversity (measured as the number of different base ecoregions within the MHT, and shown as bars in the figure), and six indices of utility value (UV, not shown in the figure). The UV indices measure the importance of the MHT in sequestering carbon and in supporting genetic diversity (with respect to trees, plants, and domestic crops and animals). From the data in the original paper we calculated one-tailed Spearman correlation coefficients (n = 11) between

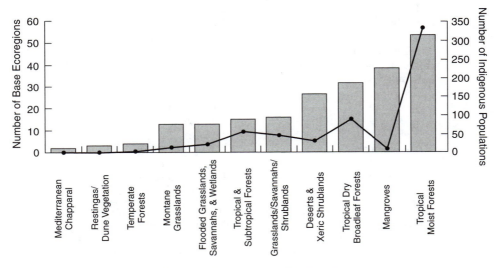

Indigenous Diversity and Biodiversity.
Plot showing the association between the number of base ecoregions (a measure of biodiversity) and number of indigenous populations across major habitat types. Courtesy of Cultural Survival.

cultural diversity and both biological diversity and the six indices of utility value. Cultural diversity is significantly associated with biodiversity ($r_s = 0.75$, $P < 0.01$, as shown in the figure, with biodiversity as a bar graph and indigenous populations as a line graph), as well as with various measures of UV (forest tree genetic resources, centers of plant diversity, and origins of important crop species). Other measures are positively but not significantly associated with cultural diversity. Studies like this one, while influential, are hard to interpret because diversity scores, for cultures or species, may be confounded by the size of the areas under consideration. It is possible that environmental variables are responsible for both cultural and biological diversity (Moore et al., 2002; see section 4.6)

TABLE 4.1

Endemism in Language and Higher Vertebrates[a]

Endemic Languages (N)	Endemic Vertebrate Species (N)
Papua New Guinea (847)	**Australia (1,346)**
Indonesia (655)	**Mexico (761)**
Nigeria (376)	**Brazil (725)**
India (309)	**Indonesia (673)**
Australia (261)	Madagascar (537)
Mexico (230)	**Philippines (437)**
Cameroon (201)	**India (373)**
Brazil (185)	**Peru (332)**
Zaire (158)	**Colombia (330)**
Philippines (153)	Ecuador (294)
United States (143)	**United States (284)**
Vanatu (105)	**China (256)**
Tanzania (101)	**Papua New Guinea (203)**
Sudan (97)	Venezuela (186)
Malaysia (92)	Argentina (168)
Ethiopia (90)	Cuba (152)
China (77)	South Africa (146)
Peru (75)	**Zaire (134)**
Chad (74)	Sri Lanka (126)
Russia (71)	New Zealand (120)
Solomon Islands (69)	**Tanzania (113)**
Nepal (68)	Japan (112)
Colombia (55)	**Cameroon (105)**
Cote d'Ivoire (51)	**Solomon Islands (101)**
Canada (47)	**Ethiopia (88)**
	Somalia (88)

[a] Comparison of the top-ranking twenty-five countries in terms of numbers of endemic languages and endemic higher vertebrates. Countries appearing in both lists are in bold. Higher vertebrates include mammals, birds, reptiles, and amphibians; reptiles not included for the United States, China, and Papua New Guinea, because of missing data.

Source: Reprinted by permission from Harmon 1996. Losing species, losing languages: Connections between biological and linguistic diversity. *Southwest Journal of Linguistics* 15(1&2):89–108.

perse ecological impacts (Alcorn and Molnar, 1996; Stevens, 1997b). Insofar as traditional communities are generally found in parts of the world exhibiting high levels of cultural diversity, cultural diversity is thus seen as responsible for maintaining high levels of biodiversity.

This view is held by many ethnobiologists and anthropologists, and is based on a suite of observations (Stevens, 1997b; reviewed in Maffi, 2001). First, most indigenous communities maintain a lifestyle and technology that makes light demands on local resources (e.g., Posey and Balée, 1989), and indeed appear to leave the natural resource base and biodiversity of their lands relatively intact. Second, traditional knowledge systems often reveal an intimate and very detailed technical understanding of the natural world, perhaps in some respects still unrivaled by modern science (Atran et al., 1999; DeWalt, 1994; Johannes, 1978; Lewis, 1989; Williams and Baines, 1993). Third, on the basis of this expertise, there often exist customary regulations that protect particular species or areas, and sanction land use regulations that limit and disperse the impacts of subsistence resource use (Alcorn, 1993; Berkes et al., 1998; Gadgil et al., 1993). Fourth, religious beliefs commonly buttress these regulations, generating elaborate cosmologies and ritual practices that would seem to assure indigenous commitment to the protection of at least some species and areas (Reichel-Dokmatoff, 1971). Finally, indigenous peoples have often fought outsiders' attempts to lay claim to their territory and natural resources, sometimes successfully (Nietschmann, 1987), and in so doing create de facto protected areas. While much of the evidence that traditional resource management systems maintain or enhance biodiversity is anecdotal, there are some very convincing cases (Sarkar, 1999). For example, the banning of grazing in the Keoladeo National Park in Rajasthan in the early 1980s has devastated this wetland, once a critical feeding ground for tens of thousands of wintering waterfowl and a refuge for other species that breed in the monsoon. Paspalum grass, kept in check by grazing, has established a stranglehold in the wetland, choking the shallow bodies of water (Gadgil and Guha, 1995).

Many of these ideas are incorporated in the notion of traditional ecological knowledge (TEK Box 4.2). Great claims have been made for these knowledge systems, for instance, as "holistic, intensely dynamic, constantly evolving through experimentation and innovation, fresh insight and external stimuli" (Posey, 2001, 382). We must beware of idealizing such traditional knowledge, however, since it is often fragile, easily lost, and somewhat less dynamic and open-ended than some claims might imply. Though TEK is by no means static (e.g., Ross and Pickering, 2002), there are, almost by definition, limits to the range of conditions to which TEK can respond (Sillitoe, 1998): chain saws can render the returns to harvesting so attractive that traditional regulations or beliefs regarding constraint are simply

overlooked (e.g., Box 5.4). Furthermore, not all traditional practice is necessarily good for biodiversity per se; when local communities espouse conservation as a goal, this may reflect a concern with ensuring the sustainability of organisms deemed useful rather than a commitment to the abstract concept of species diversity. Additionally, the consequences of traditional management are unpredictable: Kenya's coral reefs are no more diverse or undamaged when fished with traditional methods than with modern gear (McClanahan et al., 1997). A further point is that TEK is by definition local; there are some sorts of expertise that cannot be local, such as broad meteorological patterns, and in these areas local knowledge must be superseded by that of outsiders. Finally, TEK may be irrelevant to biodiversity loss; thus, dugong numbers are declining in Queensland, Australia, because these large marine mammals suffer from the degradation of sea grass bed, get caught in commercial fish nets, and are injured by recreational power boats; as such, bolstering traditional Aboriginal hunting regulations will have little if any moderating effect (Ross and Pickering, 2002). In a sense, then, we must recognize that TEK is valued by some social scientists more for its political than its ecological impact: in this former respect it has the potential to question the science behind the exclusionist and protectionist strategies discussed in chapter 2, and to reintegrate spiritual values into natural resource management in a world in which management is so often driven purely by the demands of production and accumulation. As such, it is perhaps poised to play a role parallel to that of citizen science in the West, bringing local communities as active participants to the management of nature (chapter 11).

Despite the mismatches between TEK and biodiversity conservation, the suspicions of a skeptical few that TEK is no more than tribal superstition, and its political import, it is nevertheless abundantly clear that this knowledge can be of critical ecological sig-

Box 4.2. Traditional Ecological Knowledge (TEK) and Adaptive Management

Gadgil et al. (1993, 151) define TEK as "a cumulative body of knowledge and beliefs, handed down through generations by cultural transmission, about the relationship of living beings (including humans) with one another and with their environment." TEK is not a simple compilation of facts that might yield useful leads to drug discoveries or raw materials for biotechnology and agricultural innovation. Rather, it is a body of knowledge and practice that guides decisions relating to natural resource management, nutrition, food preparation, medicinal cures, education, and community social institutions. Practices that relate directly to the environment include the management of multiple species (e.g., cultivating many different crops together; Warren and Pinkston, 1998), staggering the harvest of particular prey species (as with the James Bay Cree hunters' rotation of beaver trapping areas on a four-year cycle; Berkes, 1998), and dealing with unpredictability (as with the daily and seasonal movements of African herders in response to highly variable and unpredictable rain and grazing conditions; Niamir-Fuller, 1998). A particularly well-documented example comes from Arvika in western Sweden, where Olsson and Folke's (2001) study shows how the local fishing association manages the common-pool crayfish resource. To deal with water acidification, predation by mink, a fungal crayfish disease, and overexploitation, they have developed sophisticated techniques that protect not just the individual crayfish but the whole watershed. For example, they are liming the lake, trapping mink, exchanging crayfish among different parts of the lake to prevent inbreeding, and imposing temporary fishing bans. Anthropologists have documented many instances of how TEK can help communities deal with recurrent shocks and events, such as food shortages and pest infestations.

Many of the prescriptions of TEK are generally consistent with adaptive management, the flexible, hands-on method of ecosystem management discussed in Box 3.4 (Holling, 1986). First, TEK is for the most part adaptive, since it accumulates

Huaorani Workshop Held in 2000 at Yasuni National Park and Biosphere Reserve.
Photo courtesy of Maggie Franzen.

(continued)

(Box 4.2 continued)

incrementally through a trial-and-error learning process over generations of observers whose lives depend on the accuracy of the information. Second, it relies on social learning (between and across generations) to offer strategies for dealing with unpredictable or uncommon situations; as such, it embodies well-tested solutions to recurrent local ecological crises. For these reasons some have suggested that TEK be viewed as a less scientific analogue of adaptive management (Berkes et al., 2000).

The value of TEK for sustainable management was recognized by cognitive, economic, and other anthropologists several decades ago (Brokensha et al., 1980; Berkes, 1999). However, it was not until the mid 1980s that scholars began to voice concern over its loss, noting a concomitant decline in language diversity and, by the late 1980s, biodiversity. Initially, these parallels were drawn in a largely metaphorical sense, to evoke the comparative gravity of the different aspects of diversity under threat (Maffi, 2001). Increasingly, however, the direct value of TEK for biodiversity conservation, sustainable food production, and modern medicine became apparent, such that conservation biologists, anthropologists, ethnobotanists, and other scholars are taking renewed interest in TEK for a range of scientific, social, and economic reasons (Berkes et al., 2000). In this context workshops are increasingly held to discuss the management role of indigenous communities within protected areas (see photo). Compared to other types of knowledge, TEK is perhaps at a particularly high risk of disappearing due to its place-specific and subsistence-related nature.

nificance, for both sustainability and biodiversity conservation. As regards sustainability, TEK is intricately linked to complex agricultural and land use systems that ensure year to year production through controlling disease and insect threats, minimizing erosion, and ensuring food availability, as with the neighborhood management of pests and water on the rice terraces of Bali (Box 6.5). As regards species diversity, a classic example of the positive role of TEK are the sacred groves of India and elsewhere, where custom and religion contribute to the protection of whole suites of species at selected sites (Box 4.3). Furthermore, we should not forget that TEK contributed heavily to what we now consider modern scientific knowledge, an interesting case being the Hortus Malabaricus (Grove, 1992); this Dutch botanical classification was based on the tremendously sophisticated understanding of the medicinal properties of plants among the toddy-tapper caste of Malabar,

and was later adopted by Linnaeus. Finally, there are innovative initiatives where TEK is used to both inspire and guide management solutions, such as the controversial reliance on Aboriginal grass-burning regimes in Australia (see Box 11.4).

On the basis of these observations regarding TEK, a "conservation ethic" is often attributed to indigenous and local communities. To some authors this phrase implies simply taking no more than is needed (Johannes, 1978) or wise management (Berkes et al., 1995). To others it suggests an intentional goal of conserving resources. In fact, the geographer Nietschmann (cited in Stevens, 1997b, 2) posits a "Rule of Indigenous Environments"; by claiming that "where there are indigenous peoples with a homeland there are still biologically rich environments" this dictum posits an invariance to the coexistence of indigenous populations and biodiversity. In both the extreme and the less extreme characterizations, indige-

Box 4.3. Sacred Groves

Sacred groves are areas, ranging in size from a clump of five trees to more than 50 ha, that are set aside for religious and spiritual purposes (Gadgil and Vartak, 1976; Gadgil et al., 1998). Nowadays in some areas they protect the only remaining patches of primary vegetation (Wilson, 1993). Such refugia, comprised variously of ponds, springs, coral reef lagoons, and sacred forests, were once widespread across India, Africa, and Europe (Frazer, 1955). Historically the Greek and Roman landscape was dotted with such areas, often incorporating a temple or shrine. In western Europe they became sites for pagan resistance against Christianity, although a few were later incorporated into monastic gardens and church yards. The same kinds of sacred habitats can be found in the Americas, for example, among the Kunas of Panama (Chapin, 1991) and the Tukanos of Brazil (Chernela, 1987), in Amazonia (Shaman's gardens; Posey and Balée, 1989), and in Africa, where they protect remnant closed canopy forest (Wilson, 1993). Key to their sanctity is the outlawing of hunting and other uses of natural resources, including even the removal of dead wood. These restrictions are sanctioned by fear of religious or social retribution and by taboos. There are a variety of explicit reasons why such taboos exist: the species may have strange behavioral and morphological characteristics, be thought to be toxic, feature in creation myths, or bear religious symbolism. Nevertheless, whatever the reason for the avoidance, unintentional nature conservation is often the outcome. Gadgil et al. (1998) suggest that these practices of ecological prudence spread as part of a system of religious beliefs or social conventions without the secular function being fully realized.

Sacred groves were largely destroyed by the British in India, when they converted village-based forests to state property. The remaining groves, still under pressure from deforestation, are now of critical ecological importance. They buffer against the depletion of local species variants, serve as important recruitment areas for surrounding ecosystems, and provide corridors between dispersed protected areas. They sometimes protect keystone species (see chapter 3), such as various figs (*Ficus* spp.) that are pivotal for the maintenance of biological diversity in the tropics (Terborgh, 1992). In the Western Ghats, one of the world's twenty-five biodiversity hotspots, local farmers have taken the lead in establishing new sacred groves, as in the states of Manipur and Kerala, designating what are now called "safety forests" within social forestry programs (Gadgil, 1987; and see Box 7.5). At these sites community members are extending vigilance and protection to remaining patches of forest, now with an explicit understanding of the secular conservationist functions of these refugia.

Sacred groves support Holling's (1986) schema that human societies at different stages of economic development employ different mechanisms to promote sustainable use of biological resources. In small-scale horticultural and hunting communities the emphasis is on sacred sanctions; in agrarian societies social convention and state regulations are key, as in royal hunting reserves (Box 2.1). In industrial societies emphasis is on state regulation for scientific and recreational goals. In the emerging age of information and democracy, decentralized management may be able to create positive conservation incentives at local levels, as in the revival of sacred groves in India.

nous and native peoples are hailed as natural guardians of biodiversity (e.g., Durning, 1993; IWIGIA, 1992) or "ecosystems people" (Dasmann, 1988), a claim that provides both a moral and practical foundation for the new move toward CBC, and has added a suite of nuanced meanings to the term *biodiversity* (Table 2.3). But how accurate is the proposition that people prior to the development of large-scale agrarian nations and industrial states were so environmentally benign? What are the environmental impacts of traditional communities? The evidence that indigenous and local groups intentionally limit their use of natural resources is actually quite thin.

4.4 Ecological Impacts of Traditional Ways of Life

Paleoecological and archaeological studies show that the first humans to settle islands brought a demise to many species of indigenous fauna and flora (Kirch and Hunt, 1997). There are many classic examples of extinctions on islands in the Pacific and elsewhere. Humans seem to have driven to extinction the giant lemurs and elephant birds of Madagascar, the pygmy hippos of Cyprus, and, only 100 years after the Polynesian settlement of New Zealand, eleven species of moa weighing between 20 and 250 kgs (Holdaway and Jacomb, 2000). Perhaps most poignant of all the island extinction stories is that of Easter Island (Box 4.4).

There are many hints in the archaeological record that humans were the culprit. For example, comparing the rate of extinction on the Galápagos Islands, only settled in 1535, with Oceania (settled in prehistoric times), we see three or fewer extinctions over the 8,000 years preceding human arrival on Galápagos, compared to 800 in Oceania over the same period (Steadman, 1995). Thus, the rate of extinction in Galápagos was roughly two orders of magnitude lower than that on earlier settled islands. Other studies have provided more detailed

evidence identifying the culprit. For example, we can look at the proportion of bones of different taxa (birds, turtles, shellfish, fish, and mammals) used for food by the settlers of Polynesia, and compare these across earlier and later periods of settlement (Figure 4.1). The main pattern is for domestic mammals to increase over time as birds and turtles decline in numbers (Dye and Steadman, 1990). Note also that on islands where fish make a major contribution to the diet, the decline in birds and turtles is less precipitous. Though not shown in the figure, the swiftest declines are seen in the large-bodied ground-living birds, the taxa most directly exposed to humans and their congeners, like dogs and rats, and most likely best to eat, a result similar to that found for the Polynesian colonization of New Zealand (Anderson, 1997). All these patterns would make sense if human pressure was the culprit.

Consistent with IBT (see Figure 3.2) islands offer the clearest evidence that human foraging populations are responsible for species extinction. This is because islands protect populations that are small (and hence more easily driven to extension by random events like a big storm), offer fewer reservoirs from which to draw in the case of a local extinction, and primarily harbor species that are endemic. Events on continental land masses are more controversial. Even though Alfred Russell Wallace, a co-founder of evolutionary theory (see chapter 3), ventured over 100 years ago that "the rapidity of . . . the extinction of so many large Mammalia is actually due to man's agency" (cited in Leakey and Lewin, 1995, 172), the debate over who or what killed the New World Pleistocene megafauna is only now drawing to a close.

On the American continent after the drying and cooling that began in the Miocene 25 million years ago, there evolved many species of large grazing mammals ("megaherbivores") and their predators that persisted through the Pleistocene. These included saber-toothed cats, ground sloths,

Box 4.4. Rapa Nui (Easter Island) Extinctions

When humans first landed on these islands in A.D. 400, they were densely forested with palms and shrubs. Over time a stratified political system developed. Archaeological evidence shows that chiefs demanded huge agricultural surpluses from their followers, a phenomenon that promotes crop intensification and population growth. During this period pollen records indicate dramatic deforestation, with many species including the large *Jubea* palms going extinct. By 1500 the human population had grown to 7,000, and had developed great artistic and cultural accomplishments of which the 800 giant stone statues are one example (see photo). Archaeologists can show that deforestation caused such bad soil erosion that there was no regeneration, and that crops would not grow; also, without trees for building boats, fish dropped almost entirely out of the diet. As deforestation progressed, no more statues were built. Indeed, the archaeological record reveals that within a few generations all the stone carving stopped, and Rapa Nui was plunged into a period of endemic warfare, raiding, and general terror. The mariners on the Dutch ship that landed on the island on Easter Sunday 1722 found people who were starving, and practicing both cannibalism and slavery. The statues had been toppled and decapitated, perhaps as a result of warfare among the competing clans on the island. The people were desperate to abandon the desolate island, but could not do so without boats (Bahn and Flenley, 1992; Kirch, 1984, 2000).

Easter Island Monuments.
Photo courtesy of Peter Bellwood.

Bahn and Flenley (1992) take Rapa Nui as a model for the whole planet. They compare the environmental history of the island since Polynesian colonization with the global computer model of the Club of Rome (Meadows et al., 1972) and find disturbing congruence. Though it is clear Rapa Nui is an extreme case of anthropogenic ecological decline, in many other islands agricultural intensification was an ongoing process at the time of European contact. Had the history of these islands not been disrupted by disease, there could have been many other Easter Island stories (Kirch, 2000).

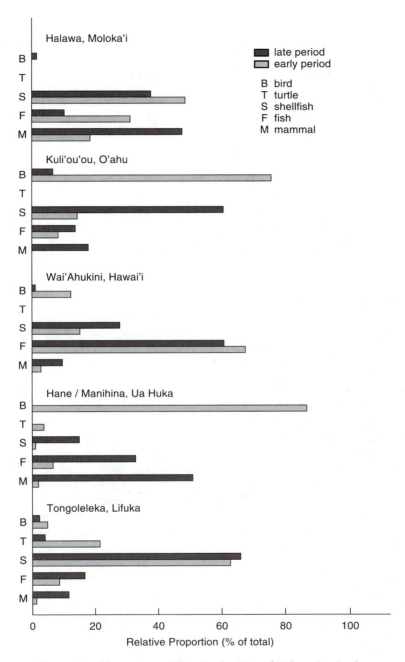

Figure 4.1. Changes over Time in the Diet of Polynesian Settlers.
The relative proportions of the main taxa used for food by the settlers of Polynesia—bird, turtle, shellfish, fish, and mammals—change over time. In the late prehistoric period, domesticated mammals account for a much larger proportion of the total meat weight at each of five sites, whereas birds and turtles account for a smaller proportion than in the early period. Data from the early period are shown here in a lighter shade for comparison. Reprinted with permission from Dye, T. and Steadman, D.W. 1990. *American Scientist* article, Polynesian Ancestors and Their Animal World. *American Scientist* 78:207–215.

mastodons, and wooly mammoths. Then quite suddenly the vast majority of these, in fact thirty-three genera of large animals, were wiped out between 11,500 and 8000 BP in North America, with yet more extinctions occurring in South America in this period. Since these events coincided closely with (most likely) the first arrival of big game hunters at the edge of the glaciated sheet in southern Canada (11,700 BP), some have argued they were anthropogenic. Mosimann and Martin (1975) model a feasible scenario, in which a small group of foragers arriving at the tip of the North American icefield decimate most of their favored prey items between Edmonton to Tierra del Fuego in 1,000 years. We can question the assumptions and estimates that go into such models, not least the original densities of prey species, but this exercise suggests that anthropogenic extinction in the Americas is at least a possibility. Furthermore, two recent models, for both the New World and Australia, provide further support for the so-called blitzkrieg hypothesis (Alroy, 2001; Roberts et al., 2001).

Empirically based studies suggest that the Pleistocene extinctions cannot be attributed exclusively to hunting pressure per se, but to the multiple ways people modify the environment, such as the burning of hunting grounds and the introduction of predatory exotics like rats. Also, the climate was warming up at the end of the Pleistocene. Humans may have sufficiently disturbed the environment in the northerly cold zones with their settlements that large mammals were no longer able to penetrate these areas for use as a retreat during periods of warmer climate. Finally, each species' disappearance changes the environment for the others remaining, potentially triggering cascades of further extinction (Owen-Smith, 1987; Terborgh et al., 1999); for example, predators like dire wolves disappear once the prey species on which they depend are hunted to extinction. In short, humans were probably the key ingredient in a complicated and fatal recipe for mass extinction (Burney, 1993).

What about contemporary foragers? Here we examine only the harvesting of animals and plants, leaving the dimension of indigenous land tenure and its role in contemporary conservation conflicts to chapters 5 and 8. The image of Plains Indian hunters, driving herds of bison over cliffs and picking only the choicest cuts while huge amounts of flesh are left to rot (Haines, 1970), is no doubt a caricature. Depletion of particular prey species may not, however, be so rare. Specific methods are available to determine whether current harvesting levels are sustainable. In northeastern Peru, Bodmer (1994) examined the sustainability of ungulate hunting. With estimates of r_{max} (the natural rate of increase in a population; see chapter 3) and animal densities he calculated animal productivity, and then compared this to hunting offtake to determine the percentage of production taken by hunters (Table 4.2). In a heavily hunted area (Tahuayo) only the productivity of tapir failed to compensate for hunted offtake, which was 160 percent of recruitment (right-hand column), suggesting that tapir are at severe risk of local extinction. In support of this interpretation, tapir densities were much higher in an adjacent lightly hunted area (96 kg/km^2) than in Tahuayo (64 kg/km^2), as were the densities of a number of primate species not shown in the table. The precarious situation of tapir and various primates in Tahuayo reflects commercial hunting for sale at a nearby market. Using another approach, current harvests can be compared to the MSY (see Box 3.2), determined as a first approximation on the basis of population density and rates of population increase (Robinson and Redford, 1991). Using this method FitzGibbon et al. (1995) showed that current harvests of elephant shrews, duikers, and squirrels in Kenya's coastal Arabuko-Sokoke Forest were sustainable, whereas those of primates and larger ungulates were not. Generally, low population density, low reproductive rates (Ginsberg and Milner-Gulland, 1994; Robinson and Redford, 1994; Peres and Terborgh, 1995),

TABLE 4.2
Sustainability of Current Hunting of Amazonian Ungulates in Tahuayo

Species	Total Productivity (Average No. of Young/ Individual/Yr.)	Density (Individuals/ km^2)	Production (Individual/ km^2)	Hunting Pressure (Individual/ km^2)	Sustainability (% of Production Taken by Hunters)
Tayassu tajacu (collared peccary)	0.55	3.3	1.83	0.27	15
T. pecari (white- lipped peccary)	0.61	1.3	0.80	0.30	38
Mazama Americana (red brocket deer)	0.33	1.8	0.60	0.13	22
M. gouzoubir a (gray brocket deer)	0.37	0.8	0.30	0.06	20
Tapirus terrestris (lowland tapir)	0.12	0.4	0.05	0.08	160

Source: From Bodmer, 1994, Table 5.5, p. 126.

as well as a variety of behavioral characteristics of prey such as being attracted to areas of burning (FitzGibbon, 1998; O'Brien and Kinnaird, 2000) are responsible for a species' susceptibility to hunting.

A compilation of recent evidence shows that in virtually all kinds of tropical forests worldwide large game animals are highly valued as prey animals, and are strongly depleted where human population densities exceed one person per square kilometer (Bennett and Robinson, 2000a), even in traditional or indigenous areas (Peres, 2000). Though a lower population size in hunted than unhunted areas is not in itself indicative of unsustainability (because of the MSY model; Box 3.2), Bennett and Robinson's review shows that very often harvested populations decline in productivity over time, constituting clear evidence against sustainability (in the sense of maintaining an "undiminished" supply; see Box 3.5). Depletion occurs largely because a major portion of the meat people harvest comes from a relatively small number of large-bodied, slow-breeding species; for example, three species of ungulates comprise 80 percent of the bio-

mass taken by rural hunters in Sarawak (Bennett et al., 2000). However, these three ungulates constitute only 22 percent of the animals hunted in Sarawak, indicative of a widespread pattern that many different species are potentially at risk as harvesters shift from one prey to the next over time. As such, anything ranging in size from an elephant to a humming bird may be at risk of local depletion or extinction. Furthermore, there are indications that even where levels of hunting are sustainable, prey populations may be so low that their ecosystem functions are almost eliminated (Terborgh et al., 1999; see also Jackson, 2001 for an application of this idea to coastal fisheries; and chapter 5).

Inevitably, these results are controversial. Some commentators are skeptical about the generality of the data suggesting that studies have been conducted where hunting was already seen as a "problem" (Schwartzman et al., 2000a,b). Others make the important point that only long-term studies of hunting activities and yields can determine the effects of subsistence hunting on local animal populations. For example, Vickers (1991)

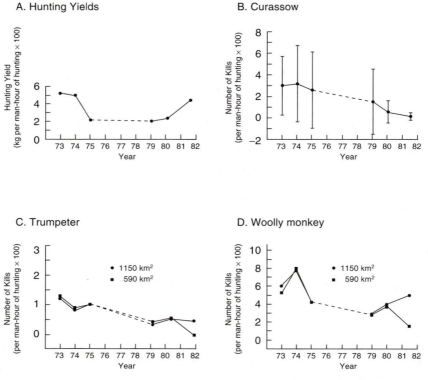

Figure 4.2. Hunting Yields of the Siona-Secoya in Northeastern Ecuador.
A. Overall hunting yields   B. Curassow   C. Trumpeter   D. Woolly monkey
The yields are calculated as mean live weight per man-hour of hunting per year. Yields are multiplied by 100 for specific species rates. Data are based on 863 man days of hunting, about 3.77 percent of the estimated total number of man days of hunting from 1973 to 1982. A. Hunting yields in the 1,150 km² Shushufindi territory from 1973 to 1982. B. Curassow (*Mitu salvini*) rates declined steadily after 1974, and suggest depletion of this large bird's population in the 1,150 km² Shushufindi territory. C. Trumpeter (*Psophia crepitans*) are being depleted when observations are restricted to the 590 km² core hunting area. D. Woolly monkey (*Lagothrix lagotricha*) is the most frequently killed primate in this area. Yields appear sustainable when the 1,150 km² territory is considered, but depletion is suggested for woolly monkeys when observations are limited to the 590 km² core area. Reprinted with permission from Vickers in *Neotropical Wildlife Use and Conservation*, ed. Robinson & Redford (Figures 5.3 and 5.12) The University of Chicago Press.

has collected longitudinal data on the hunting returns of Siona-Secoya of northeastern Ecuador. An early analysis in 1979, based on overall hunting returns, suggested hunting was leading to a depletion of game species in the area, whereas when incorporated into a full ten-year data set, it became clear that the decline was not sustained (Figure 4.2a). In addition to collecting long-term

data, it is important to pay attention to geographic scale. Again, Vickers's study shows how the presumed patterns of depletion (based on hunting returns) can vary, depending on the scale of the analysis. Whereas there is some evidence for the depletion of curassow, two other species (woolly monkeys and trumpeter birds) show only local depletion (Figure 4.2b–d). Vickers's results suggest that surveys at a broader geographic scale are needed to identify the exact impact of local depletion on a species' persistence in larger areas. Hill et al. (1997) find that locally depleted species show normal densities in areas 6 or more km away from Paraguayan Aché settlements (see Box 5.5), and Alvard (2000) calculates that the observed offtake of the Wana on the Indonesian island of Sulawesi would be sustainable if they had a larger territory. All in all sound cases for or against sustainability cannot yet be made until encounter rates and hunting returns across an appropriately broad geographic area over time are monitored. Perhaps at present the firmest conclusion is that despite abundant indications of overhunting in traditional and rural communities, especially in tropical forests (Bennett and Robinson, 2000a), overhunting is not inevitable (FitzGibbon et al., 1995; Alvard, 2000). Clearly, more research is needed to determine the potential for ecologically sustainable offtake in conservation areas set up as extractive reserves (see chapter 10).

Other important areas of investigation are the effects of changing technology on hunting practices and the impact of the market (see also chapter 5). As regards the former Alvard's (1995a) comparison of Piro and Machiguenga in Peru found that though shotguns are often more efficient than traditional technology, shotgun hunters do not necessarily obtain higher per-consumer harvests than do traditional groups, though they can devote less time to hunting. As regards markets, it is abundantly clear that where opportunities for the commercial sale of meat arise, offtake is commonly unsustainable because of the intensification of hunting effort (Lavigne et al., 1996; Freese, 1997; see chapter 5). Markets can also drive specialization on larger-bodied prey (Ayres et al., 1991). This is particularly poignant in areas like the northern Congo where a vibrant bushmeat trade follows the clearing of commercial logging roads that lead to urban centers (Auzel and Wilkie, 2000; see Box 11.3) decimating primates and other

favored species. The income from wild meat at regional and national levels can be startling—$83 million per year from Sarawak and $175 million per year from the entire Amazon Basin (figures in Bennett and Robinson, 2000b, 4). Furthermore, as Alvard (1995a) concludes, with a reason to harvest more, the impact of shotguns on harvest rates would be enormous.

As regards plant use, the story is more mixed. We already saw that within 1,000 years the Polynesian people of Easter Island turned lush forests into a barren plain. Similarly, the Mayans reduced the tropical forests of the Yucatan to scrub, and the Yuqui of Bolivia felled whole trees just to get the fruit, once they lost the labor of slaves who used to do the dangerous work of climbing for the harvest (Stearman, 1994). In a similar vein Terborgh (1999) blames the poor regeneration of Brazil nut (*Bertholletia excelsa*) trees in Amazon forests on nut overharvesting by forest people (see also Peres et al., 2003). But there is also ample evidence that gatherers and early farmers actively managed certain (generally useful) species to ensure continued productivity. Ethnobotanists studying indigenous use of plant resources have described many horticultural and gardening practices that preserve species diversity. For example, horticulturists exhibit a keen interest in the location of rare and useful species, replanting these when necessary. They often intervene in pollination and succession, thereby potentially protecting threatened species. Commonly they disperse gardens and orchards, with fruits attracting wildlife during fallow years. They may also create anthropogenic islands of forests; these are credited to the Kayapó of Central Brazil, who transplant useful species for themselves and their preferred game into an organically constructed soil base (consisting of leaf litter and termite and ant nests) laid in patches of open savanna (Anderson and Posey, 1989; but see Parker, 1992 for a different interpretation). Similar processes have been described for West Africa (Leach, 1996; see

Box 7.4) and parts of Amazonia; in fact, Balée (1989) estimates that at least 11.8 percent of *terra firma* forest in Amazonian Brazil is anthropogenic. We look at the interpretation and significance of these cases in chapter 7.

Ethnobotanists generally believe that the high levels of intra- and interspecific diversity cultivated in the gardens of native horticulturists ensure against disease and crop failure, minimize risk, help to prevent soil erosion and maintain soil fertility, as well as protect diversity at the genetic and species levels. On the other hand, there are no quantitative studies comparing species diversity within and outside of areas used by horticulturist groups. Furthermore, these management practices are not intended to do anything other than serve human interests or protect species that are valuable to humans (Johnson, 1989; Berkes et al., 1995). Is careful horticultural management of useful species really equivalent to biodiversity conservation? This is an open question, and its answer depends to a large extent on how biodiversity is defined (see chapters 1 and 3). Finally, as with hunted resources, gathered items are particularly vulnerable to local depletion and extinction when harvested for commercial ends (Vasquez and Gentry, 1989), particularly in contexts where the density of the species harvested is low, where resource tenure is open-access (see chapter 6), and where powerful people in local communities have a stake in the extraction business, as in Kalimantan (Salafsky et al., 1993).

In short, our ancestors wrought considerable changes to the species diversity in different parts of the world. In some cases ecological devastation and species extinction occurred, particularly on islands. Although limited extractive technologies, relatively low population densities, and restricted opportunities for marketing products precluded ecological damage of the extent witnessed in the developed world, the ecological impacts of our ancestors should not be underestimated. However, in generalizing these observations to the contemporary context, we need to be very careful; issues of time and geographic scale become critical. Consider this case of miscommunication: to the claim of Terborgh that forest peoples' subsistence strategies are responsible for the poor regeneration of Brazil nuts, Schwartzman et al. (2000b) reply that the Brazil nut stands themselves are the relics of plantings of now extinct Amerindian cultures, with the express implication that forest peoples are good for biodiversity. Both claims may indeed be true. Exchanges in this vein are at cross-purposes, and are not well illuminated by stereotype and generalization. Better research with comparative quantitative data into anthropogenic effects on the landscape and the conditions favoring sustainability, already initiated with tropical forest hunting (Bennett and Robinson, 2000a) and collection (Peres et al., 2003), is still needed.

4.5 The Long Shadow of an Ecologically Noble Savage

So we have some contradictory observations. On the one hand, in the contemporary world, areas of cultural diversity are those with the highest biological diversity. On the other hand, the impact of human subsistence on particular species is by no means ecologically benign. Despite these inconsistencies the rosy image of indigenous communities as prime guardians of biodiversity persists, and has been very influential in capturing the imagination of many people, including anthropologists and other social scientists (e.g., IWGIA, 1992; Durning, 1993; Lee and Daly, 1999). Among the general public, it provides something of a spiritual touchstone for the moral decay of contemporary society, offering hope through new simplicity, even primitivism (Konner, 1990).

Redford (1990) captured this romantic view of indigenous peoples as natural conservationists with the term *ecologically noble savages*. This colorful epithet has generated more heat than light, perhaps because it touches a tender nerve among academics and professional groups generally wary about indulging in caricature. In one sense Redford was right to bring under scrutiny the notion of an ecologically noble savage, since the very idea bears all the signs of a politically nuanced interpretation of a past that never was. Indeed, many of those who hold that cultural diversity enhances biodiversity would also concur with Leacock's notion (Sahlins, 1968; see also Birdsell, 1958) that it was Western capitalism, and specifically industrialism (Dobson, 2000), that upset an erstwhile harmonious balance of humans and nature.

Viewing other societies as repositories for values different from our own is of course nothing new. In many traditional societies fanciful imagination had fearsome monsters looming just beyond the horizon. In imperial Rome this fear was embodied by "barbarians," the term the Romans used for the non Latin-speaking hordes that threatened the northern frontiers of the eclipsing empire. However, these almost inhuman and despised figures joust with an alternate image, a more pristine noble character hovering on the skyline. From classical Greek times we see the first traces of such romanticism, where innocent savages are depicted as untainted by society. At Europe's encounter with the New World chroniclers expounded on this notion, depicting the native inhabitants of the Americas as free from the evils of civilization, a theme elaborated in the philosophy of Rousseau and the rich colors of Gaugin's naïve tropical scenes. In recent years this romanticism has taken a peculiarly ecological turn, priming social scientists, policy analysts, and many others to believe that indigenous people are the best natural conservators of biodiversity. The final ironic twist to this story is that indigenous activists, recognizing the divergence of their views from those of traditional Western conservationists, now debate among themselves whether they should be viewed as natural conservationists (Box

4.5; and see chapter 8). The fabulous rhetorical power of the ecologically noble savage motif seems to have stung its own tail.

In another sense, however, Redford's colorful quip triggered a virtual crusade against noble savagery within anthropology, with the names of offenders guilty of presenting romantic stereotypes appearing in strings of citations across scholarly publications. It is only fair to point out that the accused did not buy into these exotic caricatures quite as wholeheartedly as the critics would imply (as noted also by Bennett 1976), and they probably never intended to essentialize indigenous peoples (Alcorn, 1994). Also, there were many dissenters from the start (e.g., Ellen, 1986; Richards, 1992). As Stearman (1994) notes, the origins of this romantic misconception lie in a few classic ethnographic examples repeatedly cited, rhetorically embellished, and then somewhat indiscriminately applied to all indigenous peoples by those on the policy-making front. That said, the view of an ecologically noble savage, first aired in the South Pacific in the early 1980s (Chapman, 1985) and now rejoined in Amazonia (see Redford and Stearman, 1993), is undoubtedly still prevalent. While persisting only in small corners of anthropology (e.g., Lee and Daly, 1999), it continues to pervade other areas of academia, the media, and popular imagination (Figure 4.3). For example, in a generally measured and scholarly discussion of conservation practices among indigenous peoples Maffi (2001, 11) recognizes an "overall consensus" with regard to the sustainability of indigenous practices. As such, these issues continue to drive confrontations over how best to conserve biodiversity (e.g., Schwartzman et al., 2000a; Terborgh, 2000).

Since indigenous peoples are being nominated as guardians of biodiversity, at least in part because of their apparent conservation ethic, studies of the ecological impacts of early human and traditional contemporary populations are seen as inflammatory, even subversive, with respect to debates surrounding conservation and indigenous affairs

Box 4.5. Transitions in Ecological Noble Savage Thinking

"I am as free as nature first made man, 'Ere the base laws of servitude began, When wild in woods the noble savage ran." (First recorded use of the term, in John Dryden's 1672 drama, *The Conquest of Granada*, 7)

"The land belonged to all, just like the sun and water. Mine and thine, the seeds of all evils, do not exist for those people . . . They live in a golden age . . . in open gardens, without laws or books, without judges, and they naturally follow goodness . . . So in harmony with their surroundings that they all live justly and in conformity with the laws of nature." (Pietro Martire d'Anghiera, writing from the New World in 1500, cited in Hemming, 1978, 15)

"The point is that for centuries many societies had evolved some kind of accounting system whereby the number of people in their group and their age structure were thought of in relation to available natural resources. This was especially true in hunting and gathering, horticultural and pastoralist societies . . . Such socially operated cultural accounting systems, which foster local and regional social management, must be revitalized in societies around the world." (Arizpe and Velazquez, 1994, 33, who are social scientists)

"Indigenous peoples, who number around 300 million today, are the traditional guardians of the Law of Mother Earth, a code of conservation inspired by a universally held belief that the source of all life is the earth, the mother of all creation." (Martin, 1993, xvi, from the introduction to a book based on the World Congress on National Parks and Protected Areas, Caracas, 1992)

"We Africans long ago developed wildlife conservation customs compatible with sustained production, embracing soils, plants, water and animals. Resources were harvested in accordance with tribal laws that were reinforced by the authority of indigenous religious sanctions expressed, as in the West, through a combination of historical legends and charismatic myths. Taboos were an intrinsic part of life with punishments aimed at demeaning offenders." (Simbotwe, 1993, 15–16, a consultant in resource management in southern Africa)

"In this sense, then, I don't believe that you can say that indigenous peoples are conservationists as defined by ecologists. We aren't nature lovers. At no time have indigenous groups included the concepts of conservation and ecology in the traditional vocabulary. We speak, rather, of Mother Nature. Other organizations need to be clear about this before jumping in to solve some problem with the indigenous populations." (Gonzalez, 1992, 45, the indigenous leader and founding technician of the PEMASKY project, Panama; see Box 8.6)

"It is unclear whether and to what extent native peoples even understand the goal of maintaining biodiversity, or sympathize with that goal in its most extreme form. Why, they ask, should we save poisonous snakes, stinging ants, or wood ticks, when all they do is cause human misery?" (Hill, 1996 189, an anthropologist writing primarily about Paraguayan Aché; and see Table 8.1)

"Why do you white people expect us Indians to agree on how to use our forests? You don't agree among yourselves about how to protect your environment. Neither do we. We are people just like you. Some of us view nature with a great sense of stewardship whereas others must perforce destroy some of it to obtain what they need to eat and pay for expensive medical treatment and legal counsel." (Marcus Terena, the president of the Union of Indigenous Nations, in Belem, Brazil, speaking in 1990, cited in Low, 1996, 372)

Figure 4.3. The Co-opting of Chief Seattle.
"The earth does not belong to man, man belongs to the earth." These are the famous words of Chief Seattle, a great Suqamish chief born in the 1780s who as warrior, politician, and orator facilitated the inevitable transition of power from native to European. Though there are several versions of this speech extolling the sacred power of every shining pine needle, and dripping with implications for the modern ecological crisis, his real speech, written up in a newspaper article thirty years after he made it, had nothing to do with ecological awareness. Photo Courtesy of The Suqamish Museum.

(e.g., Alcorn, 1994; 1995; Puri, 1995). As a result, considerable mistrust has arisen among those who argue for and against the conservation impacts of foraging in indigenous and traditional communities. This situation is unfortunate, and derives from an inappropriate conflation of "what is" with "what ought to be" (the naturalist fallacy). In reality, conservation performance is never a valid reason for divesting any indigenous people of their land (Kaplan and Kopishke, 1992; Stearman, 1994; Vickers, 1994; Hill, 1995). Furthermore, basing policy decisions on a community's conservation performance is counterproductive for at least three reasons. First, behavior can change very quickly. The Ecuadorian Siona-Secoya, for example, had no conservation ethic under traditional conditions, but have readily become conservationists as they adjust their behavior to the novel political and economic environment (Vickers, 1994; and see chapter 11), particularly with respect to oil company ventures on their land. Second, reports on conservation performance stir emotional responses that only cloud policy judgments. This is true whether environmental practice is apparently good or bad. Unsurprisingly, reports of destructive practices may get used as a justification for removing a community from a conservation initiative, such as the evictions discussed in Box 2.3. Interestingly, however, even reports of good environmental practice can misfire, as with all the media furor over Brazil's Kayapó conservation warriors (Box 4.6). Finally, even if indigenous foraging practices cannot be described as conservationist in terms of species-specific protection, indigenous tenure may still be preferable to other forms of land management (chapter 6). This point is again well demonstrated by the Kayapó. Even though they and other indigenous groups allow selective logging and mining in their reserves, they have been effective in halting forest clearance, as seen from satellite images of land lying inside versus outside of their reserves (Schwartzman et al., 2000a).

Box 4.6. The Kayapó Controversy

The Brazilian Kayapó are celebrated in the popular media for their energetic protest against the construction of a series of seven dams along the Xingu River in the states of Para and Mato Grosso, and for finally laying claim to a 13 million ha territory. They have appeared on North American television shows, and sent delegations to multilateral development banks, government representatives, and hardwood importers all across Europe and North America, often in their traditional gear. Their most prominent leader has appeared in a concert with the British rock star Sting (see photo). Much of this they have cleverly captured on video, producing images that are useful both for internal consumption and as public relations tools for a larger audience, such as the readers of *Vogue* and *People* magazines (Turner, 1992; 1995). After the last forty years of abuse at the hands of rubber tappers, a corrupt state Indian agency (FUNAI, the National Indian Foundation), and more recently international developers of the Amazon (Fisher, 1994), the Kayapó have developed a suite of savvy political tactics, emphasizing their role as indigenous guardians of the rainforest (Conklin and Graham, 1995). They have done this in a calculated way, highlighting their traditional status by removing Western clothing and donning body paint. For the Kayapó to be environmentalists nowadays parallels being warriors in the late 1940s, when they threatened the entire Brazil rubber harvest, or brokers in the late 1970s when they dominated FUNAI by putting a Kayapó leader at its head.

But selling themselves as guardians of the rainforest may have misfired (Fisher, 1994). When reports appeared, again in the popular media, that several Kayapó leaders had granted timber companies concessions to log large virgin tracts of tropical hardwoods and were also receiving substantial royalties from the sale of gold from their reserve (Verissimo et al., 1995), conservationists became outraged, feeling deeply betrayed by the Kayapó. "What the Kayapó are doing is absurd, illegal, immoral and wrong" we find in *The Dallas Morning News* (cited in Stearman, 1994, 352), or "the savage can also be ignoble" in *The Economist* (cited in Conklin and Graham, 1995, 709). Such stories fueled moves in the Brazilian Congress to thwart the legalization of indigenous territories, and bolster the claims of the mainstream Brazilian press that Kayapó and other groups are mere pawns in the hands of Western capitalists. As Stearman asks, would this reaction have had the same repercussions if the Kayapó had not received international attention as "natural conservationists" and "keepers of the forest"? Furthermore, pushing a few tribal leaders into global

Sting with Kayapó Elder and Warrior, Chief Raoni.
Picture courtesy of Sting and Dutilleux JP (1989). *Jungle stories: the fight for the Amazon.* London: Barrie & Jenkins.

(continued)

(Box 4.6 continued)

visibility as guardians of nature renders them highly vulnerable. Thus, on the eve of the Earth Summit at Rio reports of a violent crime in connection with Payakan, a highly visible Kayapó leader who had forged alliances with international conservation agencies, totally undercut his usefulness at the UNCED Congress (Conklin and Graham, 1995, 704).

Today the Kayapó seem to have weathered the media storm. With militia-type activities they have halted incursions of logging companies onto their lands (Schwartzman et al., 2000a), most notably in a project developing economic alternatives coordinated by the NGO Conservation International do Brasil (CI-Brasil, Zimmerman et al., 2001). Population densities of vulnerable vertebrate species are relatively high in their reserves, consistent with light hunting pressure (Peres, 2000). Nevertheless, there are many different Kayapó groups, and in their attempts to deal with an increasingly complicated set of threats (Fisher, 2000) they often disagree over what to learn from each other's mistakes (Giannini, 1996). New initiatives by the Amazon Conservation Team (ACT) are addressing this problem.

4.6 Revisiting Cultural and Biological Diversity

So, what can we make of the fact, introduced at the beginning of this chapter, that the highest species richness is found in areas inhabited by members of diverse, small-scale, indigenous ethnic groups? There are three broad logical but not mutually exclusive possibilities (Smith, 2001): first, that cultural diversity directly enhances biological diversity; second, that biological diversity directly enhances cultural diversity; and third, that both are affected independently by another set of factors.

As regards the first hypothesis, the argument is that small-scale societies protect and enhance species richness because they rely on these species for their subsistence and livelihood; hence, a regional aggregate of these societies will show high levels of biodiversity. It is quite clear from the evidence of hunting, human expansion, and corollary extinctions reviewed in this chapter that humans can hardly be viewed as natural guardians of biodiversity. On the other hand, we focused primarily on the notorious fate of species favored for the pot. The impact of

traditional human communities on many plant species, let alone insects and amphibians, is largely unknown, since carefully controlled studies of traditional natural resource management have rarely been conducted. Clearly, without research comparing various indices of biodiversity between areas subject to different levels of indigenous use, the impact of cultural practices on species diversity can only be guessed at.

The second hypothesis is that biological diversity directly enhances cultural diversity. One idea is that areas of high biological diversity might provide multiple niches for cultural diversification, in a manner analogous to speciation through niche partitioning (Harmon, 1996). This simple notion is probably flawed, since people utilize a relatively limited number of species as resources and much biodiversity is either undistinguished in terms of its value as a resource or simply unusable; human niches are shaped more directly by technological, political, and economic arrangements than by the environment (Steward, 1955). In fact, quite contrary to this hypothesis, cultural heterogeneity is dramatic in places with quite low levels of species richness, as in Mali's Upper

Niger Delta, where Tuareg camel herders, Fulani pastoralists and agriculturalists, and Bozo fishermen all coexist (Moorehead, 1991). Despite little species diversity in this Sahelian zone, multiple niches are available to humans with their intense technical specializations.

The third hypothesis is that cultural and biological diversity are both enhanced (or depressed) by an independent factor, itself antecedent to both kinds of diversity. One suggestion here is that high levels of net primary productivity may permit local communities to be self-sufficient, engaging little in trade and therefore remaining culturally distinct (Harmon, 1996, 98). The strong correlation between linguistic diversity (a measure of cultural diversity) and length of the growing season (an indicator of productivity) found both in West Africa and worldwide (Nettle, 1996; 1998) supports this idea; in a similar vein, Cashdan (2001) finds high ethnic diversity in areas with predictable and constant climates. Though the factors favoring high levels of species diversity are not well understood (chapter 3; for example, species diversity is not linearly correlated with net productivity, Balmford et al., 2001), a compound environmental variable (measuring rainfall and productivity) is correlated with both species and linguistic diversity, at least in Africa (Moore et al., 2002), supporting the hypothesis that independent environmental factors are responsible for language and species richness. Since species richness is better predicted by environmental variables than it is by cultural richness, the authors of this study conclude that there is no need to invoke a direct causal link between cultural and biological diversity (the first two hypotheses; see also Sutherland, 2003).

Another likely important independent factor is inhospitability to humans, though this in part reflects environmental conditions and could be thought of as a variant of the third hypothesis, albeit with a different mechanism. Certain parts of the world, due to a medley of factors including high productivity, dense vegetation, disease, and climatic regimes, particularly in conjunction with geographic barriers and heterogeneous terrain, both prohibit intensive agriculture and foster species diversity. Since states and empires depend on agricultural surpluses to feed the nonfarming sectors of the population, such as bureaucrats and soldiers, and rely on viable systems of transport, they are often disinclined, or even averse, to spreading into such areas. Consequently, the original linguistic and cultural diversity in such areas remains intact. Many writers have elaborated on the similarities among invasive forces decimating cultural and biological diversity, respectively (e.g. Wurm, 1991), be these rats and cats in the realm of nature or television and McDonalds in the realm of culture, and Diamond (1997) reviews the broad-reaching and often decimating effects of nation-states on biodiversity. Thus, European colonial powers in East Africa took the highlands, the erstwhile "White Highlands" of Kenya, without venturing into the low-lying disease-infested Congo basin. Similarly, metropolitan powers in Central and South America are expanding to the limits of where intensive agriculture, mining, and deforestation are economically feasible (Painter and Durham, 1996). Given these dynamics, it is not surprising that cultural diversity persists mainly in the hot, wet tropics, inhospitable to intensive agriculture, afflicted with endemic diseases, and thus buffered from the advances of metropolitan society. According to this hypothesis, biological and cultural diversity are both promoted somewhat independently, albeit through a similar set of environmental variables that affect isolation, communication, settlement, and imperial expansion. The implications of this are explored in chapter 7.

In sum, the broad but not well-understood association between cultural and biological diversity that fuels the modern alliance between conservation biologists and rural communities in the developing world (chapter 8) may, at some geographical scales, be spurious, or at least reflect broader environmental variables. It is nevertheless important to recognize that biodiversity and cultural diversity are linked albeit indirectly, and that the dynamics of this tangled relationship are not well determined. In part this is because we understand so little about why some cultural groups stay small, how cultural diversity is maintained, and what causes biological speciation and radiation. Further, these links often entail complex socio-political and historical processes, examined further within the context of political ecology (chapter 7). All in all, there is still a great need for careful scrutiny of how local communities can play a role in biodiversity conservation, and how this differs across different scales of analysis; for example, the human invasion hypothesis may hold at fine scales, but fail to explain the broader latitudinal distribution of biodiversity (chapter 1).

4.7 Conclusion

Indigenous and local communities are often the principal guardians of vast undisturbed habitats that modern societies depend on to regulate water cycles, maintain climatic stability, and house species and genetic diversity. Why is this the case? In this chapter we questioned the view that indigenous or local populations are necessarily good conservationists, a conviction that implicitly dominates much recent policy development. This critique is not tantamount to denying that in most cases traditional societies have many fewer negative impacts on the environment than do developed Western societies. Nor does it undermine the value of studying traditional resource use systems. Finally, it in no way undercuts the fact that indigenous tenure may be preferable to other forms of land management. These communities can bring immense knowledge to conservation, and have often finessed land use practices that are adapted to local conditions. In short, their negative impacts on specific species may be offset by their protection of broader habitats, rendering them strong allies (as we discuss in chapter 8) for national and international conservation movements where these are needed—on the ground.

Hopefully debates over the ecologically noble savage are now on the wane, since they did much to antagonize scholars who otherwise share a commitment to the rights of local communities and indigenous peoples. The disagreement has nevertheless been useful. First, it reveals the naivety of believing that modern environmental problems are simply the result of the spread of Western culture, and would evaporate were Western science to be replaced by animism, paganism, or eastern mysticism (see Box 1.3). Second, it has exposed the dangers associated with erecting romantic stereotypes of indigenous communities, and thrown into relief the importance of distinguishing descriptive from prescriptive pronouncements. Third, and moving now to the following chapter, the debate has sparked anthropologists' and other social scientists' interest in applying models founded in evolutionary ecology and micro-economics to determine where, when, and how conservation behavior might occur (chapter 5). It is on the basis of such studies, particularly when they are situated in a broader social and political framework (chapters 6 to 9), that ultimately we will find new models for involving local people and their practices in biodiversity conservation. These debates also bring to a head the very question of what we mean by conservation, a focus of the following chapter.

CHAPTER 5

Conservation and Self-Interest

5.1 Introduction

The environmental problems facing humanity stem ultimately from human behavior—our reproduction, our consumption, and our misuse of technology. It is becoming increasingly clear that the solutions lie not in a technological fix, but in a radical restructuring of human priorities and social policy. This call to change human behavior is no modest goal. It is therefore hardly surprising that despite repeated warnings from scientists, activists, and others of an inevitable decline and collapse of the world's natural resource base, there has been no slackening in the assault on the Earth's resources. The problem is that policies and projects that ignore the wants and needs of people are doomed to failure. In this chapter we consider what we know about human behavior with respect to the conservation of natural resources, asking when and how human interests are compatible with the protection of resources for later use or enjoyment. The disciplines we call on lie in the behavioral sciences—evolutionary anthropology, economics, and evolutionary and environmental psychology. In keeping with our present concern for individual-level dynamics, we look mainly at foraging and simple horticulturist populations that have few community-wide institutions regulating resource use, and focus on the methodological individualistic approach of behavioral

ecology. In chapter 6 we broaden our focus to community institutions, and then to the wider social, economic, and political sphere contingent on national and global factors (chapters 7 to 9).

With the specter of an ecologically noble savage on one side, and unequivocal evidence of anthropogenic extinctions on the other, we turn in this chapter to look more closely at what is meant by a "good conservationist." We employ an approach popular in economics, the political sciences, and some branches of anthropology, predicated on individual self-interest (5.2). We propose three ways of detecting whether a particular method of resource use is conservationist: first, by identifying the costs and benefits of actual foraging decisions to individuals (5.3); second, by determining whether there is an intent to conserve (5.4); and third, by measuring the ecological consequences of supposedly conservationist acts (5.5). Evolutionary anthropology, with the attention it pays to the costs and benefits of behavior, offers a useful framework within which to address the broader question of when, where, and how conservation practices might arise (5.6). However, since neither the evidence that people are limiting their harvests of natural resources, nor the existence of a conservation ethic, assures true conservation outcomes, it is critical to establish an independent assessment of the ecological consequences of different resource-harvesting strategies; this will require modeling and improved field methods (5.7). Finally, we look at some of the broader policy implications of assuming that individuals for the most part behave in accord with their own self-interest (5.8), before reaching the conclusion that investigations of conservation conflicts must start by looking at individuals, and focus on what people need to survive and compete in their social context (5.9).

5.2 An Evolutionary Viewpoint

Human desires and actions are to a very large extent both the source of and the solution to conservation problems. Innovative new evolutionary ecological studies offer a promising route toward untangling how and why people sometimes nurture and sometimes deplete the natural resources on which they depend (Penn, 2003). According to this perspective, forces that have shaped the behavior of all organisms over evolutionary time have for the most part favored rapid exploitation of resources rather than prudent stewardship. Natural selection preserves the kinds of behavioral strategies that allow an individual to out-compete his or her competitors. Thus, the most efficient strategies for garnering resources in any particular environment will become the most common, at least if individuals using these strategies leave more offspring. Humans are no exception in this regard. As a result, in many circumstances people are unlikely to prioritize the common good or the distant future, but rather reap short-term gains at the expense of competitors (Heinen and Low, 1992; Wilson et al., 1998).

The situation is not, however, entirely bleak since evolutionary ecology does not rule out conservation per se (Smith and Wishnie, 2000). All organisms have to make decisions about whether to use resources now or in the future. Some mammals and birds, such as squirrels and acorn wood-

peckers, store food to eat in the season of scarcity, trading short-term costs for benefits that will be enjoyed in the future. Organisms also often show reproductive restraint. In most human societies parents do not use up all their wealth producing and raising babies, but save their worldly goods to pass on as inheritance (Rogers, 1994). All in all, an evolutionary perspective, insofar as it is predicated on self-interest, leads to the prediction that individuals will show restraint whenever the long-term benefits outweigh the short-term costs. As such, it offers a framework within which to scrutinize the conditions in which a prudent resource strategy might pay off.

Optimization theory is one of the most frequently used analytical tools in tackling such questions. With regard to how individuals use or harvest resources, optimal foraging theory was developed to determine how animals should behave when foraging to maximize their biological fitness (Krebs and Davies, 1993). Since the biological fitness contingent on a particular foraging strategy can rarely be measured, a currency such as the rate of energy captured is used as a proxy. Notably, most of the models consider only how foraging returns can be maximized over a relatively short period, such as a particular day or a specific hunt; we will return to this later. Foraging theory has been successfully applied to the behavior of humans that rely directly on natural resources for subsistence, and increasingly commercial, purposes (Kaplan and Hill, 1992). It is used to make predictions about prey choice, in other words, to identify which food items foragers should attempt to exploit and which items they should ignore in favor of continued search for more preferred foods (the optimal diet breadth model). In situations where resources are clumped together, it has also been used to determine the length of time foragers should spend in one patch before leaving for another. The first approach depends on establishing a profitability-ranking of preferred food items, whereas the second entails application of the marginal value theorem, a tool lifted from microeconomics to assess, in this case, relative returns to continued foraging in different patches.

An instructive application of the optimal diet breadth model shows how introduced technology can change hunting behavior. To apply this model, all potential prey types must be ranked in terms of their profitability; profitability, or energy return rate per encounter, is calculated as the energy obtained from each encounter with that prey type divided by the time spent in pursuit or process. The model predicts that prey yielding the highest profitability should always be pursued. Lower-ranked prey types should also be pursued (or included in the diet) until the next most profitable resource yields a lower energy return rate per encounter than could be obtained by continuing to search for and pursue the more profitable items; resources ranked below this threshold should not be pursued when encountered. Unsurprisingly, new technology can substantially affect the efficiency with which certain prey items can be pursued and processed, reducing handling times and increasing the profitability of certain items. Studying Canadian Cree foragers in the late 1970s, Winterhalder (1981) reports how the introduction of muskets, repeating rifles, nets, and steel traps increased the speed at which certain species of fish and game could be handled, and hence their profitability. As a result, prey that had previously been considered too small or too difficult to capture now fell into the optimal diet breadth as suitable foraging targets. For example, waterfowl could be quickly shot at without seriously taking time away from the more profitable pursuit of a moose. However, as more Cree gained access to modern means of transport, the search time associated with foraging decreased. A forager with a good snowmobile or motorized boat is happy to travel 30 km to a good spot for finding moose tracks, passing by evidence of other

less valued species like muskrat or beaver, whereas a forager on snowshoes or paddling a canoe would have taken these less profitable species long before 30 km were traversed. In short, using the logic of the optimal diet breadth model, a researcher can predict quantitatively how changes in the social and technological environment affect which prey items will be taken and which will be left. As such, foraging theory sheds light on many aspects of resource harvest—what species to take, what sized groups to forage in, and even the kinds of technologies to invest in. Additionally, these models can be adapted from the conventional realm of foraging to other areas of human resource use, such as fuel wood collection (Abbot and Homewood, 1999) and pastoral subsistence (Borgerhoff Mulder and Sellen, 1994; see chapter 6).

This style of analysis, predicated on the assumption that individuals act in ways that are ultimately selfish, is part of a broader approach known as methodological individualism that is popular among political scientists and a growing number of anthropologists. Methodological individualism, which arose in economics, attributes characteristics of groups, such as rules, practices, and institutions, to the objectives and actions of self-interested individuals (Arrow, 1994; Basu, 1996), and is the foundation of rational action theory (Elster, 1986; Smith and Winterhalder, 1992). Self-interest can be defined and measured in many different ways. Thus, economists view individuals as maximizing a currency loosely defined as utility (or the satisfaction derived from the consumption of a good), whereas evolutionary anthropologists focus on the concept of reproductive fitness. Both, however, view individuals as ultimately self-interested. From this perspective, most economists, political scientists, and evolutionary anthropologists are, like evolutionary biologists (Williams, 1966), suspicious of explanations for behavior, customs, and practices that are couched in terms of the group's advantage.

5.3 Design of a Conservation Act

Using this framework of foraging theory, evolutionary anthropologists developed an operational definition of conservation that allows them to test for its existence among native peoples (Smith, 1983). Hames's (1987, 1991) key idea here was to determine whether people exercise restraint in their harvests by pitting the predictions of a "conservation hypothesis" against predictions derived from optimal foraging theory, termed an "efficiency hypothesis." The efficiency hypothesis predicts that foragers should choose the prey items that offer the highest net profitability in the short term. According to this hypothesis, resources will be harvested according to their profitability ranking and entirely independent of the effect this harvesting has on resource availability in the future. In contrast, conservation entails the costly sacrifice of immediate rewards (the amount of meat in tonight's pot) for delayed benefits in the future (a healthy prey base). According to this hypothesis, highly ranked food items will be passed by if, for example, they are becoming scarce. In this sense conservation is a particular kind of self-restraint that may have the consequence of preventing, or at least mitigating, future resource depletion, species extinction, and habitat degradation (Hames, 1987; Alvard, 1995b; Smith, 1995). Because the links between short-term restraint and conservation outcomes are quite loose (as we will see later), we prefer the political economists' term *stinting* over *conservation* for this particular kind of behavioral restraint. Stinting is defined as the exercise of short-term restraint for long-term benefit.

Hames's approach stimulated a number of valuable empirical studies showing that the decisions hunters make generally maximize their efficiency, that is, the size of tonight's pot (Box 5.1). Indeed, there are now a number of studies contrasting the predictions from the efficiency and so-called conservation hypotheses (reviewed in Alvard,

1998a) that provide unambiguous support for the former, at least with respect to hunting. Some observations are nevertheless more difficult to interpret. For example, Amazonian hunters avoid hunting in game-depleted areas, and tend to move to less heavily hunted areas farther from their villages. On its own this observation is consistent both with the conservation hypothesis (sparing of the prey base) and with the efficiency hypothesis (which, according to the marginal value theorem, predicts that hunters should leave patches where the returns are low and move to richer patches; Hames, 1987). However, if hunters were really concerned with sparing the prey base, they should never take animals that they encounter on their treks to more peripheral areas, but they do (Alvard, 1994). In general, hunters appear not to eschew decisions that have major negative impacts on the population dynamics of their prey.

Useful as this approach has been, it has weaknesses. First, in situations where the human population is low and resources are plentiful, even the so-called conservation hypothesis dictates rate maximization, since no resource will be at the risk of depletion; consequently, under such conditions, stinting may be indistinguishable from (and interpreted as) efficiency. This is important, because most tests have been conducted in small-scale, low-density populations. Second, we should stress that stinting is not necessarily an alternative to efficiency maximization, once longer timeframes are considered. As we explore further in the following chapter, for a group of hunters with secure land rights limiting today's take to secure food for tomorrow (or next year) is entirely rational. Third, the assumption that foragers have as an objective the maximization of harvests may well be overly simplistic; they may, for example, be more concerned with not falling below a certain threshold; such behavior would be interpreted as stinting in this approach, whereas it may in fact reflect this different goal. Fi-

Box 5.1. Prudent Predators?

Working with the Piro Indians of Peru, Alvard (1995b) has shown how contrasting predictions can be tested regarding which individuals within a prey species should be taken. In line with the conservation hypothesis, a hunter should choose animals of low reproductive value. A hunter should specialize on young animals which, on average, have a high probability of not reaching reproductive maturity, and old animals whose reproductive potential is spent. In line with the efficiency hypothesis, hunters should chose prime aged adults, who represent the biggest return for effort spent. The age structure of Piro kills made during Alvard's fieldwork is compared against animal census data collected by staff from Cocha Cashu research station in the adjacent (unhunted) Manu National Park. For each species hunted by the Piro (see table below) the ratio of adults to juveniles is either statistically indistinguishable from that of the populations censused at Manu or else is biased toward adults. Clearly, Piro hunters do not stint when it comes to choosing which individuals to hunt. In related work looking at inter- rather than intraspecific prey choice, Alvard shows that Piro hunters do not spare prey species that are highly vulnerable (1993) or locally depleted (1994).

TABLE 5.1 BOX
Comparison of Age Profiles of Piro Kills and Censused Populations

Species	Piro kills		Censused Populations			
	Immature	Adult	Immature	Adult	X^2 adj	Sig.
Collared peccary (Tayassu tajacu)	0.27	0.73	0.31	0.69	0.37	NS
			0.24	0.76	0.19	NS
			0.45	0.55	9.15	0.0025
			0.44	0.56	8.17	0.0043
			0.26	0.74	0.01	NS
Deer (Mazama americana and M. gouazoubira)	0.18	0.82	0.39	0.61	1.85	NS
Capybara (Hydrochaeris hydrochaeris)	0.46	0.54	0.30	0.70	0.19	NS
			0.42	0.58	0.14	NS
Spider monkey (Ateles paniscus)	0.14	0.86	0.47	0.54	5.84	0.0157
			0.34	0.66	2.18	NS
Howler monkey (Alouatta seniculus)	0.14	0.86	0.51	0.49	12.09	0.0005
			0.54	0.46	13.95	0.0002

Source: Reprinted with permission from Alvard, "Intraspecific Prey Choice in Amazonia," *Current Anthropology* 36 (1995): 789–818 (Table 4). The University of Chicago Press. Bibliographic sources in the original.

nally, there is the problem of generalizing from hunting to broader environmental practices. While traditional foragers may maximize short-term game harvests, this does not mean that they lack broad longer-term management plans that enhance species productivity and even diversity (Balée, 1989; 1994). For example, the Cocamilla of eastern Peru enrich their flood plain fisheries by adding manure and garbage to the lake waters (Stocks, 1983), and contemporary lowland Mayan Indians practice agroforestry in ways that enhance species diversity, tree cover, and soil nitrates (Atran et al., 1999). Clearly conclusions from studies of how individuals in traditional communities hunt mammals cannot necessarily be generalized to their proclivities regarding broader aspects of habitat management, a point also made in chapter 4.

These limitations aside, tests of simple alternative predictions from foraging theory, predicated on individual self-interest, have the potential to clarify not only what adaptive problem a particular foraging strategy may be designed to solve, but also how this strategy might change as a result of ecological, technological, or socio-political developments. We have already seen how the prey capture of Cree Indians responded to changes in handling and search time, in predictable ways. Other very clear applications lie in tackling a particularly interesting question—how economic development (and new income) affects local people's use of their natural resources.

It is generally believed that as rural economies become linked to regional and national markets dependence on the forest for medicines, fuel wood, and construction materials declines (Bates, 1985; see also Browder, 1992), and that with new sources of wealth the value of the forest derives increasingly from timber, ecological services, amenities, and biological diversity (Godoy and Bawa, 1993; Pinedo-Vasquez et al., 1992). Optimality models, and other microeconomic tools founded on maximization, suggest how this might happen (and why in some cases it might not happen). Alternative sources of income such as wage labor, farming, or cattle-raising might lead foragers to take only the most valuable forest resources, in other words, only goods that yield a wage comparable with (or better than) what they could earn from these new alternative activities. Alternatively, economic development might tempt people to invest in better equipment for extracting, harvesting, and transporting forest goods, thereby increasing the productivity of their labor in the forest. Yet again, with cash in hand, foragers might be encouraged to substitute forest goods, such as skins, with cheaper commercial alternatives (clothes) that take less labor to obtain.

Teasing out the relative importance of these dynamics in real life produces mixed results. In some situations, for example the Sumu Indians of Nicaragua, richer villages show no evidence of specializing in the most valuable forest resources; though the very richest villages make less use of their forests, up to a certain threshold wealthier villages show greater reliance on nontimber forest resources (NTFR) than do poorer villages (Godoy et al., 1995). Similarly, in Honduras, income shows ambiguous effects on game densities, with game animals not necessarily prospering (nor adversely suffering) in the vicinity of richer villages (Denmer et al., 2002). As regards technology, in the Sumu case, as income rises, people do invest in better equipment for extracting forest goods (see also Kaplan and Kopishke, 1992 for the Peruvian Machiguenga), such that more goods are brought in from the forest, not less (Panayotou and Ashton, 1992). As regards substitution, families with more income do substitute proteins and fats from the forest with domestic animals; while this temporarily takes pressure off forest meat sources it might ultimately support larger village populations and thus more hunting pressure (as modeled by Winterhalder and Lu, 1997). In fact, any economic intervention creating new sources of income, for example, employment by an oil company, will attract immigrants to forest villages, thereby increasing harvesting pressures on plant and animal resources. Finally, the effects of increased income are difficult to disassociate from greater access to markets, insofar as markets are well-known to exacerbate hunting pressure (Freese, 1997; and see chapter 4). In sum, tropical forests are impacted (both negatively and positively) as villagers become wealthier; microeconomic models can help further specify how the marginal returns to different foraging strategies vary across socioeconomic and ecological contexts.

5.4 An Intention to Conserve

Intent to conserve, or the existence of a "conservation ethic," is commonly believed to lie at the heart of conservation behavior. Simply put, people will not conserve resources unless they want to, be this for any blend of practical and religious reasoning. In some senses this is self-evident, since the exercise of short-term restraint for long-term benefit that we have just discussed is founded on rationality, and hence intent. Furthermore, customs and practices, particularly those buttressed by spiritual sanctions or elaborate cosmologies, can facilitate sustainable resource management, and even the protection of some species, for example, through taboos (see below). As such, intent can play a key role both in promoting traditional ecological knowledge (Box 4.2) and in designing institutions that ensure cooperation over environmental challenges (see chapter 6). Limits on fish trap numbers in Lao provide a good example of restraint used with the intention to conserve (Figure 5.1).

There are, however, some problems with this line of thought. First, there are serious difficulties in establishing whether an "intent to conserve" lies behind a particular practice. A longstanding debate over the function of territories among subarctic beaver hunters reveals the ambiguities inherent in attributing intent to behavior or custom

(Box 5.2). Second, there is a danger in imputing a reverence for nature on the basis of a community's deep understanding of the natural world. Extensive traditional knowledge of nature does not in itself guarantee prudent use of natural resources; indeed, knowledge can be used to further unsustainable exploitation just as much as it can be used to ensure conservation. Finally, people do not always act in ways consistent with their beliefs, and even if they do, they are not always successful at realizing their intentions (Hames, 1991).

Hunting taboos are an interesting case here, since many (e.g., McDonald, 1977) have proposed that these mitigate the impact of hunting on selected species. A newer study, based on a cross-cultural survey of species whose use is entirely prohibited by a particular society, showed that 62 percent of avoided reptiles and 44 percent of avoided mammals appear in the IUCN red list of threatened species (Colding and Folke, 2001), the generality and causality of which warrants further investigation. Nevertheless, before accepting that these taboos ensure conservation outcomes, we need to know their generality, impact on behavior, and effects on ecological outcomes. Low's (1996) survey of 122 nonmodern societies suggests that sacred taboos are actually quite rare and are unassociated with whether resources are used in a sparing or wasteful way, as well as with levels of environmental degradation (this sample includes early chiefdoms and states, not just traditional foraging groups). In a similar vein, Aunger's (1992; 1994) work with Efe Pygmies and Lese horticulturists in Zaire shows that food taboos are often ignored, or at least beset with such extravagant exceptions, that less than 2 percent of calories are actually restricted as a result of tabooing. This differs remarkably from McDonald's (1977) early estimate that taboos can reduce the hunting of various species by 50 percent or more. Without more focused studies looking at the effects of species-specific taboos and their ecological effects on the target and related

Figure 5.1. Conservation Value of Intent: "Ou" Trap Fisheries in Lao.
Photograph of an "ou" trap. This is a fence filter fish trap. Rocks and bamboo are used to direct fish into a series of baskets placed in a row perpendicular to the direction of flowing water. The ou filter trap fishery is unique to certain areas of Khone Falls, southern Lao. The villagers have been using this technique to harvest fish sustainably for almost a century. Fishing effort is limited through an agreement to fix the number of traps in use and to exclude individuals from outside the local villages, such that the same number of traps have been set in the same location for decades. User rights remain village-based and each trap is maintained by two to eight families, though this may change from year to year, depending upon immigration and emigration of families in and out of the communities. The ou trap fishery is complemented by other measures such as no fishing zones, demonstrating the purposeful intention to conserve fish stocks in this community. Photo courtesy of Zeb Hogan.

species (or ecological communities) we cannot reach firm conclusions about the conservation impacts of hunting taboos. We must also be careful not to mistake the scrutiny with which nature is observed for evidence of an explicit "conservation ethic," or to assume that the existence of such an ethic ensures favorable environmental outcomes.

In short, it is important to recognize that the impact of traditional ecological knowledge on the conservation of specific resources or species is undoubtedly often largely unintentional, a point that the more cautious advocates of traditional ecological knowledge appreciate (Berkes et al., 1995; Colding and Folke, 2001). Where outcomes are intentional, they are extremely interesting to learn from, both with respect to how the

outcomes are attained and how the intention is preserved, as, for example, in the Lao fishery (Figure 5.1). Conversely, where the positive conservation outcome is unintentional, it becomes critical to determine what institutions or practices are responsible for this outcome, and how these might be adversely affected (or bolstered) by social and ecological changes.

5.5 Ecological Outcomes

Perhaps the trickiest question of all is how the design of behavior and the intent to conserve affect conservation outcomes (Ruttan, 1998). There are at least two reasons why stinting in a particular harvest does not

Box 5.2. Family Hunting Territories in Subarctic Canada

Contending explanations for the systems of family hunting areas among some Algonquian-speaking native peoples of the eastern subarctic Canada, of which the Cree are the major group, demonstrate how controversial claims for "intent to conserve" can be.

The first observers of the unusual family territories thought these were a traditional institution designed to provide for beaver conservation (Speck and Eisley, 1939, cited in Knight, 1965). The idea here is that every family wants beavers conserved and plays its own role in stinting, reminiscent of the ecological noble savage ideas of chapter 4. In contrast Leacock (1954) proposed that these territories emerged only recently, in response to the establishment of the Hudson's Bay Company fur trade, a proposal more in line with explanations based on individual self-interest advanced in this chapter. Specifically, Leacock argued that a shift from cooperative caribou hunting to individualized beaver trapping, and, in particular, the competition for fur resources, prompted the development of family territoriality, with families vying for entry into the lucrative fur trade. In a similar vein Knight (1965) suggests that the James Bay Cree, who had been trading with the Hudson's Bay Company since the founding of Rupert House in 1668, established beaver-trapping territories and a beaver conservation ethic only in the 1930s, when beaver quotas were allocated on a family basis. Leacock's and Knight's view prevails because it is based on evidence, though to be fair it is hard to know how good evidence to support Speck and Eisley's view might be uncovered. More generally, this debate points to the dangers in assuming that territoriality (or any other institution) reflects an "intent to conserve."

This conclusion is not, however, tantamount to a claim that "intent to conserve" cannot play a major role in shaping conservation outcomes. Analyzing a broader span of Cree environmental and institutional history Berkes (1998) intricately describes how the Cree have responded over the decades to a series of resource management crises by purposefully inventing (or redesigning) new institutions; particularly innovative was the system of beaver "bosses" whereby access to each territory is controlled. Bosses, though they are not the owners of territories, strictly regulate hunting effort, and can lose their positions if found breaking the rules. Berkes's historical study shows how an explicit intention to conserve a species can arise in a population where there was perhaps no such traditional ethic, and how important an element it can be in strategies designed to cope with novel ecological circumstances. This rationalistic approach turns much of the ecological savage logic on its head.

guarantee that the prey species will not be driven to local extinction. First, an act of stinting, such as closing a fishery for a period of time, can have exactly the opposite effect of conservation; for example, after a temporary closure, large concentrations of fish can be harvested extraordinarily efficiently, thereby accelerating a long-term decline in stock levels (Johannes, 1978). Second, stinting over a resource that is not limited by its harvest will be irrelevant to its long-term conservation. This is a point well recognized by ethnographers working in the subarctic, where natural fluctuations in prey popula-

tions are subject to unpredictable year-to-year and long-term variations, fluctuations that may have been greater than those induced by hunting (the disequilibrial systems discussed in chapter 3). Under such conditions it is hardly surprising that hunters view animals as "infinitely renewable resources whose numbers could neither be reduced by overkilling nor managed by selective hunting" (Brightman, 1993, 280). In a similar vein, grassland productivity in arid areas is primarily dependent on stochastic factors like drought; the number of cattle grazed there may affect productivity on a yearly basis, but not the longer-term health and persistence of the rangeland (Ellis and Swift, 1988; see Box 3.3). Similarly, fishery production in traditional marine systems is often more responsive to environmental factors than to fishing effort (Acheson and Wilson, 1996). Though the absence of density-dependent population limiting factors is often controversial (see chapter 3), it is clear that under such conditions stinting would actually be of no ecological consequence.

Just as stinting cannot ensure a particular conservation outcome, so conservation outcomes need not depend on stinting. There are several reasons why this might be. The most rehearsed is "epiphenomenal conservation" (Hunn, 1982). Benign ecological outcomes might arise not from the short-term restraint implied by stinting but from low human population densities, high community mobility, or a meager extractive technology; this conservation outcome bears no relationship whatsoever to the behavior of the foragers and is aptly referred to as epiphenomenal. Alternatively, prey and predators may be at equilibrium. A subsistence foraging population may not practice stinting at all, but simply avoid hunting their prey to extinction because of the density-dependent equilibrial dynamics to the predator-prey relationship (see chapter 3). In essence, predators reduce prey numbers, and then with less prey the predator population declines, allowing the prey to recover. Win-

terhalder et al. (1988) have applied these ideas to human foragers (Figure 5.2). For these two reasons, then, it is fallacious to believe that sustained harvesting is attributable to stinting, when it could be due simply to conventional density-dependent processes or entirely independent exogenous factors.

Though subsistence hunting may threaten certain prey species it can have important conservation benefits in the broader sense. There are both ecological and political grounds for this claim. Let us look first at the ecological argument. Stinting refers to the depletion of a single resource, whereas good conservation entails much more than the sustainable management of a single species, as we saw in chapter 3. Terborgh (1992) presented a theoretical model and empirical evidence (Terborgh et al., 1999) that top predators play a key role in maintaining species diversity and community structure. Though conventionally Terborgh's hypothesis is formulated with jaguars (or other species high in the food chain) as top predators, it is easy to see how the removal of a traditional band of foragers (also top predators) might also have an irreversible impact on the diversity of communities that have evolved in the presence of humans for thousands of years (Box 5.3). Second, turning now to political considerations, if local communities benefit from hunting, in terms of either food, the control of crop predators, or income, they are likely to have an interest in the continued existence of natural habitats and thus be willing to bear some of the costs of forest conservation. Thus, a ban on the hunting of selected species might remove a key incentive for forest conservation (FitzGibbon, 1998), jeopardizing alliances between local conservationists and biologists as partners in conservation efforts (chapter 8).

Perhaps the most important contribution of considering separately behavioral restraint, conservation intent, and ecological outcomes is to reveal the fragility of conservation outcomes. Consider, for example, the Waswanipi Cree Indians of the boreal

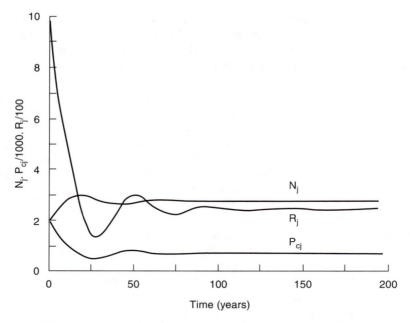

Figure 5.2. Dynamic Interactions of Foragers and Their Prey.
This is a simulation of the dynamic relation between a population of foragers and a single prey population analyzed over 200 years (Winterhalder et al., 1988). Initially the prey population (P_{cj}) shrinks, and as a result the net acquisition of energy through hunting (R_j) also declines. However, because the prey population at the start was large and the net acquisition of energy was high, the forager population (N_j) can continue to grow, even overshooting its equilibrium. Due to overexploitation, the prey density declines and foraging success falls below the basic maintenance rate of the foraging population in year 20. Consequently, the forager population shrinks, exploitation diminishes, and by year 29 the prey population begins to recover. Each subsequent oscillation is of diminished amplitude. Reprinted with permission of Academic Press/Elsevier Science from *Journal of Anthropological Archaeology* Research Vol. 7: 289–328.

forests of Canada (Feit, 1973), who rotate their hunting zones for moose and beaver every few years in order, it is argued, to avoid depleting prey populations. It is equally feasible that such rotations result from entirely utilitarian objectives (Smith, 1983). According to the marginal value theorem, a hunter concerned only with the size of tonight's dinner will do better to move from one hunting zone to another before the first is completely exhausted. According to these lines of logic, Waswanapi Cree rotations, which superficially look like intentional conservation, may actually be designed to maximize efficiency, but nevertheless have the consequence of promoting the sustain-

able harvest of the particular resource in question. These complexities highlight the need for careful studies that combine anthropological investigations of intentionality and meaning, quantitative analysis of economic decisions and their return rates, and independent monitoring of the resource (or resources) affected by foragers' many activities.

5.6 Conservation—Where, When, and Why?

Leaning heavily on the individualistic materialist framework borrowed from evolution

Box 5.3. Humans as Top Predators

Terborgh (1992) hypothesized that top predators exert top-down control of plant regeneration in tropical forests. He proposed that large felids and raptors limit herbivorous mammals and, where these predators are absent, an overabundance of herbivores alters plant regeneration. Though not all components of this hypothesis have received consistent support (Brewer et al., 1997; Wright et al., 2000), there is evidence that where humans (as poachers) reduce the abundance of herbivorous mammals this affects the regeneration of some plant species (Terborgh et al., 1999; Wright et al., 2000). Furthermore, when top predators like jaguar are removed, an overabundance of middle-sized predators can lead to rapid declines in populations of songbirds and other small vertebrates (Terborgh et al., 1999). Though these cascade effects differ in size and direction across study sites and are not well understood, the magnitude of their impacts is striking, and they are used by conservation biologists to highlight the potential ecological risks associated with local hunting, even if these patterns appear sustainable from the point of view of the particular quarry.

Most commonly Terborgh's hypothesis is invoked to highlight hidden dangers associated with losing top predators like jaguars and large preferred game such as monkeys, peccaries, and tapir, species that are often the focus of local hunters' efforts. But anthropologists have tweaked the argument. Considering somewhat longer timeframes than those concerning most ecologists, they point out that humans themselves are the top predators and thus the critical element responsible for maintaining species distributions. On the basis of human ecological studies in the Neotropics they have argued (the evidence is certainly not entirely conclusive) there were no uninhabited areas at European contact (Denevan, 1992), nor any very large uninhabited areas since that time (Steward, 1946). Since the earliest native populations spread throughout the Neotropics prior to the terminal Pleistocene extinctions, the current set of species has never existed without humans around. As such, humans have acted as top predators, competitors, and seed dispersers in the Neotropics for 12,000 years (Hill et al., 1997), and have disturbed the forest with swidden (Balée, 1998) and other activities (Bray, 2000) for at least 5,000 years. If the removal of top predators is detrimental as suggested by Terborgh's hypothesis, conservationists need to think very carefully about the effects on the community structure of evicting foragers on the basis of local or short-term prey depletion.

and economics, evolutionary anthropologists have begun a fruitful inquiry into where, when, and why strategies that sacrifice individual short-term benefits for longer-term profits might arise. We continue to refer to such strategies as stinting (in the tradition of political economists) rather than conservation, because most definitions of conservation require not just individual short-term restraint, but also appropriate levels of knowledge and a particular set of ecological outcomes. While several anthropologists have mulled over the question of whether and why people might restrain in their harvest of natural resources (e.g., Firth, 1959; Leach, 1972; Bennett, 1976; Jochim, 1981; Chapman, 1985; Johnson, 1989), evolutionary anthropologists have taken up the matter most systematically. This is probably in part because observations of restraint challenge their fundamental assumption that individuals maximize their interests (Borgerhoff Mulder and Ruttan, 2000; and see chapter 6), and in part because they are concerned with finding general explanations for human behavioral variability.

Generally, restraint will pay off to an individual only if he or she can be sure of benefiting from this restraint in the future. This will depend on three points: the intrinsic characteristics of the resource, how an individual's evaluation of the benefits of a conserved resource decay over time, and how sure an individual can be that someone else will not "steal" the resource in the interim. This last point raises the whole question of how conflicts of interest among individuals are resolved within communities, a topic explored in chapter 6 and dealt with only summarily here.

For stinting to have a benefit, the resource must have a value that will persist into the future (e.g., Alvard, 1998b). This is because people will incur the short-term cost of forgoing consumption only if this good has certain future value and no suitable substitutes (Hames, 1987). Thus, there is no benefit to be reaped from conserving a short-lived resource that will not be around when the forager returns. If I resist taking the gazelle that happened to wander into my village today, I have minimal chance of encountering it tomorrow (ungulates rarely stray into human habitations). This idea can be recast as a broader ecological prediction, that stochastic environments favor opportunistic strategies whereas systems closer to equilibrium favor greater restraint (e.g., Ellis and Swift, 1988). Where the resource base fluctuates wildly in response to unpredictable events, it is hard to see intuitively how a strategy of long-term resource management could repay the costs to an individual forager of forgoing immediate returns for uncertain future returns.

So what is the evidence? Low's (1996) comparison of 122 premodernized communities drawn from all over the world shows no clear ecological correlates to the expression of a conservation ethic. Most pertinently, people in predictable environments (measured as high productivity) are no more likely to show a conservation ethic than those in less productive environments. However, high productivity does not necessarily entail equilibrium dynamics, nor does a conservation ethic mean that resources are being conserved, so the hypothesis cannot be rigorously tested with these particular data. Empirical observations point in fact in the opposite direction. Chapman (1985) observes that conservation taboos in the South Pacific are more prominent on islands that suffer from hurricanes and droughts than on islands that are less environmentally stressed. In a similar vein, Stearman (1994; see also Beckerman and Valentine, 1996) notes that most of the original ecological noble savage stereotypes came from those areas of South America where human resource constraints are most acute, as in highly acidic blackwater "rivers of hunger" of northwest Amazonia and dry savannas of central Brazil. These observations echo the much earlier anthropological idea that only in very extreme environments do people become aware of the delicate balance of humans with nature (Bennett, 1976; Leach, 1972) and Ostrom's (1990) finding that common-pool resources are most successfully managed in harsh environments. On this evidence some suggest that people might be willing to exercise restraint in their harvesting of resources when they perceive these resources to be limited and/or unpredictable. How this relates to broader aspects of ecological stability and productivity is still uncertain, since in cases where resource abundance is only loosely linked to offtake, harvest restraint would have little effect in ensuring conservation of the resource (see discussion of nonequilibrial systems, chapter 3).

The second factor affecting whether individuals will stint in their use of resources is their time preference rate, a concept introduced from economics (Rogers, 1994). This is a tool designed to capture how a person's valuation of a good declines as a function

---

Box 5.4. Selling the Forest for Instant Returns

In 1987 Chupa Pou, a registered settlement of Aché foragers living on a reservation in eastern Paraguay, acquired legal title to about 2,000 ha. At once the Aché there began selling off valuable hardwood trees for lumber, on the informal understanding that any Aché was the "owner" of any tree they cared to claim. Consequently, local timber concessionaires contacted individual Aché about purchasing "their" trees. Within a period of a few months most of the valuable timber on the reservation was sold. The Aché lived well, and some individuals earned more than $2,000, when previously there had been practically zero opportunities for earning cash. By 1988 most trees of value had been felled, and the money all spent on parties, clothes, radios, gambling, and other luxury goods. By mid 1988 there was no evidence of the previous year's wealth, since no investments had been made in housing or living conditions, and no community projects had been funded with timber income (Hill, 1996). Aché behavior revealed very low evaluation of the future importance of their forests, at least when compared to the value of loggers' cash—in other words, a very high discounting of the future.

Such outcomes are not inevitable. In a nearby Aché settlement called Arroyo Bandera with its own forest, a community meeting was called at which it was decided that only a limited number of trees should be sold and the proceeds should fund a schoolhouse. Once a wood and brick structure with cement floors and glass windows was built, the Arroyo Bandera Aché put a moratorium on selling more hardwood from their land. With this precedent in place, when the Chupa Pou community received 6,000 ha due to the intervention of TNC in 1993, they also outlawed sales of timber by private individuals. Clearly, high discount rates can be bridled by community institutions, as we examine further in chapter 6.

---

of how long she or he has to wait for it. This decay function over time is often referred to as a discount rate. People who think little of tomorrow have a high discount rate. To the extent that the future is discounted there will always be a tendency to favor instant benefits and deferred costs, and to avoid present costs for delayed benefits. How a person discounts costs and benefits over time will be affected by many factors, a key one being probability of survival. For example, individuals will have little reason to stint if they are old or engage in a risky lifestyle, and are thus unlikely to survive to reap the benefits. An example of a very high discount rate, and a successful community effort in lowering this rate, comes from two Aché settlements in Paraguay (Box 5.4). Though discount rates are a serious disin-

centive to conservation, the one piece of good news is that humans (and other animals) do not appear to discount with an exponential function (as economists had assumed) but rather with a hyperbolic function (Kacelnik, 1997); as such, the future retains some importance (Henderson and Sutherland, 1996).

Discount rates are strongly affected by the population dynamics of the resource, and options need to be compared. In effect, when a potential harvester chooses to either stint or harvest, he or she has to compare the relative discount rates associated with both strategies. A classic example comes from Clark's (1973) work on the whaling industry. When the interest rate in the bank is higher than the reproductive rate of the whales left in the ocean, the optimal choice for a harvester is not to catch whales sustainably over time, but rather to kill them all off as quickly as possible and bank the proceeds. In short, because money accretes value faster in the bank than whales amass body size, fats, and precious oils in the ocean, extinction occurs. The very same principle applies in quite different contexts. Once isolated human communities come into the ambit of a market economy, their discount rates are affected in all sorts of ways. For instance, the deferred benefits of sustainably managing a peccary population over time may be eclipsed by a new market value for meat. If today's meat price in a local town is high enough, long-term benefits from conservation and restraint will pale by comparison. While the effects of discount rates are central to the study of community institutions (insofar as these secure future access to the resource; see chapter 6), the impact of time preferences (both individual and community-wide) on conservation outcomes in the real world has only very recently come under empirical investigation. One thing that is known is that men and women often differ in their discount rates (see below). Another is that links between peoples' discount rates and the amount of forest they cut down are not straightforward. Working with four different groups in the Bolivian lowlands Godoy et al. (2001) concluded that the lack of any clear relationship between experimentally determined individual discount rates and the amount of old growth forest cleared reflects the fact that some people cut trees to establish gardens for their children's inheritance, whereas others cut trees for immediate consumption, such as timber sales (as in the Aché community of Chupa Pou); in other words, forest clearance is evidence of a low discount rate for some, and a high discount rate for others.

The third factor affecting whether individuals will manage their environment sustainably is whether they hold secure rights to the ownership, control, and exploitation of the resource in question. A good example of how individual differences in rights to resources, even within a household, can affect the effort each will put into environmental improvements comes from a study of tree planting in Zimbabwe (Fortmann et al., 1997). Most Shona women move to their husband's household at marriage, but lose any rights to property on his farm should they become divorced. Since divorce rates are high, women have little hope that they can enjoy rights to this land for more than a few years. This is particularly relevant for the planting of trees, since most species commonly planted bear fruit only after several years. So, although women do as much agricultural work as do men, they are much less likely than men to plant trees on household land (Figure 5.3). Physical strength is not an issue (Zimbabwean women do much of the hard work), nor are there any cultural prescriptions against women's tree planting, since on communal woodlots, where women have rights irrespective of their marital status, they are just as likely to plant trees as are men. Analyses of this type raise issues of cooperation and community institutions that are addressed more directly by common-pool resource theorists, whose work we look at in chapter 6, when we move to a community-level perspective.

Viewing stinting and potential conservation acts in terms of costs and benefits to the individual as a function of resource characteristics, discount rates, and ownership offers a sound theoretical basis from which to make predictions about how people are likely to harvest and manage their natural resources. Depending on the nature of the resource, the institutional structure associated with the resource, and the particular decision maker (or group of decision makers), both incentives for conservation and major potential stumbling blocks to project implementation can be identified (chapters 10 and 11).

5.7 Expanding the Toolkit

To understand the ecological impacts of particular foraging strategies, optimal for-

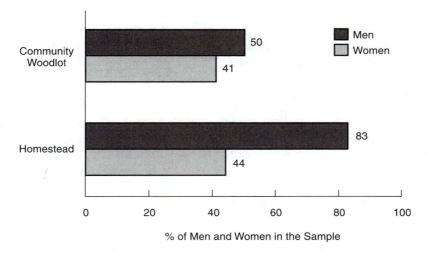

Figure 5.3. Women's Tree Planting in Zimbabwe.
The percentage of men and women who plant trees on community woodlots and homestead land. Based on interviews with 154 respondents (48 men and 106 women) in two villages (from Fortmann et al., 1997).

aging decisions can be built into broader population ecological models, in which the demographic profile of a human group and its various prey populations are tracked. Using dynamic models, it is possible to determine whether various measures of sustainability are met (chapter 3). For instance, Winterhalder and Lu (1997) found not only that forager-prey systems can stabilize without intentional conservation behavior (Figure 5.2), but also that the full suite of prey species available to foragers needs to be tracked. Their model shows that if hunters switch from one species to another when the first becomes scarce, as an optimal foraging model would predict, this can contribute to the depleted resource's persistence, at least if the forager responds to scarcity by searching in a new area or investing in a new kind of hunting technology. Another finding is that slowly reproducing prey species are particularly vulnerable if the human foragers can also rely on a quickly reproducing species; this is because the human population does not crash in response to the declining numbers of the slowly reproducing species. In the real world this might mean that if forest dwellers start keeping domestic ani-

mals (usually these reproduce quickly), this can lead to overexploitation and possible local extinction of the wild game species; the same may be true when foragers can depend on highly reliable gathered resources (Belovsky, 1988). An empirical study of how domestic cats are decimating local songbird populations in southern California vividly illustrates this point. If it were not for the daily meals that these "recreational hunters" enjoy in their owners' kitchen, cat numbers would decline in response to the increasing rarity of songbirds, giving these avian populations a chance of recovery (Crooks and Soulé, 1999).

We also need better methods to study the impact of human communities on their natural resource base (see chapter 4). As regards mammalian and avian prey, most studies determine the impact of hunting on prey numbers by simply examining hunters' encounter rates with prey at different distances from a settlement. There are several problems with this method. First, low encounter rates near villages may reflect the greater evasiveness of certain species that are heavily hunted or the departure of certain species from areas disturbed by human settlement (FitzGibbon, 1998); this is especially the case when data are collected on trails. As such, low encounter rates cannot necessarily be taken as indicative of overhunting, nor even indeed of local depletion. Second, many variables are likely to influence hunters' encounter rates with prey, other than how heavily the area is hunted (Hill et al., 1997). For example, people often establish their settlements at ecologically distinct sites, meaning that the prey base in the vicinity of settlements may be different from elsewhere. As such, multivariate analyses are needed to tease apart the effects of hunting from factors, both anthropogenic and nonanthropogenic, ideally at multiple sites controlling for differing ecologies (e.g., Peres, 2000). Third, local depletion due to hunting around settlements may not cause any overall threat to particular prey species. Harvesting is often intensified around villages and markets, yet (as discussed in chapter 3) some hunted species may be able to tolerate high local levels of exploitation if population sources are left intact and unhunted, as in the case of the culpeo foxes hunted on ranches in Patagonia (Novaro, 1995; see also Novaro et al., 2000). Innovative new research, done in collaboration with traditional foragers, assesses hunting impact with a method that provides a relatively unbiased way of isolating hunting from other factors affecting vertebrate encounter ratios, allows determination of the broader landscape effects of hunting at different distances from settlements, and requires little time input from expensive staff (Box 5.5).

Finally, we need to think much more carefully about the context in which resource harvesting occurs, and in particular whether there is a market. From the perspective of optimal foraging theory, markets reduce or eliminate the diminishing marginal returns or the value of the next item taken. Consider hunting: when a hunter is supplying only his family (or a small settlement), each animal he takes is worth less than the previous animal, because his needs are already met; in other words, the marginal value of each item declines. Markets undermine this self-regulatory phenomenon because the next animal, and the one after that, and the one after that can all be sold. In other words, each prey retains its marginal value to the hunter, whose behavior is now determined by the effectively limitless needs of the market rather than the quickly saturated needs of his household or village. There is plenty of evidence for this, as we saw in our discussion of the impact of the market on traditional hunting patterns in chapter 4 and will examine with the bushmeat trade in Box 11.3. Market access is also typically associated with development, which brings higher standards of living and levels of consumption and demand for meat. These are all issues that link the payoffs of individual stinting behavior to the broader social and economic context.

5.8 Policy Implications

Conservation is all about restraint. Scrutinizing such restraint from the purview of individual costs and benefits has suggested ways in which conservation policy and awareness initiatives can become more effective, and in a sense more acceptable.

One insight that arises from this angle of self-interest is that preferences for environmental conservation might be linked to sex, life stage, parenthood, social status, and life expectancy cues. Using evolutionary logic, such differences are predicted on the basis of how different individuals might value resources, how they discount this value over time, and how willing they are to entertain risk. In one of the first such studies Wilson et al. (1998) test the hypothesis that men will accept greater risks than women in acquiring, displaying, and consuming resources, and are therefore more likely to disregard or downplay environmental degradation. Part of the reasoning behind this hypothesis is that men, because they have a higher mortality risk at every age than do

Box 5.5. Studying Hunting with Hunters in Paraguay's Mbaracayú Reserve

The Mbaracayú Reserve was purchased by TNC in 1990 in conjunction with the Fundación Moises Bertoni for the protection of the second largest forested area of the Alto Parana formation of the Atlantic Forest. The law that created the reserve states that the resident Aché population is permitted to continue subsistence hunting and gathering using traditional methods. Because of this law, studies of Aché resource use patterns are critical to conservation planning for Mbaracayú. Though the Aché have lived at mission settlements since 1970, their economy is still centered around hunting mammalian game with bows and arrows on forest treks that last weeks or months. Meat contributes 60 percent of the calories in their diet, 95 percent of the weight of which is supplied by eight species of mammal.

To determine the factors affecting prey densities, diurnal transects were established in the reserve, and walked by a team of five local assistants and a data recorder. Each carried a VHF radio to communicate with the data recorder, who also carried a GPS unit. At each 200 m assistants recorded vegetation type, all encounters (sight or sound) with large vertebrates (direct encounters), fresh signs of large vertebrates such as spoor (indirect encounters), as well as evidence of hunters' activities. This is participatory action research (PAR, see chapter 11) insofar as all

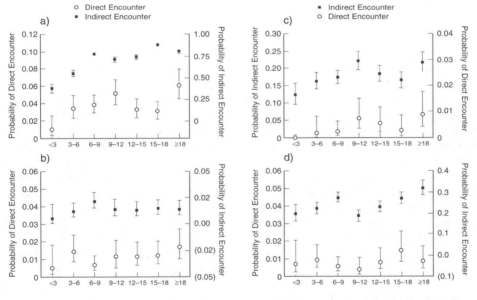

Distance (km) from Nearest Ache Access Point (DNAP)

Encountering Prey in Mbaracayú Reserve.

Encounter rates (and 95 percent confidence intervals) with (a) *Dasypus novemcintus*, (b) *Cebus apella*, (c) *Tapirus terrestris*, and (d) *Mazama* spp. as a function of distance from entrance to reserve. Reprinted with permission from Hill, K et al., *Conservation Biology* 11: 1339–1353. Blackwell Science Ltd., Publishers.

(continued)

(Box 5.5 continued)

local research assistants were Aché, experienced hunters, and trained for one week in data collection techniques.

Encounters each 200 m were analyzed by multivariate logistic regression, allowing for the isolation of multiple factors affecting the probability of an encounter, and were examined in terms of the distance from the nearest point of access to the reserve. Analysis of these encounter rates (see figure) showed that (1) no species had a lower encounter rate in hunted areas than unhunted areas without also showing a similar decrease in indirect encounter rates, suggesting that a decrease in the encounter rate with game in hunted areas is due to animal absence and not just increased wariness; and (2) locally depleted species (nine banded armadillos [*Dasypus novemcintus*], brown capuchin monkeys [*Cebus apella*], tapir [*Tapirus terrestris*] and brocket deer [*Mazama* spp.]) show normal densities at distances greater than 6 km from nearest points at which hunters can enter the reserve. The authors conclude that Aché do have a localized impact on some of their main prey. However, they note that evidence of local depletion due to human hunting is not sufficient to assert that a species will be lost in a large protected zones where humans are allowed to hunt in some areas (Hill et al., 1997).

women, discount the future more heavily than do women. Through an experimental scenario male and female students were presented with the dilemma of whether to turn the family farm over to hybrid corn production (profitable but requiring heavy chemical fertilization that renders the land unusable in sixty years) or to retain hay production (modest profits but sustainable). Men were more likely to choose the soil-degrading option than were women (Figure 5.4). Though these data support the simple hypothesis, other factors can intervene, reversing such sex differences (as we saw in Figure 5.3). With respect to policy, since time discounting may lie at the heart of why waste and inefficiency are so hard to eradicate from our behavior (Kacelnik, 1997) not only should conservation education schemes be designed differently for individuals with low and high discount rates, but conservation interventions should seek to modify high discount rates through manipulating incentives and taxes (an issue we look at further in chapter 9).

Evolutionary reasoning also accounts for the insidious ratchet of consumption. Easterlin (1995) has shown how people are happy only if they have a little bit more than everyone else, noting (in economists' terms) that our wants are relative. We not only compare ourselves to others (our fitness competitors), but we also want others to know how well we are doing. The idea that people compete with each other over what they own or procure goes back to Veblen's (1899) concept of conspicuous consumption: to demonstrate their status, high-ranking individuals, or the leisured classes, consume conspicuously. Veblen also recognized how such consumption escalates into increasingly wasteful displays: as soon as lower-status people obtain a status item normally reserved for the leisure class (or at least a cheap copy thereof), the wealthy must invest in even more costly and ostentatious items. Consumption becomes a treadmill, and advertisers can exploit our vulnerable psychological preferences for increased consumption. Frank (1999) has proposed a

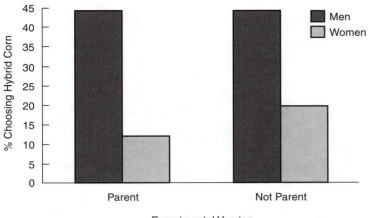

Figure 5.4. Sex Differences in Conservation Preferences.
Percentage of subjects choosing the soil degrading option (corn over hay) according to sex and parental status, when posed with a hypothetical dilemma (Wilson et al., 1998; and see text). Men were significantly more likely to choose the soil-degrading option ($X^2 = 6.6$, $p < 0.01$) than were women; neither parental status (which was distinguished in the hypothetical dilemma by the subject's status being either cast as a widowed parent with two children or left unmentioned), nor the interaction of sex of the subject by parental status, was significant by logit loglinear analysis. From *Behavioral Ecology and Conservation Biology*, edited by T. Caro, Copyright 1998 by Oxford University Press, Inc. Used by permission of Oxford University Press, Inc.

heavy tax on luxury expenditures that might reduce excess consumption without impoverishing everyone.

Another issue raised by this perspective is the very utility of environmental education as a strategy to change behavior. Educational initiatives are based on the assumption that if people knew more about the implications of their actions for the long-term health of the planet, they would reduce their consumption and fertility accordingly. This assumption is questionable if people discount the future too sharply (Penn, 2003). Though education can play a key supportive role in changing behavior (see chapter 10), it will be ineffective if there are no incentives to sacrifice short-term interests for the long-term good of humanity and the planet (Ridley and Low, 1993). As Gardner and Stern (1996) point out, people can be easily educated into the relatively cost-free practice of depositing cans into a curb-side recycle bin,

but not into environmental efforts that require greater effort and expense. E. O. Wilson recognized this too, with his claim that "the only way to make a conservation ethic work is to ground it in ultimately selfish reasoning" (1984, 131–32).

Accordingly, some evolutionists believe that environmentalists' appeals for personal sacrifices on behalf of the planet are doomed to failure. Ridley and Low (1993) suggest it will be necessary to rig economic options in such a way as to make environmentalism the most rational choice for the individual or corporation, or (in the phraseology of ecological economists) to put a true price on nature and its services, as we explore further in chapter 9. At present, good environmental practice simply does not pay for itself. If it did, laws to deter polluters or fines to punish cheaters would not be needed, and incentives would not be required to reward those who decide to bear the short-term

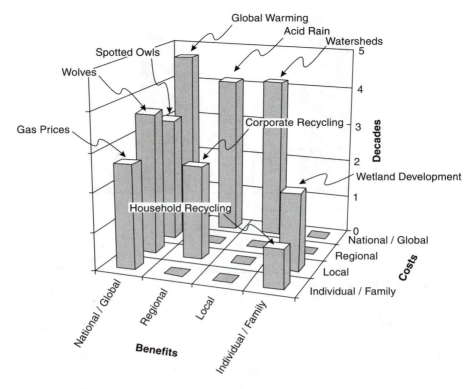

Figure 5.5 The Difficulty in Solving Environmental Problems.
Environmental problems plotted in terms of the spatial scale at which costs and benefits are experienced (horizontal axes) and the elapsed number of decades between costs being paid and benefits
being reaped (vertical axis). Reprinted with the permission of Cambridge University Press from
Heinen and Low, *Environmental Conservation*, Vol. 19, No.2 (1992): 105–116.

costs of conservation. The big question then
is who will put these incentives in place,
since politicians who raise taxes are not
popular. These broader policy issues are
tackled in chapter 9.

Without such a system of incentives and
disincentives in place, the difficulty in solving environmental problems can be viewed
as a product of two functions: how the costs
and benefits of environmental problems are
distributed across communities, be these local or global, and the period of time elapsing
between payment of cost and enjoyment of
benefit (Figure 5.5). As Heinen and Low
(1992) point out, the easiest problems are
those in which the costs are paid, and the
benefits derived, by the same individuals,
and the benefits come quickly, as, for exam-

ple, in household recycling. When costs and
benefits accrue at levels higher than the individual level, conflicts of interest arise; for
example, most people agree that too much
gasoline is being used, but no one wants to
pay the extra tax. When costs are accrued at
one level and benefits at another, resolution
can also be very difficult; for instance, in
the spotted owl controversy in the U.S.
Northwest, a moratorium on logging will
depress the regional economy (at least in
the short term, by cutting jobs) but will yield
the national (and perhaps global) benefit of
conserving old growth forest and its associated endangered species. Finally, where the
benefits of conservation are experienced
only in the distant future, these are likely to
be heavily discounted, even by individuals

with low discount rates; these, according to Heinen and Low (1992), are the hardest environmental problems to solve.

Even aesthetics is being brought into the purview of materialist and evolutionary ecological analysis, and may have policy implications. Emotional responses, both positive and negative, to the environment are not random. According to one school of thought, they may reflect our evolutionary heritage, such that we find particularly pleasing features that offer (or once offered) security and prosperity (Box 5.6). Studies such as this alert conservationists to the importance of determining the deep psychological and physiological roots of human responses to nature; effectively, such responses need to be fostered if the arguments for preserving biodiversity are to retain cogency in a world where most people look out onto concrete.

Negative emotional responses can also be harnessed to the service of conservation. As Ornstein and Ehrlich (1989) suggest, humans fail to respond appropriately to many environmental hazards, such as pollution, because the human mind bears at least some of its Pleistocene heritage. We simply do not respond to threats via our intellect as quickly or intensely as those that come via our senses. Forcing industrialists to put a nasty smell into odorless and invisible (but otherwise polluting) emissions is guaranteed to spark much more community action against the corporate enterprise than a local educational scheme that simply alerts people to the long-term statistical risks associated with the pollutant (Ridley and Low, 1993). Similar activism would surely emerge if polluted water were stained with a dye in proportion to its level of toxicity.

5.9 Conclusion

All organisms have evolved to deal with the problem of when to defer the expenditure or consumption of resources. Assuming individuals behave in ways that ultimately serve their self-interest, restraint in resource-harvesting strategies will be favored when its future benefits outweigh the present costs. This is most likely when the individual believes that the resource will still have value in the future. Thus, the emergence of so-called conservation practices within local communities, and indeed the prospects for solving all environmental problems, will depend on how costs and benefits are distributed across time. They will also be affected by whether an individual can be sure of personally reaping the benefits, raising the question of how conflicts of interest among individuals are resolved at the level of the community, a topic explored in chapter 6. Though the perspective of methodological individualism has brought rigor to anthropological analysis, it is worth emphasizing that neither stinting, nor the existence of a conservation ethic, assures true conservation outcomes. Independent assessments of the ecological consequences of different resource-harvesting strategies need to be done, work that will require population-level modeling and better field techniques.

Although chapter 4 showed the potential for ecological collapse in societies even with the most simple of technologies, the ideas introduced here suggest that this is not an inevitable aspect of human nature, but depends on many different factors, and in chapter 6 we will see that many communities have managed to prevent destruction of their resource base. The emergence and disappearance of conservation strategies is influenced by a whole range of factors specific to the ecology of the natural resources under consideration, the behavioral tendencies of individuals particularly with respect to time discounting, the institutions of the community, and the broader social, economic, and political sphere. Having dealt with individual-level analyses in this chapter, we turn to community issues in chapter 6 and to the wider regional, national, and global context in chapters 7 to 9. Only by broadening the focus from the individual can we seriously address the question of

Box 5.6. Evolutionary Aesthetics and the "Savanna Hypothesis"

How humans affect the environment depends at least in part on how we see the environment, and how we value and respond to these signals, visual or other. Orians (1998) proposes that humans have evolved emotional responses to the environment that over evolutionary timescales guided us to safety, food, and a prosperous life. As a result, our contemporary responses to nature, our habitat preferences, and

Photos of Differently Shaped Acacia Used in the Experiments.
Reprinted with permission of Standard Scientific and GH Orians.

(continued)

(Box 5.6 continued)

even our aesthetic values may reflect these past selection pressures. Within this framework, and inspired by the notion that humans have instinctive aesthetic preferences for natural environments and other species (biophilia, Wilson, 1984; and see Box 1.3), Orians explores the evolutionary-based response patterns of humans to various landscapes.

The "savanna hypothesis" yields two predictions. First, savanna-type environments should be favored over other biomes because of their critical role in the evolutionary development of our species, anatomically modern humans. Second, tree shapes that predictably signal high-quality savanna (a spreading, multilayered canopy and a trunk that branches close to the ground) should be preferred over shapes more common in low-quality savanna habitats (tall trees with narrow canopies in wet savanna and dense, shrubby-looking forms in dry savanna). A cross-cultural study of peoples' responses to different tree shapes was conducted using photographs (see photos). To control for other features of plant architecture and general habitat quality, photos of only one species of acacia (*Acacia tortilis*) were used. Trees selected for inclusion in the questionnaire varied in canopy density, canopy shape, trunk height, and branching pattern. Trees rated as most attractive by all three experimental groups (in the United States, Argentina, and Australia) are those in which canopies are moderately dense and trunks bifurcate near the ground as found in high-quality areas (Heerwagen and Orians, 1993; Orians, 1998).

Further evidence for the savanna hypothesis comes from an analysis of the drawings of a nineteenth-century British landscape designer, Humphrey Repton. In his "before and after" drawings, work we can assume was being made more attractive for potential clients, can be found the following note. "Those pleasing combinations of trees which we admire in forest scenery will often be found to consist of forked trees, or at least of trees placed so near each other that the branches intermix, and . . . the stems of the trees themselves are forced from that perpendicular direction which is always observable in trees planted at regular distances from each other" (cited in Orians and Heerwagen, 1992, 559).

A related prediction is that people will prefer environmental features that offer shelter from the physical environment, safety from predators, and vistas of their surroundings, such as caves, fires, climbable trees, promontories, and pathways. This has also received some support from the details of nineteenth-century landscape paintings, as well as by content analysis of Repton's "before and after" drawings (Heerwagen and Orians, 1993). Human environmental aesthetics is part of a growing field of environmental psychology (Gardner and Stern, 1996).

---

when and how human interests can be compatible with conservation objectives. Nevertheless, it is important to realize that any such investigation must start with individuals, seeing what they need to both stay alive and to compete within their own social context. This is critical, because the success of management programs depends on their acceptance by individuals, and it is easier to persuade individuals to accept recommendations that minimize disruption of their optimal choices. Viewing individuals as utility-maximizing agents may be overly reductionist, but it would seem to offer a firmer conceptual foundation than does the fragile assumption (explored in chapter 4) that individuals in non-Westernized communities live in harmony with their environment. In short, individual dynamics cannot be ignored, though clearly they cannot be considered in isolation. Accordingly, in the following chapter we widen our focus from the individual to the community that a person lives in.

CHAPTER 6

Rational Fools and the Commons

6.1 Introduction

Commons, or common-pool resources, have conventionally been seen as particularly vulnerable to overexploitation. In many respects this received wisdom has been overthrown in recent decades, with evidence that some communities practice deliberate restraint in their harvesting and resource use strategies. In place of the conventional wisdom now reigns the alternative view, that community-based ownership and management of a commons provides the best avenue for sustainably managing a natural resource. The reasoning for this currently popular position eludes many conservation biologists, especially those suspicious of community-based initiatives. And even those social scientists and ecologists who favor community-based approaches rarely scrutinize how and why community-based ownership and management might work.

This chapter traces the debate over the commons, starting from the assumptions of rational action theory espoused in chapter 5. Social dilemmas result whenever an individual's decision to maximize short-term self-interests leaves everyone in that person's community worse off than feasible alternatives. Social dilemmas clearly characterize the management of many common-pool resources. Everyone would like a clean, productive environment, just as they would like public goods such as safe frontiers and just government, but it is in no one's per-

sonal interest to shoulder a contribution to such initiatives. The study of how these goods are produced is therefore pivotal not just to political scientists but also to conservationists. The fact that some human communities appear to supply such goods, at least to a degree, and even to avert ecological disasters raises deep theoretical questions about cooperation and self-interest. How can we account for the quixotic tendency of human nature to alternate the pursuit of self-interest with the internalization of norms and rules that generate cooperation? And what are the features of common-property systems that potentially serve to assure sustainable use of common-pool resources?

We address these questions in this chapter. We start with the observation that common-pool resources, such as migratory pelagic fish, are at real risk of degradation when they are unmanaged (6.2). Some common-pool resources are nevertheless managed by communities, and though the ecological impacts of such regimes are rarely examined closely, apparently successful systems have persisted, sometimes over centuries (6.3). These successes necessitate a more careful scrutiny of the conditions in which individuals cooperate to overcome social dilemmas, based on game theoretical analyses and experimental economics (6.4). In fact, generations of cultural evolution may have instilled productive norms in human behavior and crafted rules that have the potential to solve social dilemmas (6.5). We then look at what has been learned about the design of common-property resource institutions (6.6), before turning to the implications of different property rights regimes for sustainable management outcomes (6.7). It is clear that community institutions for managing local resources have great potential in the design of conservation strategies, at least if they are properly understood. Conservation interventions that work on capacity-building (essentially enhancing community organization skills in the service of conservation) would do well to pay sharp attention to the institutions already in place, and how these can be strengthened and shaped for achieving new goals.

6.2 The Rational Fool Fumbles the Common Good

The last fifty years have witnessed a collapse in the productivity of the world's most important marine fisheries, most famously the Atlantic cod banks, which declined by 69 percent between 1968 and 1992, and the West Atlantic bluefin tuna stocks, which dropped by more than 80 percent between 1970 and 1993 (McGinn, 1998). Pacific and Mediterranean fisheries also show dwindling catches, and the extraction of fish worldwide, which peaked at 82 million metric tons in 1989, has now been in decline for over a decade (Watson and Pauly, 2001). Indeed, by 1997, 79 percent of the world's commercially important fish species were either fully fished, overexploited, depleted, or slowly recovering (FAO, 1997). As the top predators like cod and tuna go, fishers are switching to lower-ranked and less preferred prey, as the optimal diet breadth model would suggest (chapter 5). Furthermore, the seas are increasingly dominated by species lower in the food chain, inducing changes in the composition and abundance of marine animals and plants that endanger the healthy functioning of marine ecosystems (Jackson et al., 2001). As technology improves and demand increases, so pressure

on renewable resources grows more severe. Fishing vessels are now like floating factories, employing gear of massive dimensions, such as bag-shaped trawl nets large enough to entrap twelve jumbo jets (Safina, 1995).

There are two principal reasons why offshore fisheries are so prone to depletion, both of which stem from the physical properties of pelagic migratory fish. First, they are a common-pool resource, as are many grasslands, lakes, forests, and indeed the open seas (Berkes et al., 1989). Fisheries, like forests, are renewable common-pool resources, yet there are others that are nonrenewable (such as oil wells) or that are cultural products (public highways). Common-pool resources are defined as resources for which exclusion is difficult and joint use involves subtractability. Exclusion ranges from being excessively expensive, as in the case of large tracts of forest or grassland, to virtually impossible, as in the case of the oceans. Subtractability implies that each user's exploitation of the resource leaves less for the next user. It is because of this subtractability that individuals are fundamentally in conflict with other individuals over the use and management of the resource. Thus, if my cattle graze on the meadow, yours will find less to eat, and if you drive overweight freight on the highways, I experience pot-holed roads. Yet neither you nor I have the ability to secure total exclusivity of use, and we have to share our common-pool resources in joint use.

The second reason why the recent history of pelagic fisheries is so dire is the absence of well-defined property rights. As long recognized, oceanic fish defy clear property rights (Gordon, 1954). Prior to the twentieth century access to most pelagic resources was unregulated, free and open to everyone. In the last fifty years, however, the situation has changed. Oceanic fisheries are no longer open-access, but hedged by a web of treaties, conventions, and regulatory boards (Berrill, 1997). Nevertheless, because of the supreme difficulties in developing and enforcing a regulatory system that is acceptable to all interested parties, the system is really a free for all and open access persists. This is particularly the case with fish that migrate across international borders, as in the case of prized sockeye salmon that move from the waters off Alaska to breed in the rivers of British Columbia in the summer. When the U.S. and Canadian representatives of the Pacific Salmon Commission broke off annual negotiations over the permitted catch in 1997, Alaskan fishers took three times as many sockeye salmon as allowed under the existing treaty, desperate to hold onto their share of the beleaguered salmon population. Offshore U.S. fishers, Canadian fishers, and Indian groups living along the river systems find themselves in direct competition with each other over the dwindling stock. Without an effective regulatory system, a resource is essentially open-access. Regulations are broken and quotas exceeded, with fishermen using illegal small mesh and false log books. "Fish wars" occur worldwide at hotspots, such as in the northwest Atlantic and Southeast Asia. Furthermore, with 200 million people dependent on fish for a livelihood, it is hardly surprising policy initiatives toward regulation have been weak.

The potential fate of unmanaged (or open-access) common-pool resources such as pelagic fish, grasslands, and forests is vividly captured in the famous "tragedy of the commons" model. Hardin (1968) was concerned primarily with overpopulation, but his most definitive insight lies in his metaphorical example of how and why a resource that can be freely used by many is at such risk of depletion (Box 6.1).

Fundamental to the tragedy of the commons is a social dilemma. Social dilemmas occur whenever an individual's decision to maximize short-term self-interest leads to a situation in which all other participants in that person's community are left worse off than feasible alternatives. The study of social dilemmas, initiated fifty years ago in fisheries and wildlife management as the "fisherman's problem" (Gordon, 1954), is

pursued in many disciplines (Olson, 1965), appearing under such names as the *public goods, social trap, collective action,* or *altruism problem.* Typically they are analyzed through game theory, a tool used by economists to capture situations where the payoffs to individual strategies depend on what others do. The Prisoner's Dilemma is a game with a set of payoffs that sheds light on why many common-pool resources are overused, at least if they are open-access. The game captures the dilemma faced by two guilty accomplices held in separate cells and interviewed by the police. If both confess (defect), they both go to jail for three years. If both stay silent (cooperate) they will be sentenced to single-year sentences on a lesser charge that the police can prove. But if one confesses and the other does not, the defector (or cheat) walks free on a plea bargain whereas the cooperator, who stayed silent, gets five years. Since the prisoners cannot communicate, neither will risk staying silent. The result is that they both get the three-year sentence, two years more than had they cooperated. In the language of game theorists, individual rational strategies result in collectively irrational outcomes.

The difficulty a community faces in avoiding overgrazing, or many other environmental ills, is captured in a multiperson Prisoner's Dilemma (Figure 6.1). Should an individual cooperate in the management of the commons by stinting, limiting the number of cattle she grazes on the village commons (C-S), or should she defect (D), grazing as many cattle as is personally advantageous? It is clear from the C-S linear function that the more people who stint, the better the outcome for each of them, because the commons is well managed. Indeed, if all villagers stint, the net payoff to each individual is G. But it is also clear that because there is a personal cost to stinting, any particular individual does better to shift to the defection strategy, since this will raise her or his personal payoff; defection (D) always provides a higher payoff than cooperation (C-S). Self-interested individuals, the

Box 6.1. Freedom in the Commons Brings Ruin to All

To demonstrate the need for major policy changes regarding population and environmental pollution, Hardin (1968) proposed a metaphor. Imagine what would happen to an English village commons if each herder were to add a few animals to his herd. Individual herdsmen would find it personally profitable to graze more animals than the pasture could support, because each takes the full profit from that extra animal but bears only a fraction of the cost of overgrazing, the declining quality of forage. The inevitable eventuality is destruction of the resource on which the entire community of herders depend. This is the "tragedy of the commons." Technically Hardin's model captures a Prisoner's Dilemma game played by more than two players. Its logic is impeccable.

Hardin's classic paper is controversial for three reasons. First, the metaphor of a commons was unwisely chosen. Commons use in medieval England (and elsewhere in many parts of the world) was not unregulated. Under the aegis of the lord of the local manor, each household received specific rights to graze livestock, cut wood, dig turf, turn out pigs to eat acorns, catch fish, or take gravel, sand, or stone, subject to rules and regulations (Country Landowner's Association, cited in Ridley, 1997). In short, commons are not necessarily unmanaged, though they can be. Some persist to the present day, as in England's New Forest. The error in assuming that all common-pool resources are necessarily open-access, unmanaged, and hence subject to degradation (see text, section 6.3) was pointed out years ago by economists and others (e.g., Ciriacy-Wantrup and Bishop, 1975), and is now well recognized (Berkes et al., 1989; Feeny et al., 1990). Hardin's model is nevertheless pertinent to the question of whether individuals should follow rules and regulations (where these exist), and how they should behave if common-property institutions break down.

Second, the model assumes that individuals act independently as self-interested players, entirely unguided by exogenous expectations, rules, or institutions. In recent years evolutionary biologists, political scientists, economists, and others have tinkered with conventional rational choice models like the Prisoner's Dilemma, and shown how self-interested individuals engage in cooperation and punishment for their own good (see text, section 6.4). Furthermore, it may be that the Prisoner's Dilemma does not accurately portray the incentive structure of commoners, in which defection is always the favored outcome. If this is the case, a cooperative solution becomes more feasible (Box 6.4).

Finally, many took Hardin's model as a prescription for the privatization or nationalization of common-pool resources. In fact, it is now clear that neither privatization nor nationalization of a commons assures its sustainable use, nor indeed did Hardin believe this to be the case (see text, section 6.7). In fact, returning to the medieval English commons, historical research shows that the key factor contributing to the degradation of village common land was privatization. When feudal lords found wool production commercially profitable, they grabbed much of the commons for their own sheep. Later, as agricultural innovations increased the productivity of cultivation, originally unfenced commons areas were enclosed under the Enclosure Acts, ushering large tracts of the best land into the hands of private owners; this only intensified grazing pressure on the remaining commons. Transformation of a secure peasantry with clear rights to common land to landless laborers tied to the estates of the rich was the true "tragedy of the commons" (Ciriacy-Wantrup and Bishop, 1975), or rather the "tragedy of the non-commons" (May, 1992). Intriguingly, privatization is behind much of the contemporary tragedy of African commons as well (Galaty, 1994).

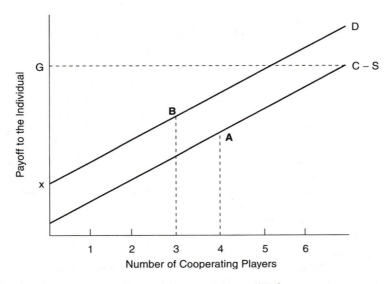

Figure 6.1. Multiperson Prisoner's Dilemma.
N individuals can either cooperate by stinting (C-S) or defect (D). If all individuals cooperate, the payoff to each is G. If, for example, four individuals cooperate (receiving payoff A), any individual who switches from cooperation to defection receives a higher payoff B, because of the avoided costs of cooperation. The predicted outcome is that no one cooperates, and all receive the payoff X.

actors in rational choice theory, will therefore push the outcome of the game to X (the overgrazing tragedy), an individual payoff that is considerably less than G. Since G was a possibility, a stark conflict exists between the outcomes associated with individual as opposed to group rationality, hence Sen's (1977) memorable term *the rational fool*, depicting the social moron who ignores the common good for his or her own private ends. The challenge posed by social dilemmas of this type is to find a way to move closer to the provision of benefit G.

Common-pool resources, precisely because exclusion is difficult and joint use entails subtractability, are just the kinds of resources on which social dilemmas turn. Accordingly, they have been studied as tragedies of the commons. Nevertheless, years of energetic debate over whether the tragedy of the commons is an appropriate metaphor for environmental abuse yields this insight: though the tragedarian imperative clearly has an inescapable logic, the model is appropriate only in situations where common-

pool resources are marginally managed, or where individuals can freely choose whether or not to follow rules and regulations. The kinds of questions that now emerge are these: Are fisheries, forests, and grasslands inevitably at risk of degradation, as in Hardin's grim predictions? Are environmental conflicts structured as harshly as Hardin's Prisoner's Dilemma scenario would suggest? Are we mere rational fools, in Sen's sense, prepared to eat the goose that lays our golden eggs? Luckily, the outlook is not quite so bleak as the fate of the pelagic fisheries would suggest.

6.3 Commons Classics

Over the last thirty years much work has gone into characterizing the nature of property rights and the implications of these for conservation outcomes. A common-pool resource such as pastureland can actually be held under a variety of different property regimes, and need not be unregulated as in

TABLE 6.1
Types of Property Rights Regimes with Owners, Rights, and Duties

Property Regime Type[a]	Owner	Owner Rights	Owner Duties
Open access (absence of enforced property rights)	None	Total	None
Common (communal or group) property	Collective (such as a village)	Control of access with right to exclude non-owners and regulate users	Maintain objectives defined by the village[b]
Private property	Individual (or corporate firm)	Control of access with right to exclude non-owners	Avoidance of socially unacceptable uses
State property	State (in name of citizens)	Determine rules of exclusion and use	Maintain objectives defined by the state[c]

[a] The term *property* is used by some to refer only to systems that include alienability (right to sell); we follow common-property theorists by using the term more broadly for control. Nevertheless, an important difference between private and common-property regimes is the ease with which individual owners can buy or sell a share of a resource.

[b] Hanna et al. (1996b) puts "maintenance; constrain rates of use." We rephrase this, since common-property regimes need not necessarily be designed to maintain sustainability.

[c] Hanna et al. (1996b) puts "maintain social objectives." We rephrase this, since the state's objectives may not necessarily be social, or in the interests of all citizens.

Source: Modified from Hanna et al. (1996b) Property rights and natural environment. From *Rights to Nature: Ecological, Economic, Cultural, and Political Principles of Institutions for the Environment.* Copyright Island Press.

open access (Table 6.1). It can be fenced off as private property, for exclusive use of the owner's livestock; in this system the owner (individual or corporate) has the right to lease out the grazing, or sell the land outright, and is obliged only to avoid infringing socially acceptable uses of pasturelands. Alternatively, the state can take control of the land; it appoints an agency with power to make decisions over how the pasture is to be used, ideally in line with the interests of its citizens. And finally, it can be owned and managed by a community as a common-property regime; under this system village members have the right to exclude nonmembers, and the obligation to manage the pasture in ways acceptable to the village.

Once property rights are distinguished from the resource per se, it becomes clear why Hardin's tragedy is predicted only when a large number of people make heavy demands on a single common-pool resource,

act independently, and have no institutions to regulate their use, in other words, when a common-pool resource is effectively open-access, as in many fisheries. Because the older term *commons* failed to differentiate between the resource type and the property rights regimes under which it is held, Hardin and many others erroneously assumed that common-pool resources were inevitably open-access and subject to degradation. Once the potential significance of common-property regimes was recognized, anthropologists, political scientists, and neoinstitutional economists embarked on field studies that examine whether and how communities manage their resources in ways that circumvent cheating, defection, and free-riding. Netting (1981) observed how Swiss farmers protected their upland grazing meadows (alps) as common-property regimes, with informal monitoring systems since at least the thirteenth century (see too McKean, 1992a for similar Japanese systems). Maass and Anderson (1978) described the 500-year-old irrigation system in Valencia (Spain), where farmers have enjoyed rights to water in return for maintenance work and monitoring the water uses of their neighbors. The *sereingueiros* (or rubber tappers) of Brazil organized a political movement that led to the legal establishment of common-property management in parts of Amazonia, ultimately as extractive reserves (Begossi, 1998; and see chapter 10), and in Thailand coastal fishers initiated their own harvesting regulations (Johnson, 2001). Many other case studies, revealing a wide range of coordinating institutions that have the potential to protect and manage common-pool resources, albeit in variously imperfect ways, subsequently appeared in a spate of edited volumes (e.g., McCay and Acheson, 1987; Bromley, 1992; Ostrom et al., 1994; Keohane and Ostrom, 1995; Hanna et al., 1996a; Berkes et al., 1998; Ostrom et al., 2002).

Common-property regimes are immensely variable, and systematic comparative studies are being initiated (e.g., Tang, 1992; Lam, 1996; Varughese and Ostrom, 1998; Meinzen-Dick et al., 2002), with the result that clear design principles can now be identified from reviews of both successful and unsuccessful cases (Table 6.2). Communities need to have distinct rights to their natural resource base, to coordinate closely among themselves over their harvesting practices, and to develop unambiguous rules about who can harvest what resources, when, and in what amount. They must also have the autonomy to modify these regulations as conditions change, both in the resource and in the community. In addition, users need to meet regularly, monitor each other's behavior, punish offenders, and have recourse to outside sources of policing should this be necessary. Most commonly these regimes prosper in small-scale communities, where relations are stable and where people communicate on a daily basis. They are more prone to collapse when trying to cope with many simultaneous changes in key exoge-

nous factors such as new immigrants, technological innovations, growing external demand for the resource, and changing relations with the central authorities (Bromley, 1992). Such collapses are not, however, inevitable. In some common-property regimes, as the condition of the natural resource base deteriorates (perhaps as a result of population growth or sedentarization), people actively restructure their property rights and management practices, as Berkes (1998) has shown in the case of the Canadian Cree (Box 5.2). Increasingly, common-property institutions are based on an amalgam of traditional (Box 4.2) and contemporary knowledge, technology, opportunities, and constraints, as with the management of exotic cash crops in parts of Nigeria (Warren and Pinkston, 1998).

Pastoralism provides a good example of how common-pool resources can be managed by communities. For many years ethnographers have described the precise institutions and norms whereby East African and other herders look after their grazing lands. They have taken pains to point out that, contrary to the received orthodoxy of thirty years ago (Brown, 1971), pastoralists do not cause inevitable environmental degradation. Careful ethnographic research has shown that although pastoralists do not own land privately, access to the pasture is by no means a free for all. Rather, it is managed as a common-property system, incorporating nested levels of user rights (that individuals enjoy as members of tribe, clan, and lineage), strictly observed seasonal grazing rotations, and protected reserves; specification of these areas is informed by detailed traditional technical knowledge (Niamir-Fuller, 1998). Conformity with the common-property system is maintained through a graduated set of sanctions, ranging from spiritual reverence to fines and (in extreme cases) excommunication imposed by institutional bodies (Gilles and Jamtgaard, 1981; Homewood and Rodgers, 1987; McCabe, 1990). Since pastoral common-property systems are easily destabi-

TABLE 6.2
Design Features of Successful Common-Property Systems[a]

1. Ongoing mechanisms to solve problems associated with exclusion and subtractability
   1.1 There is a clear definition of (i) the common-pool resource (e.g., the forest, groundwater source, etc.) and (ii) the individuals or households with rights to harvest resource products.
   1.2 There are rules that assign benefits and costs to users; these may specify harvest limits, technological constraints, labor inputs, and investment; these should be viewed as fair by all users, and be appropriate to local ecological conditions and the nature of the resource.
   1.3 Most individuals affected by the harvesting and protection rules (above) can participate in modifying these rules, such that those with most information and stake in the system have most voice; without this feedback from users, rules cannot be adapted to changing conditions over time and space.

2. Ongoing mechanisms for invoking and interpreting rules, and assigning sanctions
   2.1 There are monitors who actively audit the common-pool resource and the behavior of users; they should be accountable to the users, or be users themselves.
   2.2 Users who violate operational rules receive graduated sanctions, depending on the seriousness of the offense, from other users, or from officials accountable to users, or from both.
   2.3 Users and their officials have rapid access to low-cost, local arenas to resolve conflict among users, or between users and officials.

3. Institutional legitimacy
   3.1 The rights of users to devise their own institutions are not challenged by external governmental authorities, and users have long-term tenure rights to the resource.
   3.2 (For resources that are parts of larger systems) appropriation, provision, monitoring, enforcement, conflict resolution, and governance activities are organized in multiple layers of nested enterprises.

[a] Success is defined as resource systems and institutions enduring for a long time, albeit with collectively agreed upon modifications over time; note that this conforms primarily to social and institutional sustainability (see Box 2.5).
Source: Based on arguments in Ostrom (1990).

lized by changing social and ecological conditions, for example, when pastoral groups settle in villages around bore-holes or irrigation schemes (e.g., Niamir-Fuller, 1998), a careful look at how these institutions transform over time can teach us a lot about their adaptability and limits. Lane's (1996) account of the Barabaig response to land alienation in northern Tanzania suggests that some pastoral groups are committed to defending traditional community access rules despite severe grazing shortages. Probably more typical over the longer timeframe is Ensminger and Knight's (1997) account of how the common-property institutions of Kenyan Orma pastoralists have been modified over the last seventy years in response to internal and external factors, first becoming more exclusive and then moving toward

privatization (Box 6.2). The Orma case, and a similar story for the Boran (Hogg, 1990), reveals how the management of a common-pool resource shifts rapidly in response to both internal and external threats.

But what are the conditions that encourage people to form communal property regimes of this type in order to solve potential social dilemmas over the management of common-pool resources? We turn now to theoretical arguments developed by evolutionary biologists, economists, and political scientists for understanding cooperation.

6.4 The Cooperation Game

There are many fronts to the current effort in determining the conditions in which indi-

Box 6.2. Changes in the Management of the Kenya Orma Commons

The Kenyan Orma keep herds of cattle, goats, and sheep that are owned individually, but their grazing land was until recently managed as a common-property regime, as is typical among East African pastoralists. The first apparent challenge to the traditional system came in the 1940s and early 1950s when a small number of Galole Orma pastoralists began to settle around permanent water holes dug in dry season stream beds (Ensminger and Knight, 1997). Settlement is incompatible with pastoralism in East Africa because rainfall tends to be highly localized, and herders need to follow the new grass to find adequate grazing for their herds. As the settlements became larger over the next twenty years, the sedentary families experienced falling milk yields and reduced cattle fertility, presumably due to resource stress. To deal with this problem, the village elders declared a small area around their village off-limits in the wet season to the stock of anyone who did not live in the village, thereby protecting the grass for

Orma Herdsman from a Cattle Camp.
Courtesy of Jean Ensminger.

village use in the arid months to come (effectively creating a dry season grazing reserve). Initially these regulations caused no serious problem for the nomads, and conflict was easily resolved. But over the years the restricted area grew in size. By 1985 the villagers had to call in the police to keep nonvillagers out, whereas until that time sanctions had been informal and never entailed the law.

The second major shock to the traditional system came in the 1980s, and was intensified in the 1990s, with the incursion of Degodia Somali into Orma territory. By imposing new (this time external) pressure on the traditional grazing land, the Somali inadvertently furthered the cause of the sedentarists. In 1994 elders of the Galole Orma formally petitioned the government for recognition of a cooperative ranch, a demand very much in keeping with the general move toward privatization in Kenya's land tenure policies at the time. Although membership of the ranch would be open to all villagers, it entailed a fee, and as the experience in Kenya has repeatedly shown, group ranches tend to be the thin end of a familiar wedge: privatized farms for the rich, disenfranchisement for the poor, and ecological degradation for the remaining commons (Galaty, 1994). However, as Ensminger and Knight (1997) point out, many Orma, even the poor, believe that group ranches are their best option for securing legal control of their land and keeping out would-be settlers, Somali and others.

viduals cooperate to overcome social dilemmas. Work in these areas consists partly of game theoretical modeling, partly of social experiments in laboratory settings, and partly of observations in the natural world, and it lies across the disciplines of evolutionary biology, economics, political science, and psychology. Darwin himself was foxed by the extraordinary levels of cooperation he saw in ants, bees, and wasps. Why should organisms shaped by natural selection behave in ways that appear to compromise their self-interest and, ultimately, their reproduction? Over the last thirty years evolutionary biologists have successfully coaxed conventional rational choice models into generating cooperation and collective action to determine the conditions in which cooperation might pay (Dugatkin, 1998). Biologists have not been alone in this enterprise. Economists, who had since Adam Smith founded their whole discipline on the question "What's in it for the individual?" discovered from their experimental work that people appear to be motivated by something other than self-interest (Box 6.3).

In evolutionary biology several theoretical paths have been used to explain the emergence of cooperation, the principal of which are reciprocity, kin selection, and asymmetrical altruism; the fourth (cultural group selection) is addressed in the following section. Most attention has focused on reciprocity, both between two individuals and within groups. Trivers (1971) suggested that cooperative (and seemingly altruistic) behavior might arise if individuals exchange or reciprocate cooperative acts. The small cost you pay to help me could be easily compensated for if, sometime in the future, I return a favor when you need it. But there is a problem. How can the initial altruist ensure that others reciprocate? As Trivers realized, the system is open to the defection we saw in Figure 6.1, and selection will act against indiscriminate cooperators. Game theory is often used by biologists, economists, political scientists, and others to explore the dynamics of how cooperation

might arise through reciprocity. The Prisoner's Dilemma model has been their favorite toy. Though a variety of different strategies can produce stable cooperative outcomes (reviewed in Dugatkin, 1998), cooperation is dependent on there being a high probability of future interactions among participants (Axelrod and Hamilton, 1981), small groups (Boyd and Richerson, 1988), and extensive opportunities for monitoring the behavior of others (Axelrod and Hamilton, 1981; Ostrom et al., 1994). The opportunities for cooperation are enormously enhanced if there is a possibility of punishing nonreciprocators (Schelling, 1960; Boyd and Richerson, 1992; Clutton-Brock and Parker, 1995), though this raises the tricky "second-order problem" of who will pay the cost of punishment. Nevertheless, relatively large cooperative groups can emerge when group members punish nonpunishers as well as defectors (a "moralistic" strategy; Boyd and Richerson, 1992). Interestingly, in experiments with human subjects, Ostrom et al. (1994) find that individual players are willing to pay fees to fine defectors; in fact, players overuse fines, spending more than they should on fining than if they were acting in rational self-interest.

One assumption of the Prisoner's Dilemma model, that players cannot communicate before or during play, is clearly inappropriate when applied to humans. In experimental work with human subjects, there is a consistent, strong, and replicable finding that when individuals are allowed to communicate they are more likely to solve social dilemmas (Sally, 1995). Interestingly, the face-to-face component is important; electronic messages are not effective. And so, although in principle talk (at least when it is cheap) should make no difference (you could convince others to reciprocate and then chose defection for yourself, as depicted in Figure 6.1), this does not appear to be the case. Ostrom (1998) suggests that communication is so important because it allows exchanges of commitment, increased trust, reinforcement of norms, and group

Box 6.3. Experimental Games in Economics

Laboratory settings are used to present individual actors with choices concerning the production of public goods and the management of common-pool resources, set in a variety of differently structured games. A human subject, usually a college student, playing a game, perhaps for money, is not an ideal replicate of a man, woman, or child making decisions over how to allocate their energies in real life. Nevertheless, participants in these games are assumed to draw on values and decision-making processes that they have learned over their lives, thus responding to diverse incentive structures in meaningful ways. Game settings offer wonderful possibilities for testing hypotheses under controlled conditions, in which costs, benefits, risks, information, and numerous other important dimensions structuring the dilemma can be varied. In addition, experiments can be replicated and tested across different kinds of situational framing (Ostrom and Walker, 1997).

One predominant finding is that human subjects in controlled game setting experiments do not always follow the selfish strategy, something that economists see

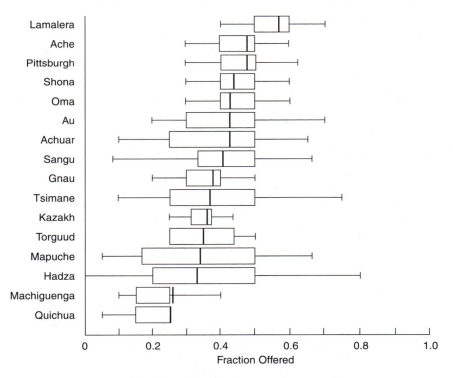

A Modified Box Plot of Ultimatum Game Offers.
Horizontal lines denote ranges, boxes interquartile ranges, and vertical lines within each box show mean offers within each society. (For the Machiguenga the mean offer lies outside the interquartile range and is represented by the vertical line just right of the box.) Figure used with permission of Joe Henrich.

(continued)

(Box 6.3 continued)

as an "anomaly" (Thaler, 1992). In fact, high levels of initial cooperation are found in many kinds of social dilemmas, albeit at levels lower than optimal for the group. Initially experimental economists put this down to the lack of intelligence of the players, but over the years many consistent findings call into question the validity of the assumptions of rationality and complete information. Subsequent empirical work has shown that cooperation is particularly high when interactants know each other (Bohnet and Frey, 1999), when they are allowed to communicate with each other, if they belong to the same artificially created groups (Dawes et al., 1988), and when the punishment of noncooperators occurs (Fehr and Gachter, 2000a; 2000b; see also Boyd and Richerson, 1992).

A good example of this kind of work comes from the Ultimatum Game, where two players split a pot of money. The first player (proposer) suggests a split with the responder. If the latter refuses her or his share, neither player gets anything. Under the rational choice assumption that players try to maximize their own incomes, the unique Nash equilibrium is that the proposer offers the minimal possible amount and the responder accepts anything. Evidence from a great number of laboratory experiments shows that proposers routinely offer 50 percent of the sum, even when the stakes are as high as three months' earnings, and responders rarely accept offers of less than 40 percent of the pot (Roth, 1995). A cross-cultural study of the Ultimatum Game (see figure) shows that offers range between 26 and 58 percent of the stake, confirming the apparent initial generosity of people but pointing to considerable intercultural variability (Henrich et al., 2001).

All in all, Nash equilibrium strategies are not good predictors of individual behavior in the kinds of games that economists and social psychologists have explored (Ostrom, 1998; Fehr et al., 2002; Gintis, 2000b). These results imply a tendency for humans to cooperate in ways that perhaps seem to be unique to our species. They also suggest that humans have an innate taste for costly punishment, a distaste for unfairness, or both, though this may well extend only to other members of our species. This is good news for conservationists, in that it suggests that cooperation over natural resource management schemes is feasible.

identity. Communication does not, however, guarantee collective action; communication becomes less efficacious when the stakes in a game increase, when the cost of providing a common good rise, and when it is difficult to monitor each individual's contribution (Ostrom et al., 1994).

Kin selection is the second path toward explaining cooperation (Hamilton, 1964; Maynard Smith, 1964). In a nutshell, kin may be less inclined to free-ride on the efforts of close relatives and at the same time be more tolerant of family members who do so. This is because an individual can enhance his or her own fitness by helping kin in proportion to their genetic relatedness. Frequently, cooperation over resources is based on family ties, not only in traditional societies but also in developed nations (Palmer, 1991). Nevertheless, kin selection, like reciprocal altruism, is less likely to promote cooperation as the size of the group increases and the degree of relatedness declines.

Explanations based on these first two

paths, reciprocity and kin selection, high-light a constellation of factors that favor cooperation: small groups of closely related individuals who maintain face-to-face communication with one another often over long periods of time, who have ample opportunities to monitor each other's behavior, and who possess some means of punishing wrongdoers. These conditions loosely characterize many, though by no means all, communities with successful common-property management systems (Bromley, 1992). For example, among the Orma (discussed in Box 6.2), most of the herders know each other, and many are probably related through various kinship ties. However, there are also successful common-property systems that exist at much larger scales. In Valencia the community of Spanish farmers sharing water rights has grown from eighty-four irrigators at the system's inception in 1435 to over 15,000 users (Maass and Anderson, 1978), and in southern California the common-property management of ground water reserves can involve the interests of whole counties (Blomquist, 1994). Furthermore, a comparison of self-governed irrigation systems shows no relationship between the number of users and various performance variables (Tang, 1992). It may even be that larger communities are better positioned to pay the transaction costs of monitoring and management, as suggested by Agrawal and Gibson's (1999) comparison of the effects of community size on forest reserves in India (see also Meinzen-Dick et al., 2002). In general, depending on the nature of both the resource, for instance, its subtractability and its speed of regeneration, and the form of collective action (labor inputs, sanctioning, attendance of meetings), we might well expect group size to encourage cooperation in some cases and not others.

A third path toward explaining cooperation has considerable potential for explaining cooperative solutions to resource dilemmas in larger and more complex communities. Inspired by Olson's (1965) theory of privileged groups, economists are exploring the impact of asymmetries among players in generating cooperation. Most models of cooperation assume that any given strategy yields the same payoff to all players. But what if some forms of cooperation are less costly for some players than others or, alternatively, more beneficial? Biologists have been interested in this possibility, modeling situations where a minority of individuals is willing to incur costs of enforcement in return for an extra share of the benefits (e.g., Clutton-Brock and Parker, 1995). In fact, a certain amount of heterogeneity seems to facilitate the solution to social dilemmas insofar as members of a small privileged group may find it individually advantageous to support the costs of collective action (Olson, 1965; McKean, 1992b; Ruttan, 1998).

This has interesting implications for the study of common-property management. A game theoretical analysis of why some pastoralists reserve pastures for later use in the year, especially when these lie temptingly close to their homesteads, shows that wealth differences among herders shape the payoff structure in such a way that men with large herds favor the cooperative (or conserving) strategy more than men with small herds (Box 6.4). A finding like this does not mean that cooperation is always in the interests of the rich. Rather, it suggests that in some situations particular sectors of the population may be more predisposed to engage in solving social dilemmas than other sectors. It also carries a more sobering message. Because high levels of heterogeneity can encourage punishment, common-property institutions that appear to work well may in fact conceal quite brutal levels of enforcement; as such, issues of equity within the community may require special scrutiny (Ruttan and Borgerhoff Mulder, 1999). The broader topic of environmental justice is discussed in chapter 7. The point here is simply to stress that common-property institutions that appear to be successful in sustainable resource management may not necessarily be equitable, as is so often assumed.

Box 6.4. Asymmetries among Herders: The Barabaig Case

Barabaig livestock herders live on the semi-arid grasslands of Tanzania's Hanang District, maintaining fixed homesteads throughout the year on the Barabaig Plains. At the start of the wet season young men and women set up temporary camps for the family herds on the distant Basotu Plains (Lane, 1996). There are no permanent water sources at these wet season pastures (WSPs), so when the rains end the herds are driven back down the Rift escarpment to the Barabaig Plains. Throughout the next four to eight months grazing becomes increasingly scarce on the dry season reserve (DSR), such that in years of normal rainfall animals lose condition, and in years of drought they are forced to enter more densely wooded areas with associated elevated risks of disease and predation. Critically, herders need to preserve the DSR grasslands surrounding their permanent homesteads; if they fail, all households suffer. As Lane concludes, Barabaig favor seasonal grazing rotations because they know that "not only are these areas [Basotu] more productive at these times, but . . . pastures (nearer home) need time to recover if they are to be fully productive at a later date" (1996, 108). Since Barabaig elders punish herders who make incursions into seasonally protected areas, we assume that for some herders at least there are costs associated with driving the herds to Basotu. These are likely to be costs in labor and time.

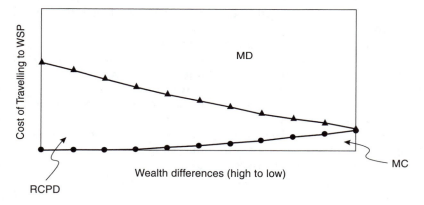

Thresholds for Cooperation and Defection among Herdsmen.
Both wealth differences and travel costs affect whether or not herders cooperate by traveling to wet season pastures. Reprinted with the permission of Cambridge University Press.

Because restraint (or stinting) on the part of some individuals opens up the possibility for cheating on the part of others, the decision to leave the DSR can be posed as a frequency dependent game (Ruttan and Borgerhoff Mulder, 1999). The decision analyzed is whether or not a herder should "cooperate" by moving his livestock to the WSP during the rainy season (thereby protecting the DSR), or whether he should "defect" and remain in the DSR all year round. Furthermore, since pastoral communities characteristically are made up of herders with different sized herds, this heterogeneity is included in the game. Analysis of the game (see figure) reveals three evolutionary stable spaces. Above both lines is mutual defection (MD). Below

(continued)

(Box 6.4 continued)

both lines it is to the advantage of both parties to cooperate (MC), and there is no temptation to defect. Between the two lines is an area "Rich Cooperate, Poor Defect" (RCPD), where the herder with more animals should always cooperate while the herder with fewer animals should defect. Two of these patterns are quite predictable: no one cooperates when the cost of using the WSP is very high, and everyone cooperates (MC) when the costs are low (and when wealth inequalities are mild). The interesting RCPD outcome emerges when the costs of using the WSP are moderately high and the inequalities in wealth are sharp, in other words, where some individuals with very large herds are primed to benefit from the WSP. In short, the protection of dry season reserves is encouraged by wealth asymmetries (Borgerhoff Mulder and Ruttan, 2000). Given that the rich herders tend to be elders, they have the traditional rights to impose fines on herders that fail to observe grazing regulations. Once the punishment of rule breakers is incorporated into the model (Ruttan and Borgerhoff Mulder, 1999), the social dilemma of preserving the DSR is even more easily solved.

An intriguing analysis of Nepalese small-scale irrigation systems shows how subtly these asymmetries might work (Lam, 1996). In traditional irrigation systems there is a lot of maintenance work that needs to be done to keep the system working, both at the headworks (e.g., a dam) and throughout the canal system. Because of the laws of gravity, farmers at the higher end of the canal (known as "headenders") have natural advantages over farmers near the bottom of the canal ("tailenders"). They tend to be less careful with their water, and to overlook fixing leaky parts of the canal, especially if these lie downslope of their fields. These asymmetries between headenders and tailenders potentially cause problems in organizing a fair and effective irrigation system, since headenders will take advantage of their comparatively privileged position, and will ignore tailenders' water claims. However, and this is the key, this is only the case if the amount of labor headenders need to get water is less than they can provide on their own, in other words, if they do not depend on tailenders' labor for their water. It turns out that in many irrigation systems the maintenance needs at the head-

works are so great that headenders must rely on tailenders' labor, in return for which they spare tailenders their fair share of water. Lam detected these dynamics as a result of a Nepalese government initiative to upgrade farmers' water supplies by putting in concrete headworks and lining the canals. After these "improvements," village productivity quite unexpectedly declined.

According to Lam's analysis, permanent headworks result in a substantial drop in the amount of labor allocated to system maintenance, because they critically alter the bargaining power among headenders and tailenders; headenders are no longer willing to agree to the water claims of tailenders, once the labor contribution of tailenders is not essential to the headenders' water supply. Consequently, only a small proportion of the village fields received adequate water. A case like this reveals the fragility of village-level cooperation in the face of externally induced changes.

6.5 Culture, Norms, and Cooperation

Most social scientists, unlike economists, long ago rejected the notion that individual behavior is dictated purely by self-interest, stressing rather the extent to which it is constrained by institutions, expectations, values, and norms, the very stuff of culture. They argue that narrow self-interest is in practice brought in check by a variety of social norms and institutions. Our everyday experiences suggest that this is a sensible point of view. An awful lot of human behavior can be construed as trying to get what we want within a reasonable interpretation of the rules, expectations, and dictates of the communities we belong to. But how do these norms arise, and what about cheating, the more skeptical biologist or economist might ask?

Across the disciplines intriguing theoretical advances on these questions are beginning to emerge. Runge (1981; 1986) argues that in communities where people are poor and resources unpredictable, individuals are so highly dependent on one another that defection is exceptionally costly (see also Hyden, 1990). For example, a man will not run an extra animal on the commons if, as a result of community tradition, this disrupts future beneficial interactions with neighbors; the man who does may, for instance, face problems in settling a dispute or in finding a wife for his son. This line of argument is particularly popular among ethnographers who deeply appreciate the intricate interdependence among members of small-scale communities (e.g., Lane, 1996; Peters, 1987). If such community issues affect people's reasoning, the commons problem is best seen not as a Prisoner's Dilemma but a game of "Assurance" (Sen, 1977), in which the payoff to cheating when others cooperate is lower than that of cooperating. Village norms, kinship obligations, religious traditions, and charismatic leadership might provide this assurance, by coordinating people's behavior (Alvard and Nolin, 2002). An intriguing case of coordination among farmers' organizations (*subaks*) mediated through religious structures has been documented for Bali (Box 6.5).

Others are looking further into the mechanisms that might stabilize cooperation in a population of highly interdependent actors. One important consideration is the role of reputation (Nowak and Sigmund, 1993; 1998; Leimar and Hammerstein, 2001). If cheating robs you of respectability, it may not be worth it. Another fundamental issue is the role of the emotions. Emotions, or moral sentiments (as Adam Smith called them), can be thought of as problem-solving devices designed to enable highly social creatures to resolve the conflict between short-term expediency and long-term prudence in favor of the latter (Frank, 1988; see also Sen, 1977). Finally, it may be that humans are not entirely rational in everything they do. Ostrom (1998) offers a useful framework whereby we can think of rationality as a family of models, ranging from complete to quite restricted (or "bounded") rationality, depending on the circumstances. From this perspective, the complete rationality implied by conventional rational choice models is merely a limiting case, applicable when social institutions and cultural constraints break down. Of interest here is Grabowski's (1988) point, unfortunately all too true, that when resources become scarce individual selfishness is particularly apt to rear its ugly head. Furthermore, even where traditional norms exist, people still have the choice of whether to conform, with either course of action potentially being in their own best personal interest.

The fourth path (not examined in the previous section) toward explaining cooperation, cultural group selection, is consistent with much of this recent work on common-property management emphasizing the role of tradition and morality. It also provides a theoretical justification for how such norms might arise or persist. Essentially, the idea is that a norm ensuring cooperation over common-pool resource management, which makes a group more productive than a group with a norm sanctioning unlimited exploitation, will persist over time, since the non-

Box 6.5. Coordinating the *Subaks* of Bali

In Bali rice is grown in paddy fields fed by irrigation systems dependent on rainfall. Weirs divert streams through a series of tunnels and canals to terraced hillsides, where households are organized into different *subaks* (or farmers' organizations). The flow of water not only irrigates the rice crop, but controls soil pH, stimulates the activity of microorganisms, circulates mineral nutrients, stabilizes soil temperature, and prevents erosion. Most important, at a larger scale the simultaneous flooding and draining of blocks of terraces controls rice pest populations by temporarily depriving these pests of their habitat by removing all standing water. How long these drying out periods should be depends on the species of rice pest, but if too many farmers synchronize their cultivation cycles, peak irrigation water demand for the rice is not met. Water sharing and pest control are therefore opposing constraints. How are cropping patterns among *subaks* coordinated? Empirical research (Lansing, 1987) and simulations (Lansing and Kremer, 1993) suggest an answer: the Goddess of Waters, and her intricately linked network of shrines and temples. At these holy sites representatives from each *subak* regularly convene for ritual purposes and to coordinate the scheduling of water (and thus cropping patterns) between different *subaks*.

Using a simulation model based on the rainfall, river flow, irrigation demand, rice growth stage, and pest levels for 172 *subaks* in two watersheds, Lansing and

Maps Showing Simulated and Real Data.
(a) The run of the simulation model showing highest average harvest, and (b) the actual distribution of synchronized cropping patterns in the traditional system of water temple networks. The different symbols represent distinct synchronized cropping patterns (i.e., *subaks*). Reproduced by permission of the American Anthropological Association from American Anthropologist 95(1). Not for sale or further reproduction.

(continued)

(Box 6.5 continued)

Kremen (1993) show that when *subaks* act independently there is high pest damage, and when cultivation is entirely synchronized across *subaks* rice productivity declines because of water stress. Peak productivity is achieved with local-level coordination among *subaks*. Further simulations show that this level of coordination can be achieved by a simple rule, that after each annual cropping cycle, each of 172 *subaks* copies the cropping pattern of its most productive neighbor. The distribution of cropping patterns found in the simulation run yielding the highest harvest closely resembles the distribution of temple systems across the two watersheds (maps in box). This suggests that the Balinese traditional water temple system may provide the means whereby optimally coordinated decisions are made. Indeed, the temple system itself may have arisen as a result of continuous trial-and-error adjustment by *subaks*.

The events of the Green Revolution support this interpretation. In the mid 1970s in many parts of Bali native Balinese rice was replaced with high-yielding varieties. Farmers were instructed to ignore temple scheduling, and to plant as often as possible. In these areas productivity dropped significantly and pests became a major problem (Lansing, 1987). It is quite possible that disruption of the traditional coordinating institutions was responsible for these changes.

competitive group will disappear along with its norm. While genetic group selection is controversial in biology, the conditions favoring cultural group selection are much less strict; this is because people tend to conform to the cultural expectations of the group that they live in and nonconformists are disciplined (Boyd and Richerson, 1985). Specifically, if socialization entails conformity (even docility, Simon, 1990; see also Gintis, 2000a), then individuals who join the group will adopt cooperative attitudes along with other locally adapted skills. In this way social norms that allow a community to persist may be favored by cultural evolutionary process, regardless of whether these norms are consistent with or antagonistic to an individual's self-interest.

In short, the last two sections have shown that humans behave neither as utterly autonomous individuals (as assumed in chapter 5), nor as entirely·conformist beings. Recent developments in game theory show that cooperation in large groups is a logical possibility, even when people have evolved

to look out largely for themselves. These insights into the conditions favoring cooperation can be of immense value in guiding the design of institutions to produce common goods, like conservation.

6.6 The Study of Common-Property Institutions

Research into common-property regimes has been unusually well organized. By 1985, when the National Research Council organized an international conference on common property, researchers already had a conceptual framework that could be used to collect and analyze information about common-pool resources and their management (Oakerson, 1992). As a consequence, researchers all focus on certain key elements to these systems (Figure 6.2). A simple framework like this has the merit of clarifying the key components of a common-property system.

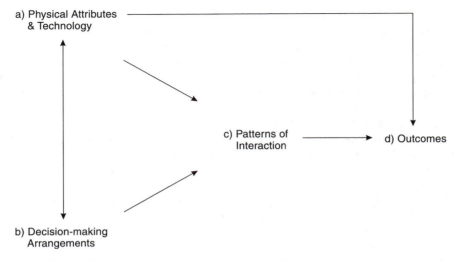

Figure 6.2. A Framework for Analyzing the Commons.
(a) The physical resource (and the available technology for its harvest) affect (i) its subtractability (its capacity for supporting multiple users without diminishing the aggregate yield available to all the group), and (ii) its excludability (the extent to which individuals can be prevented access to the resource; see text). It is important to identify the physical attributes of the common-pool resources and the technology used to appropriate its yield, since all common-pool resource problems are rooted in these critical constraints in nature and available technology. (b) The decision-making arrangements refer to the authority relationships that specify who decides what in relation to whom. They will include rules about how the commons is used, about how these rules are established, and about how regulations of the broader society can impinge on commons management. Patterns of interaction refer to the lacunae between rules and observed behavior, in other words, the unobserved mental calculations of individuals who make choices. Given (a) and (b) individuals make choices from which emerge patterns of interaction. These decisions can be independent (the free rider strategy) or cooperative, an issue addressed in section 6.5. (d) Outcomes refer to the physical outcomes produced by the patterns of interaction, in effect, the environmental impact (Oakerson, 1992); though not explicitly recognized by common-property theorists, such outcomes may result also from biological and ecological interactions, contingent on human acts. Figure from *Making the Commons Work*, published by Institute for Contemporary Studies, Oakland, CA 94612.

First, it is critically important to distinguish between the resource itself, for instance, the fish harvested from a lake, and the property rights regime under which the resource is held. It was because the older term *commons* failed to distinguish between a certain resource type and its management that Hardin proposed (and many still believe) that common-pool resources like fisheries and forests are unmanaged and hence inevitably subject to degradation, as we emphasized earlier. This distinction granted, the intrinsic nature of the resource will in-

evitably affect the kind of property rights regimes with which it can be managed. As such, a resource needs to be examined in terms of its size, spatial and temporal availability and predictability, mobility, speed of regeneration, and whether or not there exist valid and reliable indicators of its condition. A good example of how physical constraints imposed by a resource shape common-property institutions and sustainable management comes from the northeastern United States' Maine lobster fisheries, where the geography of the coastline affects the defensibility of the lobsters. Here there is a system of territoriality that has been in place since the 1870s, with fishermen affiliating themselves informally with specific "harbor gangs" whereby they gain permission to set lobster pots (Acheson, 1988). Effectively, where territories are defended around islands the lobsters are much better managed than where territories are focused around mainland fishing ports. In the latter, uncontrolled mixed gang fishing is taking place on the periphery, and there are severe signs of overexploitation (Berrill, 1997). Conversely, out in the bay fishermen patrol sharply defined boundaries, observe self-imposed trap limits, and are effectively creating a conservation area (Acheson, 1998). Key to the success of parts of the fishery is the geography of the natural resource.

Second, this framework highlights the importance of determining what exactly constitutes the decision-making forces. Is there any coordination among the people who consume, harvest, or produce from the common-pool resource? Is there an appropriator organization, a group of users who recognize established procedures to deal with the resource? Do they meet regularly, discuss their common problems, and recognize some commonly accepted rules for who has access to the resource and under what conditions? Do they have ways to settle their disagreements? Do they recognize who is, and who is not, a member? Appropriator organizations can vary enormously: in some cases, a local oligarch may dictate harvesting rules; in others, matters may be put to an open democratic forum; and in yet others, regulation may be highly ephemeral. As such, user organizations can be hard to identify and study. Nevertheless, only with such organizations does the possibility exist that individuals will switch from independent strategies for exploiting a common-pool resource to the more costly, coordinated strategies that assure conservation and sustainability, as we saw in the Balian water temple systems.

Another value of a framework like this is that it throws into relief the diversity of "outcomes" that might be measured. Different investigators, reflecting personal, political, and disciplinary biases, are concerned with different outcomes, such as the state of the resource, the incidence of rule breaking (poaching), the institutional persistence of the decision-making body, livelihood security, or local empowerment. Even if analysts agree on a common criterion to evaluate success (such as efficiency; e.g., Oakerson, 1992), these abstract concepts need to be converted into operational measures, which themselves may measure different things. In one case efficiency, technically defined as the ratio of benefits to costs, might be deemed low if at least one person could harvest more of the resource without anyone else harvesting less, whereas in other cases efficiency might be judged low if the harvest is not sustainable over time, if policing is expensive, or if economic inequities escalate in the community (see also Berkes, 1996). Another commonly used criterion, sustainability (e.g., Feeny et al., 1990), has its own problems since ecological, social, and economic dimensions are quite distinct (as we saw in Box 2.5) and often mutually incompatible. Furthermore, a set of institutions designed to manage a single resource sustainably may be inappropriate for the management of the full suite of resources implied in the term *biodiversity conservation*, as we saw first with Leopold's dilemma on the Kaibab Plateau in chapter 1 and encountered again several times when tracing the broadening agenda of conservation in chapters 2 and 3. In sum, the definition of success is often vague and, ironically, often narrowly operationalized.

Perhaps most important, this framework is useful as a diagnostic tool. Initial inquiry focuses on outcomes. What is happening to the resource and its community of users? To answer this question we can work backward (Oakerson, 1992), by tracking interactions among users. Maybe the rules are not being followed, maybe there are asymmetries that allow some users to raid the resource and then move on? There are many possibilities here. If nothing problematic is uncovered, we need then to look at whether the decision-making arrangements are appropriate to the suite of resources given the particular technology. Are the quotas correctly set? Is the level of policing and monitoring appropriate? Beyond this, we need to examine whether broader features of the state neutralize, or render inappropriate, the common-property regime. For example, do perverse incentives exist whereby the state subsidizes unsustainable harvests (chapter 7)?

6.7 Property Rights, Management, and Sustainable Outcomes

The "tragedy of the commons" model is heavily criticized for its apparent political implications. If commons are inevitably destroyed, what alternatives are there other than privatization and nationalization? Hardin is commonly read as promoting state or private management, and for this reason many social scientists see the popularity of the model as "one of the greatest tragedies of recent times . . . the degree to which Hardin's tragedy of the commons has been accepted," an adage voiced in Furze et al. (1996, 80). In fact, Hardin was well aware of the dangers associated with these latter two courses. His favored solution, "mutual coercion mutually agreed upon," refers not to the arbitrary coercion by irresponsible bureaucrats but coercion agreed upon by the majority of people affected (Hardin, 1968, 1247; though in later years he appears to have leaned more toward privatization, Hardin, 1993). These highly charged issues have stimulated a burgeoning literature on property rights, showing that overuse can occur under any regime.

In principle, private property should allow for successful exclusion of illegal users

and sound management for subtractability; this is because diffuse costs of failing to stint (in Hardin's metaphor) are now suffered personally, not shared with the community. In practice, private property is only partially successful in excluding illegal users and effecting sound management, for several reasons. First, private property ensures economic but not necessarily ecological sustainability. This is because private proprietors can externalize the costs of environmental damage; externalize in this sense means allowing the costs to be borne by others, not the owner. For instance, as Hardin himself noted, the owner of a factory can pollute the stream that runs through his property since the current will simply carry away the effluent; similarly, a farmer can grow crops until the best soil has blown off with the wind, sell the produce, and buy virgin land elsewhere to start again. The costs are transferred to the less privileged, or to nature, substituting a tragedy of the commons with a tragedy of the sole proprietor. Under these conditions, the impact of private ownership on sustainable practices and conservation outcomes will likely depend on the availability of alternative sources of investment with higher rates of return, bringing us back to the discount issues discussed in chapter 5. Second, private owners can find it very expensive to keep out illegal users; the fishing and hunting rights of medieval lords were notoriously violated by the likes of poachers such as Robin Hood. Third, it is clear from research in the developing world that there are enormous difficulties in moving from common to private property regimes, with excluded communities balking at the legitimacy of private claims; these difficulties are often referred to as transaction costs. In Kenya, for example, government initiatives promoting privatization have been unraveling for some years because of local resistance (Haugerud, 1989; Ensminger, 1996; see also Place et al., 1994), and offer no easy solutions to common-pool resource management (Lane, 1998).

State ownership regimes may be unavoidable in situations where other property rights cannot be relied on, but often they also do a poor job both in regulating harvests and in excluding illegal use. Administrative law is rightly feared for ancient reasons, and "*Quis custodiet ipsos custodes?*" or "Who shall watch the watchers?" captures Hardin's fears in this respect. It is well known that state owners easily succumb to political pressures exerted by interest groups, and thus are prone to sanction destructive short-term use (as with the excision of protected areas in Peninsular Malaysia, Box 2.2). In fact, governments often claim natural resources precisely because they wish to avail themselves of new wealth. The forests of India are a good example. Throughout the latter part of the nineteenth century village forestlands were converted into de jure state property (e.g., Guha, 1989a). The motivation was simple: build an extensive rail network on a bed of wooden sleepers, and stuff the colonial coffers with hardwood timber revenues. Over time, at different rates in different areas, villagers' rights to the forests they had formerly managed as common-property systems were eroded. Furthermore, even if sustainable management had been the state's intention, its national budget was not used to achieve this goal. Even more seriously, state appropriations were often signed in a faraway capital, with minimal accommodation for the customary rights of local populations, generating only more bitterness (Panayotou and Ashton, 1992). Guha (1989a, 55) describes this widespread process as an "alienation of humans from nature." Use of the forest became individualized and uncoordinated, effectively open-access. Increased pressure on the few remaining forested zones left to villagers could not be handled by the old common-property institutions, just as the English commons crumbled when the best land was privately fenced off (see Box 6.1 again). These same processes have been studied in Java (Peluso, 1992), and more recently in Nepal, with the interesting twist suggesting that if the state reinvigorates common-property institutions deforestation abates (Box 6.6). The implications of cases like this are now leading to a reversal in national policies in some countries, with governments aiming to re-create common-property regimes for both grassland (Lane, 1998) and forest management (Poffenberger, 1994; see social forestry below).

Common-property regimes have their own inherent tendencies toward resource degradation that can be linked to imperfections in the design principles outlined in Table 6.2. One persistent problem is lack of legal recognition of the local users' rights (e.g., Sporrong, 1998) and often a concomitant insecurity of tenure rights. Comparative studies of deforestation at the international (Bohn and Deacon, 2000) and interregional (Alston et al., 2000) level show that tenure insecurity generally increases deforestation (unless the clearing entails large capital investment, as in the case of oil extraction; an oil company would not invest heavily in an area from which it risked being evicted). At the national level evidence is mixed, but instructive; communities consisting of squatters in Peru and Brazil with no security of tenure deforest twice as much as do legal owners; however, once families obtain land titles (effectively enhancing tenure security to the level of private property) they are strongly induced to clear forest, because they now have the chance of intensifying cash crop production (Bedoya, 1996; Alston et al., 2000; and see the dangers with private property outlined above). Another frequently noted problem for common-property regimes is the impact of changing technology, immigration, and new markets. In this chapter alone we have seen the difficulties arising from immigration for the Orma, from new rice varieties for the Balinese, and from concrete dams for the Nepalese; in chapter 10 we will consider the bushmeat trade. On the other hand, Berkes (1998) takes a more optimistic stance, arguing that external jolts to a system can trigger the invention of new management institutions (see also Dove,

Box 6.6. Reviving Traditions in Sagarmatha National Park

Sagarmatha was set up as a national park in 1976 and as an early UNESCO World Heritage Site (WHS). This was in response to escalating deforestation resulting from an unmanaged trekking and mountaineering industry, and was spurred by Sir Edmund Hillary's environmental activism regarding deforestation in the area. Right from the start there was an effort to include in the plans the interests of the Sherpas, the indigenous communities made famous by their feats on Mount Everest, which they call Sagarmatha. The design of the park kept village islands inside the park, and allowed limited timber harvesting, yak grazing, and agriculture. Sherpas were incorporated into the advisory council for the park. Conflicts nevertheless soon arose over timber rights. The Sherpas wanted to cut more, so they could build and cook for the trekkers staying at over 150 tourist lodges in the national park. The park authorities responded by relaxing the rules slightly, but a decade later the relations between park and people were at rock bottom, with incidents of outright violent conflict between Sherpas and national park officials sent to police timber use. In the words of an erstwhile friend of Hillary's, "Hillary first brought sugar to the lips of the Sherpas, but now he is throwing salt in their eyes" (Hillary 1982, cited in Nepal, 2002, 753).

According to Stevens's (1997c) analysis, what really riled the Sherpas was that their own resource management institutions had been ignored. Traditionally the Sherpas had an array of institutions controlling how much forest could be cut, where, and under what conditions, regulated by *shinggi nawas* (forest guards) elected by villagers. Conventional regulations included no wood being harvested in the sight of a temple or on steep slopes. When the forests of Nepal were nationalized in the 1960s the *shinggi nawa* began to lose prominence, but the coup de grace for these local dignitaries came with the national park, when a state property regime was imposed practically overnight. Since the villagers felt the forests were no longer theirs, timber cutting escalated. In effect, a common-property regime had been converted into state property (as a park), and because the national park guards were not recognized the land became open-access.

In a move of extraordinary perspicacity the members of one village (Kunde) decided to reinvest their traditional forest guards with authority. When Mingma Norbu Sherpa, a resident of Kunde, became the first Sherpa administrator for the national park in 1981, the initiative taken first in his home village was encouraged throughout the area. *Shinggi nawas* did such a good job bringing timber-cutting under control that the national park arranged to pay them honoraria from the Himalayan Trust; the park also returned tree-felling revenues to the local communities. By the early 1990s villagers' distrust of the national park had virtually dissipated. In Stevens's opinion the success of the *shinggi nawas* lay in their local support. Herein, unfortunately, too lay the seeds of disaster, since the government department under which national parks administration fell was not comfortable with this degree of local management. In the 1990s the traditional system was undercut and tree-felling permits again had to be obtained from the national park. Sherpas complained, apparently with some justification, that each national park warden makes his own rules.

In recent years local institutions, such as the Tengboche Monastery, have begun to address social and ecological problems in the area arising from protectionism and tourism, since the national park authority has few community-orientated activities (Nepal, 2002). Furthermore, recent political changes have encouraged local people to express their grievances against park-imposed regulations. With the demise of the traditional institutions, the conflicting goals of the park and the local communities remain unresolved.

1995). Noting that there is no good ethno-historical evidence for any intentional conservation among the Canadian Cree prior to the fur trade (see Box 5.2), Berkes proposes that effective management of beaver populations began only after the serious depletion caused by the fur trapping industry. In other words, people can respond to exploitation by inventing (or redesigning) their resource institutions.

While no single type of property rights regime works efficiently, fairly, and sustainably in relation to all common-pool resources, different regimes are particularly suited for different kinds of resources. Based on satellite imagery, which shows the scarring of the earth where thin topsoil has been exposed or covered with drifting sand, it seems quite clear that grasslands are better conserved under common-property regimes (Mongolia) than private (China) or state management (China and Russia, Sneath, 1998). Mongolia has allowed its pastoralists to continue their traditional group-property institutions, which involve large-scale movements between seasonal pastures, while on their grasslands both Russia and China have imposed state-owned agriculture and permanent settlements. Similar effects of property rights regimes on savannas are seen in East Africa: land conversion and declines in dominant grazing species numbers are far more prevalent in areas with private land tenure and good market access than in areas with communal (and even state) property regimes (Homewood et al., 2001). For irrigation systems farmer-owned and community-managed regimes function more efficiently in terms of crop productivity than do government-managed systems, as shown by quantitative work in Nepal (Lam, 1996) and Bali (Lansing, 1987). In the Swiss Alps Stevenson (1991) has compared the economic performance of private and common-property regimes, and found that private regimes are more efficient at lower productive elevations, whereas in the higher and more remote zones communal systems performed well. Studies such as these have alerted policy development specialists to the fact that property regimes must fit well with the economic, geographic, ecological, and cultural milieu in which they are to function. They should also be interpreted with care since different factors are important at different geographic scales. For example, Homewood et al.'s (2001) demonstration of the significance of private property and market access in driving land conversion and wild animal densities at the transnational scale should not suggest that human and livestock populations are unimportant factors locally.

Four conclusions can be reached. First, whatever property rights regime is in place, it must be clearly defined and effective. Many of the worst excesses of degradation have occurred precisely because stakeholders were unsure of their rights or had no power to enforce them. Poorly defined property rights have an insidious way of reverting to open access, as we saw in the case of Sagarmatha. It is in this context that so much emphasis has rightly been placed on ensuring security of land tenure for indigenous, local, and resident communities. According to the arguments presented in chapters 4 and 5, the expectation that local communities will work toward conserving their natural resources is mere pie in the sky without clearly defined and secure rights of tenure.

Second, categorizations of property rights must reflect the way people think about and relate to their environment. A schema like Table 6.1 is useful for didactic and organizational purposes, but the real world, particularly beyond the reaches of Roman law, is more complicated, and for many reasons. First, property rights rarely come in secure bundles. Indeed, the whole concept of ownership, or tenure, needs to be examined according to duration (how long it lasts), breadth (what rights are entailed), and assurance (the certainty with which the right is held; Place et al., 1994). Second, property rights are commonly attached to a lineage or household and are rarely enjoyed equally by everyone falling into that grouping. Innovative work on gendered property rights shows, for example, the fragility of many African women's property rights, since these are commonly tied to household membership and hence the stability of marriage, as in the Zimbabwean tree planting study reported in Figure 5.3. Finally, nature disarticulates in numerous ways, and different property owning regimes cut her at quite different joints. In elaborating the concept of "tenure niches" Bruce et al. (1993) point out how different aspects of a resource can be subject to distinct property rights. Thus, a field may be held privately before harvest and as a common-property system for grazing when in stubble. In a swidden system, rights vary as a function of how much labor has been invested in the garden. And even a single tree can offer exclusive honey collecting rights to one person, fruits and bark to another, and firewood for all the village, even though it grows in the privately owned field of yet someone else. With such property regimes, where tenure niches are not only complex but rarely static, it is quite clear that the categories shown in Table 6.1 are only guides to an intellectual map.

Third, and critically, there must be trust between the body granting property rights and the recipients of these rights. The tremendously innovative "social forestry" programs, initiated in India and elsewhere across Asia to halt the deforestation attendant on state ownership (outlined above), demonstrate how important such trust can be. Villagers abjure cooperation if they are not confident that the benefits of the forests will ultimately be theirs (Box 6.7).

Finally, there is a danger of coming away from the common-property literature believing that local-level institutions are a panacea for all. Local-level institutions, by definition, are limited in their ability to deal with broader regional problems (Wyckoff-Baird et al., 2000; and see chapter 11). Herein lies the need to provide some over-

arching organization to coordinate the efforts of local organization, from the regional to the international scale. Sometimes, this is achieved by traditional institutions that ensure coordination (though still at a relatively local scale). The success of the farmers of Bali in distributing water among *subaks* in such a way as to both irrigate the fields and control major pest infestations is a good example (see Box 6.5 again).

It is here that outside agencies have a key role to play. While they must be careful not to dismantle currently effective local institutions, they also have to forge critical links beyond the local level: in assuring legitimacy of local users, introducing new technology and training where necessary, settling disputes that cannot be resolved locally, monitoring resources at a broader scale than just the local project area, and buffering local common-property institutions from destabilizing events, such as a market collapse, migration, or warfare. Indeed, national (and international) institutions, such as NGOs, can pick up transaction costs more effectively and cheaply than can local communities. In the contemporary world of integrated markets and global processes state protection of community-based tenurial systems becomes a necessary condition for success, as Alcorn and Toledo (1998) show for Mexican forests. Debate nevertheless revolves around how these umbrella organizations should be structured so as to protect the integrity and function of local level organizations (Ostrom et al., 1999). Clearly, if new centralized institutions are designed to provide a single, comprehensive organization that will integrate and rationalize local organizations, then we are back to all the problems associated with state management and its disruption of local governance. An alternative approach, which holds the promise of supporting local-level institutions and addressing their limits, is to nest local organizations within a larger institutional environment that facilitates coordination among them and can deal with broader-scale prob-

Box 6.7. Social Forestry: Bihar and Beyond

Since the mid nineteenth century, large areas of forests across the Indian subcontinent were declared state lands. These lands were managed for production, displacing an estimated 300 million rural resident users by the 1990s and, for reasons outlined more fully in the text, left less than 10 percent of these areas under good forest cover (Poffenberger, 1994).

To counter the shocking trend in deforestation social forestry was introduced in the 1960s and 1970s, marking a shift from commercial timber production to the sustainable production of nontimber forest products (NTFPs), in line with the World Conservation Strategy and the Brundtland Report (see chapter 2). Implemented in fourteen of twenty-five Indian states (Krishnaswamy, 1995) and across many other Asian countries (Dove, 1995), the social forestry program emphasized tree planting, specifically putting species with commercial value into individual holdings (farm forestry), developing village woodlots (often for firewood), and reforesting degraded village or state forests (which often entails controlling livestock grazing or wood collection in areas that have the potential to grow back from stumps and coppices). Policies were put in place in states such as West Bengal, Bihar, and Orissa to encourage such cooperative forest management systems (Poffenberger, 1994), with seemingly remarkable results. In Bihar, for example, the number of trees planted increased from 31.8 million in 1981 to 256.1 million in 1989 (Krishnaswamy, 1995).

For the most part these projects subsequently collapsed, largely because ownership and access issues had not been addressed (Chjatterjee, 1995). Most of the trees planted in Bihar did not reach maturity. Villagers chose to harvest them instead, on the suspicion the plantings had been for the exclusive benefit of the forest department. The fact these trees were often planted on state land and consisted primarily of exotic and commercially important species (such as eucalyptus, when villagers had wanted traditional timber species and fruit trees) reinforced this belief. Subsequently, new initiatives were experimented with, under different names, such as *joint forest management*. The focus here was explicitly on government land, with local people setting up village forest committees to oversee the management of these forests, but again participation was low and bitter mistrust prevailed. In a perspicacious critique of community forest management developments in Bihar, Krishnaswamy (1995) argues that these failed because they were centrally managed, implemented by international donor agencies, and ineffective in increasing forest resources to village communities. And in an equally negative review of similar initiatives across Bangladesh, China, India, Indonesia, the Philippines, and Thailand, Dove (1995) observes that benefits from community forest management did not trickle down to the poor because of institutional obstacles, namely, the state and its continued vested interest in forests. As Dove bitterly notes, in an atmosphere of mistrust it is only when forests lose value that decentralized management becomes an option.

There are, however, exceptions, and these successes are revealing. These are where forest-dependent communities faced with growing hardships have themselves decided to protect nearby forests from further degradation. Here there are no issues of trust, and villagers themselves monitor tree-cutting and illegal cattle-grazing in the forest. These self-initiated projects are spreading across Bihar's Chotanagpur's Plateau (Krishnaswamy, 1995) and have had other successes in Orissa (Poffenberger, 1994). The success of these schemes derives in part from the villagers' initiative, and in part from both the declining value of forest land and the emphasis of international aid donors on community control. There remains, however, the challenge of how to deal with management as forests regain value. As always, at the heart of these conflicts lie property considerations.

lems that local organizations share in common (see Table 7.1; and Little, 1994; Niamir-Fuller, 1998). This is a theme we return to under the label of co-management (chapters 7 and 11).

6.8 Conclusion

This chapter opened with the observation that common-pool resources are easily depleted, and with a simple model (predicated on rational action theory) that accounts for such depletion. Empirical evidence nevertheless suggests that common-pool resources are not necessarily open-access, as Hardin's model presumed. Common-property regimes have existed in various communities, and in places continue to exist. These institutions are not anachronistic relics of mere anthropological interest, but flexible devices with potential value for addressing current environmental problems if they are properly understood. It seems that generations of culture and experience have instilled productive norms in human behavior and crafted rules that have the potential to steer us away from commons tragedies.

There are, however, some provisos. First, common-property institutions are generally designed to manage a single resource, and their effectiveness in preserving multiple species as implied by biodiversity conservation is not well established. Second, local-level management is not a panacea for all evils (see chapter 11). To be sure, local communities command intricate environmental knowledge (chapter 4), and local communities can reach solutions unattainable at higher levels, where communication and coordination are a big problem, as reviewed here. But, on the other hand, many contemporary environmental problems stem from circumstances that transcend individual communities, such as immigration, new technology, increased market demand, changing expectations, national policies, and global environmental forces. Problems of this scale require a more coordinated set of responses than can be achieved at the level of the community. As a result, the links between a local community and its resource base must be studied in the context of a broader range of economic, cultural, historical, and global influences. It is to this scale, the emerging field of political ecology, that we now turn.

CHAPTER 7

The Bigger Picture

7.1 Introduction

Throughout the 1980s and 1990s Hardin's model stimulated an array of studies showing how ecological degradation can often be traced far beyond individual actors and local communities to the broader political power structure. An English folk rhyme of the late eighteenth century captures this:

The law locks up both man and woman
That steals the goose from off the
     common
But lets the greater felon loose
That steals the common from the goose

Recognizing such dynamics in England's seventeenth- to nineteenth-century enclosure movement, May (1992) relabeled Hardin's "tragedy of the commons" a "tragedy of non-commons" (or better, "tragedy of the enclosure"), noting that as formerly communally owned land falls into the hands of a few manorial lords, the increased pressure on the remaining commons is actually a tragedy of privatization (see also Ciriacy-Wantrup and Bishop, 1975). In a similar vein many studies in the developing world now document how colonial powers destroyed common-property systems, turning them variously into open-access, private, or state property (Berkes, 1996). These observations ushered in a new approach, political ecology, which deals with the broader institutional, cultural, and political aspects of human/environmental interactions. Through the lens of political ecology we now have a kaleidoscopic image of the vast array of social and political causes of environmental destruction. At the root of political ecology is the axiom that it is not just technology that determines human impact on the environment, but a combination of technology with economic markets, ethical standards, political ideologies, religious conventions, and (inter)national law. More pointedly, the question of how communities use (and abuse) their natural resources is recast as how forces external to local communities stimulate resource depletion. Attention is therefore drawn to framing local conflicts within the broader question of who has power and why, realities that rarely fall within the analytical ambit of a conservation project (even though they often impinge on its day-to-day pursuits). Political ecology also serves to raise concerns about human rights and the moral dilemmas entailed in designing conservation strategy.

The insights of political ecology bring social and economic realities home to conservation biologists, and ultimately suggest conservation strategies (such as co-management) that are more likely to be sustainable in ecological and institutional terms. We therefore start this chapter by introducing the principal objectives and defining characteristics of political ecology (7.2). We then look at its application to the emotive population-environment debate (7.3) as an illustration of the scale within which new conservation initiatives must be implemented. We go on to examine from a political ecological stand-point two cherished terms in modern conservation thinking, biodiversity (7.4) and community (7.5). Finally, we examine the consequences of emphasizing politics over ecology (7.6), before reaching the conclusion that the invaluable broad scope that political ecology brings to resource management and conservation comes with a heavy dose of ideology (7.7).

7.2 What is Political Ecology?

As social scientists groped to get a handle on the environmental crisis in the developing world, a new field emerged—political ecology, which can be loosely defined as the politics of environmental change, or alternatively, the political dimension of human/environment interactions. Most fundamentally political ecology springs out of a new engagement of the social sciences with human rights (Johnston, 1997), and is linked closely to the influential field of environmental justice. Environmental justice itself emerged initially from concerns with environmental hazards for impoverished neighborhoods in the developed world (Bullard, 1994); soon, however, it had brought broader aspects of health and food security within its ambit, before spreading into the literature on the developing world, where demands for justice arose in struggles over forests, rivers, lakes, and grasslands at threat from large-scale corporate interests in mining, plantations, and irrigation (Lane, 1996; Johnston, 1997).

Historically the field of political ecology can be traced to more tractable intellectual developments. First, social scientists were appalled by the influence of authors such as Ehrlich (1968; see also Ehrlich et al., 1995) and Hardin (1993), both of whose work had been interpreted, albeit in overly simplistic terms, as claiming that poverty and over-population lie at the root of environmental damage. While there is an inescapable truth to the claim that poverty is a fundamental problem, focus on the marginalized peasant who sets fire to yet another acre of forest to feed her six children may not be the most productive avenue for either exposing the underlying causes of undesirable environmental changes or averting their occurrence in the future. Second, political ecology flourished within anthropology, feeding on a reaction against an earlier school (cultural ecology) that had been concerned almost exclusively with the homeostatic processes internal to the community under study and had ignored broader political processes (Bryant and Bailey, 1997).

The intellectual roots of political ecology lie with a group of radical geographers who began to question whether "natural" hazards and disasters (like famines and landslides) were really natural (Watts, 1983). They would focus on a particular environmental problem (like soil erosion) that occurs in so many places across the world and at different junctures in history, to see what generalizations could be drawn (see Bryant and Bailey, 1997 for a review). For example, an early political ecology classic showed that the 1970s Saharan famine was in some sense not a famine at all (Franke and Chasin, 1992). The problem was not so much a shortage of food as a failure on the part of political and economic structures with regard to food production and distribution. Similar perspectives were applied to land degradation (Blaikie and Brookfield, 1987). Stimulated by these penetrating analyses, geographers began to dip into the fields of history, sociology, and anthropology, as what had once been strictly geographic subjects became increasingly contextualized within broader cultural, regional, and ultimately global developments (Vayda, 1983).

From the very start, this nascent field was highly political. Initially it embraced Marxism, though the ardor of this affair waxed and waned over the years (Bryant and Bailey, 1997). Political ecology subsequently became almost synonymous with the study of resource extraction (mining, logging, fishing, cash cropping, etc.) in a capitalist world, often betraying a somewhat morbid fascination with the destructive worldwide spread of capitalism in the nineteenth century. Colonial states are portrayed as wielding power over their subjects through the use of "tools of empire" (Headrick, 1981), notably quinine (see chapter 8), the machine gun, the steamboat, the railway, and the telegraph. The reading is compelling (Peluso, 1992) and the logic often convincing (Hecht and Cockburn, 1990).

Now that colonialism has been worked so hard, political ecologists draw on a more eclectic range of theoretical sources for determining how power mediates human/environmental interactions. Avenues include gender studies, poststructuralism, social movements theory, and rural sociology, and the work is published in the journals of many different disciplines, such as environmental history, anthropology, sociology, geography, and economics. Despite their differences, political ecologists can probably all agree that environmental problems threatening the developing world are not simply a reflection of the failure of specific policies or markets, remediable through technical "fixes" (as the World Bank tends to see it; Bryant, 1997). Rather, these problems arise from broad political and economic forces that operate both within local communities and at regional, national, and international levels, constraints that are not easily changed.

Five particularly productive themes now characterize political ecological approaches to the study of conservation and protected areas. First, there is a strong emphasis on moving away from case studies that are purely local toward a consideration of all actors, local to global, and all intervening levels, from peasants, local businesses, district administrators, and national governments to international organizations. Thus, Neumann (1998) traces the current challenges facing Tanzania's Arusha National Park through a web of connected themes—a London-based conservation organization that flourished under the British Empire, German

and British land policies during the colonial period, a succession of colonial district administrators with different briefs and sensibilities, the International Monetary Fund's structural adjustment program, peasant agricultural strategies, behavior of park guards with conflicting loyalties to their employers and their local communities, and escalating contemporary land pressure on Mount Meru. Through the lens of political ecology the picture of villagers encroaching on Arusha National Park is retold as one of powerful outsiders, in many guises, encroaching on the lands of villagers and the villagers' counter-strategies (see Figure 2.3). Wells (1992) encapsulates this multilevel approach with respect to the environmental politics of protected areas. He shows how the economic benefits from protected areas are graded across geographic and socio-political scales; though difficult to measure and varying from site to site, the benefits are usually very limited at the local scale, increase at the regional/national level, and are potentially substantial for the transnational/global community, whereas the economic costs follow an opposite trend (Table 7.1; see also Balmford and Whitten, 2003 for a global application). A detailed empirical study of the economic incentives for rainforest conservation in Madagascar highlights how useful it can be to disaggregate by scale (Kremen et al., 2000; see Table 9.3). It shows that while conservation generates significant financial benefits over logging and agriculture both locally (ecotourism, NTFPs, etc.) and globally (carbon value), at the national level the financial benefits from industrial logging far outweigh those contingent on conservation.

A second defining characteristic of political ecology is that it is actor-based, focusing on all the people and organizations that are involved in conflicts over environmental issues. Who are the stakeholders, the people whose interests are directly challenged by changes in the natural environment? What are the political consequences of a new natural resource management scheme, and who

TABLE 7.1
Comparing Protected Area Benefits and Costs at Three Spatial Scales

	Benefits[a]	Costs[b]
Local Scale	Consumptive benefits (if permitted)	Damage caused by protected wildlife to humans and their property Forgone harvests Lost economic potential of the land if put to other uses
Region/National Scale	Recreation/tourism Watershed values	Budget outlays by governments for both acquiring and operating the area to be protected Lost economic potential for regional and national development
International/Global Scale	Biological diversity Nonconsumptive benefits Ecological processes Education and research	International aid[c]

[a,b] Potentially the most significant benefits and costs associated with conservation, using as an example the establishment of a protected area, considered at different scales.

[c] International support to biodiversity conservation in developing nations is, for the most part, a very small percentage of developed nations' budgets (Balmford and Whitten, 2003).

Source: Modified from Wells (1992, 237). Reproduced by permission of The Royal Swedish Academy of Sciences from *Ambio 21.*

is particularly vulnerable? In relation to this latter question, political ecologists have been concerned that a community's most poverty-stricken members suffer not just from environmental degradation such as soil erosion, but also from conservation policies, as we saw in the critique on protectionism in chapter 2. Rather alarmingly, impoverished populations are also shown to be at risk from certain economic "improvements," as, for example, when eucalyptus plantations established on apparently barren lands in an Indian aforestation program deprived the poor of weeds they used for fuel (Bryant, 1992). In some of the most sophisticated political ecological analyses, changing land use and conservation practices can be traced back to the intersections of the interests of different constituencies within a community, as with Ensminger and Knight's (1997) analysis of the rise and fall of common prop-

erty among Orma pastoralists in northern Kenya (see Box 6.2).

The third clear characteristic of political ecology lies in its deep commitment to a historical approach. Regional histories are proving to be highly illuminating. At one level they tell sad, but often highly readable tales of environmental degradation in a particular region, as with Schroeder's (1999) depiction of the chaos caused by a succession of different development policies on both the environment and the social fabric of a Gambian village. Regional histories make for stirring reading, as for Java (Peluso, 1992), India (Guha, 1989a; Gadgil and Guha, 1992), and Amazonia (Hecht and Cockburn, 1990), and allow for useful analyses of how natural resources are differently viewed and valued by social actors at each level of analysis, from peasant to president. Bringing history to the evaluation of a contemporary conservation conflict can shed light on which development initiatives work and why (see the Sagarmatha example, Box 6.6). In fact, it would be a good idea for people working in the fields of conservation and development to elicit local histories for each community incorporated in the project area, insofar as even within a small and relatively homogeneous area different communities respond very differently to development initiatives, and in ways that are often interpretable only in terms of very specific ecological challenges, economic opportunities, or political players (Walters et al., 1999).

Fourth, political ecological analyses pay special attention to the suite of socioeconomic factors implicated in environmental destruction, most commonly class, gender, and race. Within this framework feminist political ecology has arisen as a distinct subfield (Box 7.1). While some of the debates among ecofeminists have become somewhat academic and philosophical, empirical research labeled as feminist political ecology (or related terms), designed to be sensitive to local gender differences, has been extremely rewarding. For example, Fortmann et al.'s (1997) study of a tree-planting project in Zimbabwe (see Figure 5.3) showed how it is property rights, marital security, and gender identity that hamper conservation initiatives, not the usual candidates—poverty and overpopulation. Gender differences rooted in different patterns of time discounting are also implicated in women's involvement in activist groups, such as the Chipko resistance movement against deforestation in the Himalayan foothills (see Box 8.1); in one instance, a potato-seed farm was to be established by cutting down a tract of oak forest and was supported by the local men because it would bring cash income. Women argued that the scheme would take away an area they used for fuel and fodder, thus adding 5 km to their fuel-collecting journeys. Knowing that cash in the men's pockets would not necessarily benefit them or their children, the women protested, and were successful (Agarwal, 1992).

Finally, political ecology attends to the social construction of natural resources by social actors at each level, from village headman to international institutions. Particular attention is given to how different stakeholders value and culturally define land and other natural resources (Fairhead and Leach, 1996), and the recognition that such concepts are never innocent of political import (Bryant and Bailey, 1997), as we discuss further below.

In short, political ecology spans the local to global continuum, revolves around stakeholders with often competing interests, is grounded in history, and pays particular attention to issues of race, class, gender, and meaning. In the next section we look at a political ecological perspective on population and deforestation in Latin America, before turning in subsequent sections to how the tools of political ecology can be used to examine issues that are both pivotal to contemporary conservation initiatives and that are deeply disputed.

7.3 Tropical Forest Destruction and Population Growth

Writings on the role of population in resource depletion have a long history (Malthus, 1959; Davis, 1963; Boserup, 1965) and are highly polarized. The central axis to the debate runs loosely along the divide between ecologists and social scientists, though there are mavericks in each camp. Ecologists attribute environmental degradation in a relatively straightforward way to population growth (Durning, 1989; Low and Heinen, 1993; Meffe et al., 1993; Pimentel et al., 1994), evoking a "population bomb" (Ehrlich and Ehrlich, 1990; Wilson, 1992). The stance of social scientists, led primarily by economists, is less well known. They argue variously that the impact of population is more limited (Lappé and Schurman, 1989; Simon, 1990; Tiffen and Mortimore, 1994), that "overpopulation" is a fiction, that inequity and exploitation are largely to blame for environmental damage,

---

Box 7.1. The Hen Has Started Crowing

If a woman speaks out at a village meeting for forestry management in Nepal, men say, "The hen has started crowing" (Hobley, 1991, 148). Ecofeminism focuses on women's relationship to the environment and the particular stake they have in environmental protection. While comprised of many philosophical and activist strands, ecofeminists identify women with nature at a very fundamental level. They stress parallels between the oppression of women and the exploitation of nature which, in patriarchal thought, leads to women and nature being viewed as inferior to men and culture, and thus legitimate objects of domination (Merchant, 1982; Sturgeon, 1997; and see Box 1.4).

In the 1980s ecofeminism was firmly endorsed by the policy-orientated "women and development" literature. More recently, however, it has come under attack for reifying a biological or natural essence of women, for treating women worldwide as a unitary category undifferentiated by race, class, and power, and in some cases for taking a very Western and intellectual perspective on women's issues (Jackson, 1993). As a consequence, there have developed new fields variously known as "feminist environmentalism" (Agarwal, 1992), "feminist political ecology" (Rochleau et al., 1996), or gender and development (Green et al., 1998). These approaches differ from ecofeminism in looking at how environmental change affects women's lives as a function of their political, social, and cultural position in society, not simply of their being women (Jackson, 1993). Typically studies in this area focus on the real-life situations of poor peasant and tribal women, who are so often responsible for cultivation as well as for fetching fuel, fodder, and water, and yet who, despite their knowledge of species varieties and processes of natural regeneration in their local environment, have few rights regarding the management of local resources. According to this view, gender differences in both knowledge of and susceptibility to environmental processes are not inherent in femininity; rather they arise from the division of labor that characterizes household production in any particular place, as well as the differential access rural men and women typically have to subsistence resources such as land, technology, labor, food, health care, cash, and community government.

---

and that people's labor can always be put to use with intensified production and improved technology. The debate is often polemical, fired more by politics and ideology than science. It founders in part on the confounding of causes with proposed remedies. Our planet is indeed very full. If this were not true, we would not face many of today's environmental challenges. On the other hand, interventions that control population directly, such as family-planning programs, cannot in themselves stem environmental degradation. Given that the world's twenty-five biological hotspots, identified by biolo-

gists as especially rich in endemic species (Myers et al., 2000), are characterized by especially high population density and growth rates (Cincotta et al., 2000), human population trends remain a crucial factor in the development of conservation strategy (Figure 7.1).

A particularly ingenious way of thinking about the impact of population is the IPAT equation (Box 7.2). Though sometimes portrayed in an over-simplistic way, this is not a model that specifies overpopulation as *the* cause of all our ills, but rather links it into a multiplicative model built on indices of

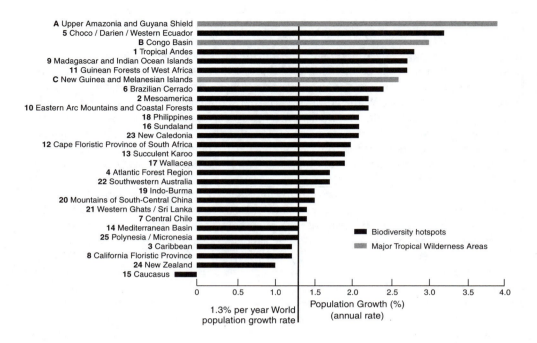

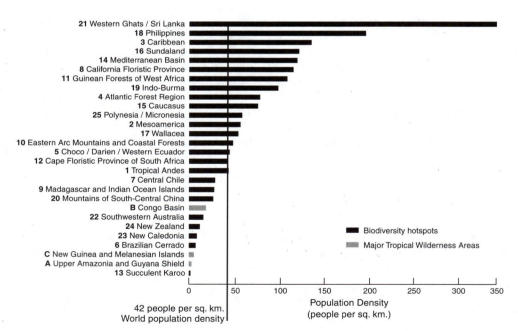

Figure 7.1. Biological Hotspots and Human Population Growth.
Human population growth (above) and density (below) in the twenty-five hotspots and major tropical wilderness areas. Since hotspots were identified in part on the basis of threat (Myers et al., 2000), this congruence is not surprising. It nevertheless highlights the demographics to be faced in areas of high remaining biodiversity. Reprinted from Population Action International. From Cincotta and Engelman: *Nature's Place—Human population and the future of biological diversity.* Copyright 2000.

Box 7.2. IPAT

Ehrlich and Ehrlich summarize IPAT as follows (1990, 58). "The impact of any human group on the environment can be usefully viewed as the product of three different factors. The first is the number of people. The second is some measure of the average person's consumption of resources (which is also an index of affluence). Finally, the product of these two factors, population and its per-capita consumption, is multiplied by an index of the environmental disruptiveness of the technologies that produce the goods consumed. The last factor can also be viewed as the environmental impact per quantity of consumption." In short:

$$\text{Impact} = \text{Population} \times \text{Affluence} \times \text{Technology, or } I = PAT$$

Since the model is multiplicative, the effects of interacting factors can be incorporated. It is now widely accepted among ecologists that although human population growth is a major cause of the current precipitous decline in biological diversity, it is not human numbers per se that matter (Meffe and Carroll, 1997), but rather the pattern of where they live, the extent they consume, and the technology used, in a word, IPAT.

The most radical alternative to this position, the argument that environmental degradation or losses to biodiversity are entirely independent of human population (e.g., Bell, 1987, 89), is hard to sustain since at the margins there is clearly a critical level of human population at which ecological systems can no longer function. Nevertheless, the whole formulation of IPAT bothers social scientists who point to all the complexities in the process of environmental degradation (see text).

---

population (P), affluence (A), and technology (T) (Ehrlich and Ehrlich, 1990). Statistically the IPAT model is sound; so long as the averages of P, A, and T are correctly defined and measured, their product must equal I. But when the equation is used to represent the processes involved in environmental deterioration, the formula is highly misleading. Furthermore, in various interpretations IPAT rhetoric insinuates that the places where most people live are the principal sources of environmental degradation, or worse, that it is poverty itself that pollutes and degrades the environment (Arizpe and Velazquez, 1994; Bryant, 1997).

To exemplify the shortcomings of the IPAT formulation Durham (1996) presents a graphical model of the social causes of environmental destruction in Latin America (Figure 7.2) based on numerous studies including those published in Painter and Dur-

ham (1996). The message that comes from this critique is that the impact of humans on the environment is mediated by cultural, political, and economic forces that cannot simply be added or multiplied if they are to be properly understood. Critically important are the social relations within and between populations that, because of the institutional history of Latin America, guarantee inequitable access to resources.

Two pathways lead to deforestation in this schematic figure. The top feedback loop represents capital accumulation. Commercial forestry expands into an area of virgin forest, fueled by domestic and foreign demand and often facilitated by weak land laws, lenient timber concessions, and bribery. Capitalizing on these subsidies from nature (treating of nature's services as free; see chapter 9), commercial logging companies get richer, improve their technology,

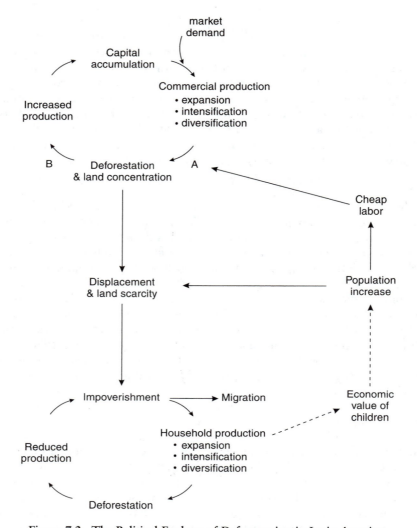

Figure 7.2. The Political Ecology of Deforestation in Latin America.
Simplified sketch emphasizing two positive feedback loops that promote deforestation, one corresponding to capital accumulation and the other to impoverishment. The loops are linked together by interdependent effects on population, resources, and environment. "A" refers to low input costs, specifically the use of cheap labor by logging operators; "B" refers to production subsidies from nature and the state, specifically the fact that logging operators can externalize the costs of their production onto nature, sometimes even with incentives from the state (see text). The figure assumes market involvement. Painter and Durham, *The Social Causes of Environmental Destruction in Latin America*, Ann Arbor: The University of Michigan Press Copyright 1996.

perhaps diversify into ranching and commercial crop production, and finally stimulate the positive feedback by starting to log in other regions, ever more distant and expensive, but now (with all this new capital) commercially feasible. In the Latin American context, the bottom feedback loop, impoverishment, is intrinsically linked to the top. This is because as forests get logged, land for household production becomes more scarce. Peasants, forest inhabitants, and others become displaced, with the inevitable consequence of lowered living standards and income. What do displaced peoples do? They may try to improve the situation by migrating and increasing household production on new lands. This can entail more forest clearance, more intensive cultivation techniques, and the adoption of cash crops and fertilizers, in any combination. When this happens in marginal areas, the outcome is a loss of soil, a buildup of pesticides, and more deforestation, leading to ever-declining yields and a further mutually reinforcing cycle spawning yet more migration. At the same time labor, and particularly children's labor, becomes more valuable. This drives increases in fertility, low school attendance, and ultimately (as land becomes increasingly scarce) a burgeoning market of cheap labor available to the operations of national and multinational corporations, perhaps even the logging companies we started out with. Note here that the causality between population growth and environmental deterioration is reversed.

So how does the IPAT formulation help in unraveling this situation? How would we define levels of technology (T) and affluence (A)? It is not the overall levels of technology and affluence that cause the high levels of deforestation in Latin America. Rather, it is the differences in technology and consumption that not only distinguish but actually drive the two positive feedback systems which, according to Durham's schema, are primarily responsible for deforestation. Finding an average national level of consumption and technology and multiplying the two together would tell us little about what is going on. Similarly, how can we deal with population (P)? First, we saw that population itself is not entirely an independent variable, but partly a product of the value of household labor. Second, who counts in P? In a nation like Panama, forests are converted to cattle pastures to feed an exponentially growing herd of cattle, some of which is exported to the world market. Should all the world's hamburger eaters be included in P? Durham's point is that too much is

left out of the IPAT equation to make it useful for anything other than pure forecasting. Furthermore, it has the danger of generating the false conclusion that we can simply tweak P, A, or T, and all will be well, thereby diverting attention away from the real cause of the problem, policies and institutions (Bryant, 1997). The "vicious circle" type model espoused by Durham is increasingly used by economists as they move away from simple IPAT formulations (O'Neill et al., 2001, 96; Geist and Lambin, 2002), and make much better use of fine quantitative work showing an array of different contextual variables that determine how and if population pressure causes environmental degradation (Bilsborrow, 1987; Abler et al., 1998). Indeed, in recent years ecologists have moved toward emphasizing policy change, institutions for equity, and regulation, with Daily and Ehrlich (1996, 991) offering the memorable quip that focusing on the biophysical as opposed to political dimensions of carrying capacity is analogous to a "drunk searching for lost keys under a lamppost rather than where they were lost 'because the light is better' there."

This evaluation of IPAT brings us to the heart of the political ecological critique, the fundamental failure of policy to both protect the environment and ensure fair distribution of environmental goods, the international elements of which will be examined in chapters 8 and 9. Within the national arena, the eye of the political ecologist is on regulations and practices that contribute to land use conversion, ecological degradation, and loss of biodiversity, albeit through complex routes, as we saw in Durham's (1996) study. A commonly used term is *perverse incentives* (Myers, 1998). These derive from programs put in place by the national (or regional) administration that encourage, through the use of government subsidies, land use practices that are beneficial for the state in the short term, often reflecting political or security exigencies. A classic example comes from Brazil, where citizens can only get credit from a bank if they have collateral, and land is only counted as collateral if it is deforested. Indeed, in much of South America landowners must show evidence of land use to retain their title. Such policies serve the government well, in that they harness private individuals to the clearing and eventual development of regions that are much too expensive for the government even to secure, let alone develop. At the same time these subsidies keep resource prices below market levels, thus encouraging overexploitation of the natural environment. As such, they reveal the deep paradox in the state's dual role as developer and steward of the natural environment (Box 7.3).

Political ecologists have usefully pointed to the critical part the state must play in resolving conservation problems; furthermore, at the Earth Summit in Rio in 1992 it was national governments that were ultimately charged with implementation of the CBD (article 3—Principle). If conservation goals are to be reached, national governments will need to find ways to remove perverse economic incentives for habitat destruction, to implement legal changes that protect conservation initiatives, particularly relating to land tenure and security (see chapters 6 and 8), and to generate specific funds for conservation actions (chapter 9).

7.4 Biodiversity and its Human Dimensions

Conservation biology emerged in the United States as an organized academic discipline in the mid 1980s (chapter 3). Its focus became biodiversity. As a shorthand for *biological diversity*, this term came into popular use after the appearance of Wilson and Peter's (1988) edited volume. It passed almost instantaneously from a hopeful neologism to the buzzword for the conservation movement and is now eagerly bandied about by politicians and NGOs. Having looked at how ecologists try to measure biodiversity (chapter 3) and how policy makers have

used the term (chapter 2), we turn now to social scientists' scrutiny of the concept.

While biodiversity is technically defined in terms of genes, species, and ecosystems (see chapter 1), there is in fact no discussion within international policy documents, such as The World Conservation Strategy (IUCN/UNEP/WWF, 1980) or Caring for the Earth (IUCN/UNEP/WWF, 1991), of how much conservation is enough, or of what levels of diversity are required. The implication is simply that more is better. There are also few guidelines on how to deal with evolutionary change and anthropogenic influence. As it stands, many conservation projects that emphasize protectionism still enshrine the unstated, and perhaps counterproductive, objective of freezing nature in its present condition or restoring it to some former state, despite Meffe and Carroll's (1997) clear statements that this is not conservation biology's goal (see chapter 3). As such, decisions as to how much to conserve, and in what state, are essentially arbitrary or aesthetic.

The question of "which biodiversity to preserve" is especially problematic when biodiversity is itself a product of human activity. If human practices are responsible for fostering some of the biodiversity we now treasure, inevitably these must be maintained if biodiversity is to be conserved. In a planet so dominated by humans it would be surprising if some key aspects of biodiversity were not anthropogenic. There are now many assaults, so to speak, on the notion of the pristine, in all parts of the world. Instructive and very different examples come from studies on the two flanks of Africa.

Early East African colonial administrators and sportsmen found the East African savannas teeming with game (Meinertzhagen, 1983). The Eden they strove to preserve, however, was not a pristine wilderness but rather a product of local historical factors, specifically depopulation. In the late nineteenth century East Africa had suffered a series of ecological shocks—an epidemic

Box 7.3. Leviathan Rules

Without a state social life would be in chaos. This was the insight of the philosopher Thomas Hobbes, who in 1651 claimed that with no governance the "life of man" could only be "solitary, poore, nasty, brutish, and short" (Hobbes, 1996, 89); for the price of conceding some liberty, citizens establishing a state gain peace and prosperity. From this perspective the state's primary role is to articulate and enforce collective interest and, insofar as the environment is a collective good, the state is also envisaged as the protector of the environment. But since most states also strive for economic development, is it even feasible to expect them to steward the environment? The inherent logical contradictions between conservation and economic development were explored in chapter 2. Here we look at how political ecologists contest the state's role in resolving environmental problems (Bryant and Bailey, 1997).

First, since most states focus primarily on providing economic development, few are willing to eschew the benefits of resource extraction. Indeed, environmental departments are often tucked away in larger ministries that include mining, agriculture, fisheries, forest, and water, against which the rights of nature compete poorly and from which very little environmental policy can be implemented. In the context of an increasingly competitive global capitalist economy, the outcome is a global tragedy of the commons, with each state acting as a herder on the global commons. This unstinting quest for economic development is restricted neither to colonial governments (see chapter 1) nor postcolonial capitalist states (such as Malaysia; Box 2.2), but also characterizes the socialist era of countries like China (Smil, 1984); indeed, in some cases socialist states may have assumed an even larger role in environmental degradation than their capitalist counterparts.

Second, given the scale of ecological problems, the state can be either "too small" to manage regional and global environmental problems, or "too big" to deal with local environmental problems (Hurrell, 1994). At the local level state apparatus tends to hinder environmental grassroots actions (as we saw with state management of common pool resources in chapter 6), whereas at the international level any particular state may act as a selfish herder, blocking agreements that it finds economically damaging, as we address further in chapter 9.

The fundamental paradox to the challenge states face in both generating development and protecting the environment is exacerbated by third world debt. The International Monetary Fund's structural adjustment program, originally designed to allow Mexico to shed its huge debt, encourages underdeveloped states to operate on comparative advantage (or do what they are best at), in other words, to continue to supply natural resources cheaply to the developed world. Then, as commodity prices fall (as they inevitably do as production increases), there is still heavier extraction of minerals, timber, and other natural resources, less environmental impact assessment, and more internal migration within developing countries, more of the vortex captured in Figure 7.2.

It is not surprising, then, that political ecologists paint such a bleak picture of the state as an environmental manager in the developing world. Possible remedies lie with grassroots political resistance allied with international environmental NGOs (to which we turn in chapter 8), or with the growth of democracy in underdeveloped nations, a topic beyond the scope of this book.

of a viral disease of ruminants (rinderpest) in 1889 that spread over the area and killed more than 95 percent of the domestic livestock, followed by outbreaks of sleeping sickness and smallpox, interspersed with years of famine (Ford, 1971; Kjekshus, 1977). These depopulating events fell on the heels of slave trading across the interior, and were exacerbated by enforced labor on European plantations. East Africa was indeed an empty place at the turn of the last century. For example, Lamprey and Waller (1990) describe how after the rinderpest epidemic much of southwestern Kenya and northern Tanzania was covered with thick bush, habitat very unsuitable to most grazing ungulates. It was actually Maasai herders who at the beginning of the twentieth century, in their struggle to build up their livestock and human numbers, cleared the bush with burning and grazing. It was the Maasai who opened grasslands which, like the Serengeti Plains and Masai Mara, are now the epitome of an African park, rich with grazing ungulates and their predators. This African Eden, when it was "discovered" by early colonialists, had a population perhaps lower than at any time since the Iron Age revolution 2,000 years before. It was gradually converted into hunting reserves, and subsequently fully protected areas, in complete ignorance of its anthropogenic character. Furthermore, as Homewood and Rodgers (1991) show for the Serengeti area, the removal of all domestic livestock from the area to be protected as a park caused an explosion of wildebeest numbers, possibly creating new problems for other grazing wildlife species. In short, according to this view, since the special mix of biodiversity on the African plains is anthropogenic in origin, management through human exclusion might be problematic. Outside of Africa there are other examples of apparently pristine habitats being maintained by anthropogenic forces (Nelson and Serafin, 1992). For example, in North America Native Americans frequently set burns to control the seasonal distribution of bison. This

practice contributed to the grassland character and species composition of the pre-European landscape. Now that there are controls on the use of fire, most of these grasslands are increasingly lost to shrub and woody vegetation (Denevan, 1992). Similarly for South America, some argue there were few uninhabited areas at European contact, and therefore that the current set of Neotropical species has probably never existed as a natural community without human activities (Sponsel et al., 1996; and see Box 5.3).

A second (and more controversial) example of the anthropogenic nature of biodiversity comes from West Africa's guinea savanna. In this vegetational zone lying between the coastal rainforest and sudanic savanna, there are dense patches of deciduous trees typical of the southern rainforests that stand out in striking relief against an otherwise open woodland savanna. Noting that these forest patches are often adjacent to villages Fairhead and Leach (1996) set out to determine whether they are relics of a once more widely distributed tropical forest (now depleted by human activities), the product of a mosaic of differences in soil and drainage conditions, or the result of human activities that encourage forest formation. In the first two scenarios, endorsed variously by colonial administrators, geographers, and contemporary conservationists, forest patches are viewed as "natural" (either as relics or part of an edaphic mosaic). In the last scenario, the argument is that the forest patches owe their existence to human activity. Clearly, if the goal of conservation efforts in the area were to protect forest patches, we would want to know what is responsible for these forests. Though the first two hypotheses were not directly tested, ethnographic observations and historical sources lend considerable though controversial support to the third viewpoint (Box 7.4).

Why is all this work on the human dimensions of biodiversity important? First, efforts to conserve biodiversity that focus

Box 7.4. Forest Islands in Guinea: Are They Man-Made?

In trying to protect their villages, animals, and fields from annual fires that burn across the savannas, the people of West Africa's guinea savanna foster moist forest groves in the vicinity of their settlements. Accordingly, around villages they cut the high grass, practice early burning, build up organic deposits (through cultivation mounding and mulching; adding domestic manure, ash, and cooking waste), and tether livestock beyond the reach of the trees to keep down the grass. In addition, villagers intentionally transplant first fire-resistant pioneer trees that permit the establishment of subsequent trees in their shade, and then economically useful forest species. Villagers elaborate on the importance of these groves, mentioning the benefits of shade, fire protection, sources of forest produce (such as oil palms), and, in former years, protection and self-sufficiency when villages were at war with each other.

Fairhead and Leach (1996) compare aerial photographs taken since the 1950s and show that forests adjacent to villages are increasing, and indeed that there is more forest now than there was at the turn of the twentieth century. The thrust of the analysis is that colonial authorities and international environmental experts have their story of human-caused ecological degradation backwards. Fairhead and Leach's (1998, xxiv) argument is that "independent assessments of forest cover change [derive] from methods, theories, deductive reasoning, disciplinary authority and institutional affiliation [which] reflect a colonially-rooted genealogy." In other words, that experts who are ignorant of the anthropogenic nature of the forest see only what they are looking for—human-caused degradation. This argument, promoted as a new reading of the landscape, has now been generalized across many parts of West Africa (Fairhead and Leach, 1998).

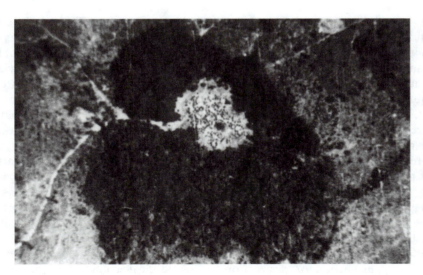

Forest Island in the Savanna, Showing Village and Paths (white)
Surrounded by Forest (black)
(Yomadou, northern Kissidougou, 1952).
As reproduced in J. Fairhead and M. Leach, 1996, *Misreading the African Landscape,* Cambridge University Press.

(continued)

(Box 7.4 continued)

The case, however, is controversial. Let us look more closely at the Guinea savanna. Comparable and somewhat more rigorous quantitative work on similar kinds of forest islands at a site 200 km. away in Sierra Leone shows how difficult it is to determine the real sequence of events (Nyerges and Green, 2000). There, careful spatial analysis reveals that at least one-third of the total forested area occurs near streams. A plausible argument can be made that because villagers value forest products they site new settlements at places where both water and forest can be found. Furthermore, new villages are often located in proximity to old village sites because of heritable rights to farming land. These abandoned sites possess fertile soils conducive to tree growth, and hence become forested. Nyerges and Green's digital analyses of satellite data reveal that the Kissidougou model of anthropogenic, peri-village forest growth is not applicable at their nearby study site in Sierra Leone, where other environmental changes (deforestation) are under way. Additionally, without rigorous species inventories it is impossible to tell from aerial photography whether secondary growth is being mistaken for primary forest (Peterson, 2000). Incidentally, very similar debates concern the so-called anthropogenic forests in Brazil (Anderson and Posey, 1989; Parker, 1992; 1993).

What do we learn from this debate? Just as we should not assume that humans are an inevitable cause of biodiversity depletion, so we should not jump too quickly to the contrary view that forest loss is a mere unscientific vestige of a colonial (and neocolonial) mentality.

---

primarily on biological inventory, assessment, and monitoring are likely to do little more than document the disappearance of species unless we have a full understanding of how humans affected biodiversity in the past, both positively and negatively, as well as how they shape biodiversity now and in the future (Nelson and Serafin, 1992). This has been recognized in India, where a tradition of research that examines the significance of traditional knowledge and institutions in contributing to maintenance of biodiversity has flourished in policy circles under the label of social ecology (Box 7.5). Second, there is the issue of respect. Without a clear understanding of a community's role in biodiversity protection or destruction, how can external conservation agents generate a relationship of trust? For example, the inhabitants of Kissingdogou react with hostility and suspicion to externally imposed measures to prevent deforestation, since they believe that forest patches are

their own creation. Third, political ecologists' appeal to portray humans as the creators rather than the decimators of biological diversity has fostered the view that people, specifically local people, are part of the solution to the biodiversity crisis, not part of its problem. While the balance of negative and positive impacts of local communities on biodiversity will vary, this recognition that communities are part of the solution is good reason for conservation biologists to examine the positive as well as the negative human effects, to work closely with political ecologists, and to integrate their own methods with those of social scientists.

A political ecological perspective on biodiversity shows how ideas and concepts are never innocent of political import (Bryant and Bailey, 1997). While technically biodiversity can be defined along a number of dimensions it is clear, as both social (Guyer and Richards, 1996) and natural (Bell,

Box 7.5. Social Ecology

In the words of Gadgil and Thapar (1990, 209), "Indian society is a mosaic of fishermen and shifting cultivators, toddy tappers and tea estate owners, nomadic shepherds and coal miners, village blacksmiths and computer programmers." They live in the slums of Calcutta and the skyscrapers of Bombay, the beach resorts of Goa, and pilgrimage heartlands of the high Himalayas. Exploring this diversity, Indian environmentalists have uncovered a huge array of belief systems, practices, and regulations governing the utilization of natural resources. Social ecology, a term coined many years before by an Indian sociologist (Mukerjee, 1942), emerged in the 1980s among these environmentalists as a particularly Indian approach to the science of biological conservation. It sprang from the recognition that to conserve biodiversity it is critical to understand a society's interactions with the environment (Guha, 1994). Social ecologists maintain that social relations among individuals and communities lie at the root of how the subcontinent's natural resources are used and abused, and more specifically that it is the interdependencies and inequities among these individuals and groups, in conjunction with their beliefs and values, that govern natural resource usage; in this sense it is strict political ecology, deeply tinged with social and environmental justice.

Social ecologists tell India's environmental history thus. Original hunter-gatherer and shifting cultivator tribal groups were made up largely of kin and were rooted to certain territories. Their natural resources were generally well managed, partly because it was in everyone's self-interest to assure productivity of economically important resources over time, partly because deviance was heavily sanctioned; most critically, resources were consumed and produced by the same group of people. With the transition to agriculture, small multicaste villages continued to manage their natural resource base in a sustainable way, largely because of their high levels of interdependency. Things changed, however, with urbanization. Urban settlements encouraged a flow of natural resources at much larger scales, eclipsing the smaller closed production-consumption cycles of earlier tribal territories and multicaste villages. City elites tapped surpluses by force or guile, and natural resources turned into commodities because of new demand and trade opportunities. European expansion and the subsequent colonialism of India only exacerbated these processes, extracting gems, elephants, spices, and subsequently cash crops such as cotton, jute, indigo, teak, tea, and coffee that were grown in newly cleared forests. With India's Independence in 1947, state-subsidized efforts to bring water, vegetation, minerals, and energy sources from the countryside to urban-industrial centers intensified (see figure).

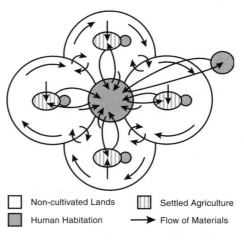

| □ Non-cultivated Lands | ▥ Settled Agriculture |
| ▨ Human Habitation | → Flow of Materials |

Social Ecology's Perspective on Goods Flows in Modern Indian Society.

The urban-industrial sector at the center taps both agricultural production and the wealth of noncultivated lands to meet its requirements, with these flows partially compensated for by material such as fertilizers that emanate from the urban-industrial sector. Other urban-industrial centers are linked. Reprinted from Gadgil, M., and Thapar, R., *Interdisciplinary Science Reviews* 15:209–223. Maney Publishing, Copyright 1990.

(continued)

(Box 7.5 continued)

Contemplating this somewhat idealized history, social ecologists, with Gadgil and his collaborators (Gadgil and Guha, 1992, 1995; Gadgil et al., 1998) in the forefront, are attempting to steer natural resource management back to semiautonomous local communities. They have identified pockets of good resource management under communal control that have persisted through colonial times, and are using these as models for the reassertion of communal control (Gadgil and Thapar, 1990). They encouraged various Indian governments to experiment with decentralizing development decisions, and in particular with social forestry that emerged as a result of local resistance to centralized exploitation of forest resources. Management of natural resources by local communities faces problems (since there is no such thing as real autonomy within a globalized world; see Box 6.7), and it bears repeating (see chapter 3) that conserving an economically important resource is not identical with conserving general biological diversity. Nevertheless, the successes of some of these schemes highlight the importance of reinvesting traditional roles of authority over natural resources (see the case of Sagarmatha, in Box 6.6) and of rekindling spiritual attitudes to nature (as with the sacred groves, Box 4.3). Just as popular concern with extinctions helped conservation biology focus on small populations in the United States (chapter 3), so social unrest with state appropriation of India's forests contributed to the emergence of social ecology as India's principal conservation movement (Sarkar, 1998).

1987) scientists have pointed out, that biodiversity has a vagueness that makes it susceptible to a number of legitimate and potentially beneficial reinterpretations (see, e.g., Table 2.3); indeed, in the view of Bowman (1993, cited in K. Brown, 1998, 75) biodiversity is "little more than a brilliant piece of wordsmithing." In the eyes of some this ambiguity has been good insofar as it generates active brokerage between the political and scientific communities over the appropriate means of conserving biodiversity. On the other hand, it leads to cross-purposed arguments: thus the debate between Redford and Stearman (1993) and Alcorn (1993), about whether and how forest-dwelling peoples of Amazonia should be involved in conservation projects, turns ultimately on whether protection of biodiversity or traditional forest subsistence systems is the objective. As we discuss in chapter 11, clearly defined goals are critical. In very similar respects the debate between John

Muir's preservationists and Gifford Pinchot's supporters of "conservation through use" (see chapter 1) turned on philosophical premises rather than any real difference in knowledge or understanding.

7.5 Community as a Casualty

With the rethinking of neocolonial protectionist strategies and the emergence of community-based conservation initiatives, grassroots participation in natural resource management became a central pillar within conservation strategies of the late 1980s and the 1990s (chapter 2), and were built right into the CBD (chapter 8). Indeed, the terms *indigenous people, local communities*, and *traditional knowledge and practice* emerged in the last decade of the twentieth century almost as mantras. As a consequence, working with the community (indigenous or other) became de rigueur, guiding a new philoso-

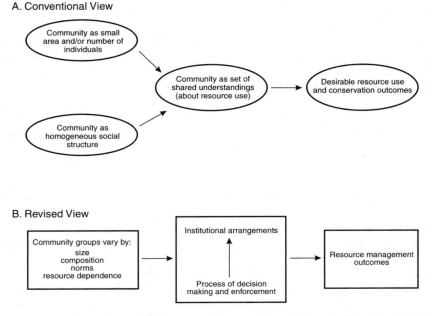

Figure 7.3. The Relationship between Community and Conservation Outcomes:
Conventional and Revised Views.
A. Conventional view and B. Revised view. Reprinted from Agrawal, A. and Gibson, C. C., *World Development* 27:629–649. Elsevier Science, Publishers.

phy in conservation policy. International agencies such as the World Bank, WWF, and others embrace the term *community* (Agrawal and Gibson, 1999), using it heavily (albeit loosely) in their promotion campaigns (Brosius et al., 1998); without the term, and its implications for participation and grassroots action, social approval (and hence funding) is unlikely.

The vision of small, integrated communities using locally evolved norms and rules to manage resources sustainably and equitably is a powerful myth, but as Agrawal and Gibson (1999) pithily note, tragically flawed. At the heart of their critique is the question: what is a community? The most influential early publication on CBC (Western and Wright, 1994) purposely eschews a definition, preferring to assume implicitly the usual suite of traits, namely, isolation, a shared location, small size, stability, social homogeneity, and shared norms (Figure 7.3A). Under these conditions it would

seem, according to the conventional view, that communities might facilitate collective action over the management of scarce natural resources.

As a result of political ecological analyses it has become clear that communities vary in size, in the extent to which they are internally differentiated, in the institutions whereby they govern their internal and external relationships, and the extent to which they share values pertaining to conservation (Gibson and Koontz, 1998; Leach et al., 1999; Barrow and Murphree, 2001). Many of these points had already been made by Little (1994), Murphree (1994), and others in an early benchmark CBC publication (Western and Wright, 1994), as well as by consultants for international agencies (IIED, 1994) and at least by some hands-on conservationists grappling with community issues (e.g., Snelson, 1995). The political ecologists' message then was simple, if not entirely new, and began to accrete nuances as it was brought to bear directly on the evaluation of conservation projects. For example, in assessing the ICD project at Madagascar's Ranomafana National Park, Kottak (1999; and see Box 10.4) points out that with regard to community benefits only landowners on the rich valley bottoms profit from project's agricultural inputs, such as dams and irrigation; horticulturist hill farming families are hurt by restrictions on using the forest, and the landless lose their primary source of subsistence. In a similar vein Songorwa (1999, 2067) shows how in Tanzania it is the wealthier and more educated sectors of the community who are most disenchanted by a community-based project designed to ameliorate the costs of protectionism in the Selous Game Reserve; this is because they were the best positioned to benefit from initial promises of game meat sales and job opportunities, and therefore the most disappointed when these promises failed to materialize.

This more rigorous and critical approach to the notion of community has stimulated innovative studies of why some communities do so much better than others with regard to the management of natural resources (Agrawal, 2002). Small communities would seem to have the edge over larger ones in successfully controlling their own use of natural resources, but are weak in the face of external or even state interventions. Economically or socially homogeneous communities, where individuals share many of the same concerns and interests, may be particularly effective in regulating the behavior of community members, yet heterogeneity can usefully spur certain groups into incurring the costs of policing (see the Barabaig herders in Box 6.4). Shared norms that promote or prohibit actions may enable solutions that are to the common good, but norms can easily break down, and break down differentially, as conditions, scarcity, markets, and other things change; furthermore, some community norms can be counterproductive for conservation ends, such as the prestige killing of large charismatic species like lions in parts of East Africa (Wilson, 1953). All in all, community (as a generic term) can be good or bad for conservation. Furthermore, even if a community is small, homogeneous, and characterized by shared norms, conservation is not a necessary outcome. Valorizing the term only papers over the fact that communities will differ enormously in the extent to which there are incentives for both defection and collective action (chapter 5).

And so the romantic view of community (Figure 7.3A) has given way to a more dynamic model (Figure 7.3B), a model that bears the clear imprint of common-property theorists (see Figure 6.2). The emphasis is increasingly on community institutions, and institution building (at the community level and above) is emerging as the new focus for conservation policy (Agrawal and Gibson, 1999; Leach et al., 1999). Increasingly conservation biologists and planners are interested in how communities work (rather than whether they simply exist), and show particular concern for five issues. First, are local officials elected? Gibson and Marks (1995) comment on this problem in Zambia's community-based natural resource management (ADMADE) scheme, where power over wildlife management was devolved to the local chiefs, who immediately subdelegated to their own family members, with the result that less privileged members of the community felt more disenfranchised than ever (and were subject to a new swath of fines and penalties). One way to correct this problem is to ensure that local officials are elected in a fair, free, and transparent process, ensuring genuine representation. As Ribot (1996) has shown for forest reserves in Burkino Faso and Mali, without elections local devolution can be nothing more than another form of privatization; it is this recognition that lies behind a new initiative among development workers for democratic decentralization (Ribot, 2002). A second area of concern is whether communities can self-evaluate, since unchecked authority at any level inevitably leads to perverse outcomes, for people and resources. Agrawal (1994) provides a good example of how communities monitor themselves in Kumaon, India, where the forest Panchayat (government) watches the guards, who watch the users, who watch the Panchayat. A third concern is whether there are institutional ways in which small and relatively powerless local communities can network with one another, to bolster their interests in the face of the state or other outside interests, as well as to adjudicate intercommunity conflicts (initiatives often facilitated by indigenous activist groups and NGOs, and examined more closely in chapter 8).

A fourth question is whether a community has appropriate institutions to manage the use of natural resources. Decentralization depends ultimately on the nature of the institutions to which power is devolved. A clear example comes from Bolivia's tropical lowland regions where, at various junctures over the last twenty years, the state has given municipal governments unprecedented formal authority over local forest resources, including the right to benefit from commercial timber concessions. Under these agreements local municipal authorities keep 25 percent of timber royalties, in return for ensuring timber companies comply with forestry regulations and passing on the majority of the money to local community groups. Wyckoff-Baird et al.'s (2000) evaluation of Bolivia's forestry reform shows that while some communities benefited from these new opportunities, others failed to organize effectively to consolidate their potential gains. In these latter cases, local elites of ranchers or city dwellers seized yet more power and profited from the decentralization initiative, either through monopolizing government posts or through allying themselves with the logging companies and turning a blind eye to unsustainable practices. Interestingly, the different outcomes of decentralization nationally reflect particular

histories of social activism: Beni Department had been a center of activism since the early 1980s, and had a history of using revenues for local community services, whereas in Santa Cruz Department (with less of a history of organized protest) there was more scope for political manipulation by local power holders. Even in communities under indigenous governance, the consequences of decentralization for sustainable forestry varied: some sold their logging rights with little concern for sustainable production, while others organized patrols to deter illegal encroachment by logging companies, ranchers, or agricultural colonists. Wyckoff-Baird et al.'s study highlights the critical role to be played by political ecologists in providing a historical understanding of the power structure in any locality, and how various interests might intersect with government initiatives or policy reforms. More generally, it shows that decentralization is by no means a magic bullet for conservation, nor even for equitable social relations; rather, its success depends on already existing power alignments and institutions.

Finally, there is the question of whether community institutions are authorized or undercut by municipal, regional, and national authorities. As political ecologists like Brosius (1999; see also Brosius et al., 1998) have noted with alarm, community-based approaches to resource management are easily appropriated by state structures, with the state ending up controlling resources for its own interests and the idea of participation relegated to mere rhetoric (see, for example, the fate of social forestry; Box 6.7). Accordingly, social scientists have been drawn toward co-management (or cooperative management) as an antidote to the dangers associated with the more politically blinkered community-based approach. With co-management local communities are granted shared management rights, thereby preempting the substantial opposition from more powerful local (or regional) policy-making elites to their involvement. While we consider this strategy more formally in chapter 11, we look here at political ecology in action, with the creation of a co-management scheme for caribou and brown bear by the Yup'ik of Western Alaska (Box 7.6). Indeed, in arctic North America co-management emerged as the dominant strategy for resolving resource conflicts and building partnerships in conservation and management throughout the 1990s, in many cases successfully giving local groups a political edge over outside forces.

In sum, political ecology has exposed the wooliness of the notion of community as used by international conservation organizations. This critique was motivated in part by some devastating recent reviews of apparently successful CBC schemes (e.g., Gibson and Marks, 1995; Hulme and Murphree, 2001), and perhaps in part by policy makers' trespassing on turf precious to anthropologists. Just as in chapter 4 where we questioned romantic ideas about human nature, here we explode idealized views of human community. The first casualty of this intellectual assault could be CBC (and its related CBRNM). However, the solution lies not in throwing the baby out with the bathwater, so to speak, but in heeding the progressive studies of behavioral ecologists (chapter 5) and political scientists (chapter 6) that tease out the situations in which sustainable patterns of community resource management become possible, in exploring new avenues for democratic decentralization in the highly critical manner advocated by Ribot (2002), and in pursuing more closely the strategy of co-management (chapter 11). Conservation biologists who consider the full political, historical, and philosophical context of their work within a particular nation are clearly still in the minority (e.g., Rozzi et al., 2000; and see chapter 10 for more on Chile's Chiloé project). In this context it is appropriate to note two points: the dangers for wildlife or natural resource authorities getting too involved in rural development, since they have neither the comparative advantage nor necessarily the legit-

imate interest (Snelson, 1995), and the need for interdisciplinary work.

7.6 Where's the Ecology?

Political ecology provides an exciting synthesis of many disciplines but at the same time, and perhaps inevitably, serves too many masters. Ecologists, and ecologically focused anthropologists, are concerned with the dearth of real ecology in political ecology, and also with the increasing recourse to rhetoric, narrative, discourse, and interpretation. In talking about reading the landscape as text, political ecologists reveal the strong bias in the social sciences toward critical theory and postmodernism. Indeed, though we have not dwelled on it in this book, much of the literature in the social sciences is currently engrossed with an approach that deconstructs meaning rather than analyzes causality and empirical outcomes. In this vein, social scientists, including political ecologists, are often more interested in what nature means, how it is constituted, and how this constitution reflects the social sphere than in ecological systems and their interactions. As Williams observes, "the idea of nature contains, though often unnoticed, an extraordinary amount of human history" (quoted in Escobar, 1995, 1). It is precisely this history that absorbs political ecologists' gaze, and which leads to a loss of focus on ecological processes and, perhaps more insidiously, to an assault on the legitimacy of scientific expertise and knowledge. Though postmodern political narratives do shed light on why we can all look at the same situation and see very different things, and can to some extent be incorporated into conservation biology (Robertson and Hull, 2001), there are costs entailed in adopting this perspective, particularly the loss of ecology and scientific method.

In a penetrating critique Vayda and Walters (1999, 168) suggest that what developed as a reaction against three decades

Box 7.6. Co-Management in a Landscape of Resistance: The Case of Alaska's Yup'ik

After Alaska was purchased from Russia in 1867, most wildlands were designated as public trust lands to be managed by the U.S. government, but because of limited management personnel and funding indigenous land use and occupancy persisted. Only in 1971 were the legal land rights of the Aboriginal peoples formally extinguished in about 90 percent of the state. This dire situation was modified in part by the passing in 1980 of the Alaska National Interest Lands Conservation Act (ANILCA), which guaranteed rural native users access to federal lands for "customary and traditional" subsistence activities. These subsistence rights were to have priority over those of commercial fishing and sport hunting but, and here lies the rub, were subject to game laws established by state and federal agencies. In 1990 the State of Alaska's Supreme Court determined that the ANILCA subsistence laws violated the state constitution. In short, while local communities could legally harvest wildlife, the practical exercise of these rights was attenuated by federal and state game laws, regulations crafted in distant agencies and enforced by nonlocal wardens. In this fraught environment emerged remarkable co-management schemes (Notzke, 1995; Spaeder, 2003).

In the remote Yukon Delta National Wildlife Refuge live 16,000 Yup'ik, for whom harvest of wildlife resources (bear, fish, caribou, etc.) still plays a central role in subsistence and culture. For many years the Yup'ik, like many other politically disenfranchised groups around the world, had simply disregarded game laws, harvested in stealth, avoided government personnel in the field, and ignored the mandatory filing of game permits, a kind of passive resistance that becomes, in itself, customary or traditional. From this stalemate Spaeder (2003) recounts how the Yup'ik developed a co-management scheme for a nonmigratory caribou herd in Yukon Delta lowlands.

In the mid 1980s, the Alaska Department for Fish and Game (ADF&M) estimated this herd to be in severe decline and banned all caribou hunting in the area. The Yup'ik, by contrast, believed caribou numbers were healthy and growing, and they continued to hunt. Rather intriguingly, the conflict over the caribou echoed the ongoing scientific debate over population regulation. While ADF&M adhered to density-dependent harvesting models (such as MSY; Box 3.2) the Yup'ik, according to their traditions, believed wild animal populations were regulated by random factors (disequilibrial dynamics; see chapter 3). Tired of a dearth of game meat and violent confrontations over the hunting ban, and with each party refusing to believe what the other said, a group of villages decided to change their strategy: submit a proposal to the Alaska Board of Game (ABG) to reopen caribou hunting. When this was denied, they filed suit in federal court. The result was a permit to hunt fifty caribou, and a harsh critique of ABG's erratic and unpredictable dealings with Yup'ik. On the heels of this ruling, the next two years saw the emergence of a cooperative caribou management plan negotiated at every step with village traditional councils. Native hunters agreed to a low harvest quota (5 percent of the herd) because they saw this as the most efficacious route to their goal: legalized hunting. They also became involved in aerial surveys of the caribou, with the result that knowledge and trust were shared.

Once native representatives had secured seats at the management table, negotiation and cooperation replaced litigation and charges of injustice. Accordingly, agreement was more easily reached over a subsequent conflict that emerged over bears, in which villagers successfully threatened to withdraw from all the co-management agreements in the region (not only for caribou, but waterfowl and salmon). Clearly, co-management succeeds not by valorizing local devolution, but by addressing and working through the inevitable historical de facto interdependencies between local users and state-level authorities.

of "ecology without politics" within anthropology has now turned into "politics without ecology." Essentially, the field has become so politicized that fundamental ecological factors are overlooked. Interestingly, this kind of development was anticipated several years ago, when Bryant (1992) spotted a similar problem with resorting exclusively to economic determinism; in his view, in the excitement of attributing desertification to economic forces, social scientists were ignoring critical climatic and other abiotic changes.

There are several strands to this "Where's the ecology" lament. First, ecosystems are active agents, and cannot be simply considered as passive objects transformed by human actors (Peterson, 2000). Just as political dynamics shape ecology (as we saw in the analysis of Latin American deforestation; see Figure 7.2), so ecology provides constraints that, because of the dynamic and variable nature of ecosystems, are themselves fluid. Thus, climates change, species compositions shift, populations fluctuate, and diseases evolve. An account that fails to look at ecological dynamics per se is likely to be flawed, as we will see below. Second, political ecology presents narratives rather than actual tests of ecological, or even political ecological, hypotheses. As such, they make for stirring reading, but can ultimately be misleading. The difference here goes beyond the matter of analytical style, with political ecologists espousing an interpretational stance and ecologists demanding conventional science. The fact is that these narratives have a tendency to overlook simple ecological realities; for example, in their widely acclaimed study of forest islands in Kissidougou (Box 7.4), Fairhead and Leach appear to have equated fast-growing tree species, including non-natives, with the far more biologically diverse old-growth forest. Third, political agenda and a priori judgment can hamper clear and unbiased analysis of ecological and social processes. For example, many political ecologists subscribe to the belief that community management

will solve problems of habitat conversion or over-harvesting. This conviction derives from the assumption that because international capitalist forces are the root of resource depletion, devolving control to local communities will mitigate destructive practices. The implicit idea here is not only that local communities can manage resources sustainably (by no means a certainty, as we saw in chapter 6), but also that devolution or decentralization buffers influences from the wider political economic system. This too is palpably false, as we know how hard it is for local communities to withstand external commercial or governmental pressures: for example, with outside pressure commons disintegrate into open access (as in Sagarmatha, Box 6.6), and loggers collude with hunter-gatherers (as in the Aché forest, Box 5.4).

So how can an approach be developed that reintegrates ecology and avoids doctrinaire bias? Vayda and Walters (1999) adopt a more focused treatment that starts by looking at an ecological issue and then working outward to its broader causes and consequences. The merit of the approach is that it starts with concrete observations and avoids privileging certain kinds of restrictive questions favored by the investigator. They present as an example their study of the loss of mangrove forests in the Philippines, a pressing environmental problem because of the importance of mangrove in providing fishery and wildlife habitat and in protecting shorelines from erosion. Conventional political ecological wisdom is that elite families, with the support of export-orientated national policies, have destroyed vast tracts of mangroves for the development of capital-intensive prawn farms. Because of this analysis, coastal resources policies have been changed to devolve greater management authority to local communities, with the view that this will lead to more sustainable management.

Vayda and Walters's own study suggests very different dynamics, at least for their specific study site. First, they find that mangrove-cutting well predates the capital-intensive and export-orientated economy of prawn farming. Second, it turns out that the elites invest in the planting and protecting of mangrove forests on the perimeters of their properties in order to shield their fish ponds and land from wave action. Third, poor families and fisherfolk also protect stretches of mangrove, but ultimately this is often motivated by the wish to secure tenure rights; landless families will claim mangrove areas, and later clear them to establish fishponds and house sites that are subsequently sold. Fourth, poor families also maintain mangrove for the production of posts for fish traps, fish corrals, and home construction. However, only one of the fifteen species of mangrove existing in the area is favored for these uses, and closer inspection shows managed mangrove stands to be a virtual monoculture. In short, a politically more open-minded approach exposes a picture that is more complicated than that presented in the earlier political ecological analysis, and suggests different interventions. These should still be designed to protect the interests of landless families, but must be more carefully thought through. Interestingly, too, it appears that the major factor limiting mangrove spread in the area is storms and pest infestations, reminiscent of the nonequilibrial models discussed in chapter 3. If this is the case, any amount of government effort to foster mangrove planting by poor families along exposed coastlines will fail.

7.7 Conclusion

How do contemporary developments in the social sciences bear on the conservation of biodiversity and environmental protection? They have been extremely useful in framing local conflicts within the broader reality of who has power and why. Political ecological analysis provides a helpful framework for looking at the differential access individuals have to land, forests, water, terrestrial and marine wildlife, and minerals. It also asks who has the power to pollute the environment, adversely affecting the livelihoods of others. Additionally, it investigates how environmental processes are perceived, how they are interpreted, what environmental narratives arise, and ultimately what kinds of environmental objectives and projects are prioritized. On top of all this, political ecological analysis is deeply linked with concern for human rights, as inequities (at both local and global levels) become increasingly apparent. As such, political ecology can suggest more specific avenues for policy interventions. For example, attributing deforestation not to overpopulation per se, but to the chain of processes outlined in Figure 7.2, allows for the development of specific initiatives regarding not just commercial logging access but local land tenure rights, alternative sources of income to rural populations, and school and health initiatives. In short, political ecology has the potential to bring new scope to understanding resource management and conservation, yet it comes with some problems.

First, the approach exhibits the somewhat monolithic agenda of pointing accusing fingers at the rich, the state, and transnational organizations in defense of the poor and the helpless. Vayda and Walters's study of mangrove destruction in the Philippines shows how easily political agendas have come to override careful ecological and social science. Second, and relatedly, the field has become so excessively politicized that the dynamics of ecological systems have been overlooked. Third, political ecology offers such a broad approach that it is difficult to stop adding new analytical dimensions. Accounts that meander across so much material (relating to many different disciplines) render comparison across cases extremely difficult, with authors sometimes seeming to revel in the complexity of their own case material. Fourth, unsurprisingly given the scale of its agenda, political ecology is generally vague on alternative courses of action, other than very naïve ones. The challenge now lies in incorporating political ecological insights into conservation initiatives. When we turn to evaluating conservation solutions (in chapters 10 and 11) we will see how important it is to pay attention to internal community fissures, to the different access various community factions have to broader national institutions, to the importance of history and the critical role of national and international institutions in shaping various stakeholders' options. Though there are undoubtedly projects that attend to one or more of these matters, we have been surprised by how few conservation biologists show cognizance of the political ecological literature.

In this chapter we have been wading again across the troubled waters between the social and biological sciences. Political ecology has accrued strong elements of postmodernism and critical theory, the dominant perspective in the social sciences. The difference here is not just that social scientists are more concerned with people (and the role of people in constituting "nature") than they are with nature, but that they have a different view about the role of science, the power of information, and the ethics of interventionism (issues we pick up in connection with citizen science; chapter 11). Insofar as science is suspect, because of its avowal of objectivity (an impossibility in the view of postmodernists), its methods of hypothesis testing and systematic data collection are avoided, even abhorred. For some political ecologists this means that any interventions based on science alone are unethical; for others ethical criteria alone should be the foundation of intervention. These differences should not, however, distract conservation biologists from incorporating the scale and complexity of the perspective of political ecology into their analyses. In the following chapter, we start to look at what happens when conservation hits the political realities of markets, global capitalism, and indigenous activism.

CHAPTER 8

Local People and International Conservation

8.1 Introduction

The globalization of conservation has brought local communities into the hub of a conflagration of novel practices, policies, and institutions. At the heart of these developments were indigenous peoples, communities that were in the early 1960s beginning to emerge from their marginalized position in nation-states by organizing politically. Increasingly, as the philosophical and political ideas regarding the incorporation of local communities into conservation strategy evolved over the 1970s (chapter 2), outsiders have been challenged with how to work together with these groups, and specifically how to put lofty ideals into actual practice. Despite persistent mistrust, with local communities anxious that those speaking for conservation interests merely want to evict them, and conservationists nervous that local communities are more concerned with feeding their families (and perhaps even political autonomy) than biodiversity conservation, there are common interests that can be harnessed into productive alliances. The safest of these focus on protecting traditionally owned land and water from unrestrained infusions of the market and immigrant settlers, but other options lie in finding economic incentives for saving nature, such as pharmaceutical prospecting and green marketing.

This chapter retains the overall political ecological focus of the previous one, but examines more specifically some of the prominent initiatives for working with local communities to slow biodiversity loss. We look first at the emergence of indigenous movements within the new international framework of conservation and development (8.2), turning to a closer evaluation of the potential dangers that stalk the alliances of local people with international conservationists (8.3). From there we examine pharmaceutical prospecting, a highly publicized strategy for incorporating local communities as partners in conservation (8.4). This brings us to the international marketing of "ecologically friendly" products as a conservation strategy (8.5) and a reaction against this approach, conservation through self-determination (8.6). We end by looking at the evolving views of people in the developing world on Western environmentalism (8.7), before concluding that despite pitfalls for local-environmentalist alliances, there are promising developments.

8.2 Indigenous Movements and Conservationists

From their common situation of disenfranchisement, marginality, and dependency in the late 1960s, indigenous peoples began to emerge as an influential force on the world stage, often through political organization. In some cases they achieved worldwide attention for their genuine grassroots origins and coordinated resistance, as with the Chipko movement of Uttar Pradesh (Box 8.1). Central to these movements is the control of resources. Thus, in Chipko villagers fought to defend their traditional forest rights against commercial forestry interests infiltrating the Garhwal and Kumaon Himalaya, with the ultimate goal of sustaining a rural livelihood. In other cases resource control is viewed as a key step toward self-determination, which can mean anything from autonomy within a national state to full sovereign independence (Clay, 1984). Often indigenous NGOs form to help organize and promote such struggles, as with the Coordinating Body of Indigenous Organizations of the Amazon Basin (COICA), that since the late 1980s have been campaigning for indigenous land rights across Latin America. Indigenous movements have additionally been helped by such international advocacy organizations as INDIGENA and Cultural Survival (CS) based in the United States (Figure 8.1), Survival International (S.I) based in the United Kingdom, and the International Work Group for Indigenous Affairs (IWGIA) in Denmark, (Furze et al., 1996; R. M. Wright, 1988). These organizations have played key roles in facilitating and documenting indigenous meetings and in coordinating campaigns against megadevelopment projects, such as hydroelectric schemes in Brazil and logging operations in Southeast Asia. In short, in the context of new international protocols (discussed below), indigenous peoples are increasingly making their positions known, either through autonomous movements or in conjunction with local or international advocacy groups.

Some of the most visible indigenous movements have focused on resistance to national parks and protectionism. This resistance can take many forms. In prominent cases it entails armed combat, as when the Bodo drove park guards out of the Save the Tiger program in India. In other cases highly prized wildlife species are killed; this may be motivated by spite (as with the spearing of rhinos in Kenya's Amboseli National Park; Lindsay, 1987), or by the need to provide funds for armaments in insurgency movements (as with smuggling of rhino horn from Kanas National Park in Assam; Kumar, 1993). Increasingly today, because violence has so often failed in the past, indigenous and local peoples are relying on legal structures and negotiation, basing their claims on a moral right to live in, exploit, and manage areas set aside for protectionism that were once their homelands (see chapter 4). Examples of such claims can be found in an anthology of indigenous (and activist) writers' statements on the historical legitimacy of "first peoples" to manage their resources (Kemf, 1993). Thus, Njiforti and Tchamba (1993, 177) lament that although Bénoué National Park in Cameroon used to be the private hunting ground of a very powerful local chief who forbade any hunting in the area without his permission, the current situation seems to them entirely opaque; in their words, "since the national park came, who knows what is happening?" The economic, cultural, and spiritual loss associated with the extinction of traditional rights under protectionist conservation strategies is a persistent rallying cry of indigenous activism.

But conservationists are not always at odds with local communities. In fact, in some respects they are natural partners. Local goals such as territorial sovereignty, livelihood improvement, and protection from predatory interests of outsiders often line up closely with conservationists' concerns about deforestation, global markets in threatened species, and regional migration. It makes sense for conservationists to work with these communities, and this can be achieved without promoting locals as "ecosystems" people (chapter 4); the alternative, to cast them as enemies of nature and then evict them from their homelands, is a sure route to social conflict and political instability, as examined in chapter 2. In other words, though Caitlan's dream of a

Box 8.1. Chipko: Grassroots Environmentalism with a Sting in its Tail

Inspired by a history of resistance against state interventions in villagers' control of their forests (Guha, 1989a), a wave of social movements swept through the hills of Uttar Pradesh during the early 1970s. The best known of these by far was Chipko, a peasant initiative against commercial forestry. The movement was precipitated by a severe monsoon in 1970 that caused the most devastating floods in living memory, with roads, bridges, villages, and farms washed away over a 100 km^2 area. This was a turning point in the ecological history of the region, as villagers (who bore the brunt of the damage) began to perceive hitherto tenuous links between deforestation, landslides, and villages as real (Bandyopadhyay, 1992). Demonstrations in the district town of Chamoli called for a restoration of village rights to the forest, but with continued harassment at the hands of commercial foresters and local government throughout 1973 the idea of embracing trees and thus "Chipko" (to cling) was born, as villagers resolved to hug trees "even if the axes split open their stomachs" (Guha, 1989a, 157).

Throughout the 1970s logging operations were foiled. The movement gained massive public support, influencing policies throughout the Garhwal and Kumaon Himalaya. Commercial forestry slowed significantly, declining from 62,000 to 40,000 m^3 per annum (1971–80; Guha, 1989a), leading to a complete ban on felling in 1980 for fifteen years. During that decade Chipko evolved from a peasants' movement against forest felling into a global environmental philosophy for sustainability and justice, accreting a moral stance attributed to Mahatma Gandhi: "The earth has enough to satisfy everyone's needs but it has got too little to satisfy everyone's greeds." As such, Chipko acquired an intellectual niche among scholars, planners, and activists across North America, Europe, Asia, and Australia, sometimes more as a myth than a movement.

As an *environmental* movement Chipko collapsed in the early 1980s, due partly to its success in attaining the logging moratorium, partly to its eclipse by a nascent political struggle (Uttarakhand) for hill districts' statehood, and partly to philosophical schisms among leaders loyal to different regions and tolerant of different degrees of Marxism (Guha, 1989a). Somewhat ironically, the characterization of Chipko as an ecological or environmental movement may have played a part in its demise (Mawdsley, 1998). To the question "what did you get out of Chipko?" a woman named Gayatri Devi (one of the original activists) replied: "I don't know . . . The road to our village is yet to be constructed and water is still a problem . . . Now they tell me that because of Chipko the road cannot be built because everything has become *paryavaran* [environment] oriented nowadays. Chipko has given us nothing. We cannot even get wood to build a house because the forest guards keep us out. Our rights have been snatched away."

Clearly, she had expected more than a cessation of logging. Indeed, right from the first protests in 1971 Chipko had called for an end to underdevelopment and exploitation in the hill districts, and for the abolition of the illegal liquor trade, untouchability, and poor governance, all calls reflecting the mistrust of the hill peoples toward the politics, business, and custom of outsiders. Hugging trees to reclaim the forests and evict commercial foresters was simply one strand, albeit a

(continued)

(Box 8.1 continued)

very important one, in what was a much broader movement, a movement that in the eyes of some has failed. But the broader lesson for us is that despite all the publicity surrounding Chipko as an environmental (and feminist) movement, and all the inspiration it gave environmental and social justice activists worldwide throughout the 1970s and 1980s, we should not overlook the fact that tree-hugging was a strategic move to control the key local resource, the forest. This does not mean it was not genuine, or that the readings of the Gita and renditions of folk-songs were not sincere, only that it was really about the right to survive in a world where livelihood depends so strongly on the forest, not about conservation.

"magnificent park, where the world could see for ages to come the native Indian in his classic attire" (Catlin, 1990, 35; see chapter 1) is now seen as paternalistic, the basic concept stripped of its romanticism may well offer the securest route to biodiversity conservation in many though not all circumstances.

Much (by some estimates over 85 percent) of the world's protected areas are either inhabited or claimed by indigenous peoples (sources in Colchester, 2000), and

their rights to this land are increasingly protected by international declarations (Mauro and Hardison, 2000). But these declarations are weak. The UN International Labor Organization (ILO) (Convention 169) recognizes indigenous peoples' rights, *inter alia*, to the use, ownership, management, and control of their lands and territories, but is signed only by fourteen countries. The UN Human Rights Committee established indigenous rights to self-determination through the Working Group on Indigenous Popula-

Figure 8.1. Aboriginal Children with Cultural Survival Logos.
Photo courtesy of *Cultural Survival Quarterly.*

tions (WGIP), but WGIP's Draft Declaration, begun in 1982 and more radical than the 1989 International Labor Organization convention, remains stalled in the General Assembly of the UN, despite having succeeded in establishing a Permanent Forum on Indigenous Issues. (This delay arises in part because of contentious debates over the definition of indigenous in the context of nation-state fears of secession in indigenous areas with great biological or mineral riches; McIntosh, 2001.) The Convention on Biological Diversity (CBD), signed in 1992 and now ratified by 171 countries, emphasizes the need to protect customary use of biological resources, although the actual interpretation of the phraseology used in article 8 (*In Situ* Conservation) remains highly fraught, particularly with respect to the ownership of local knowledge and indigenous participation in decision making (Mauro and Hardison, 2000; and see below). These instruments, limited as they are, have nevertheless stimulated a host of initiatives in other organizations, often in the form of nonratified global conventions or other types of "soft" law. Thus, after the 1992 Fourth World Parks Congress in Caracas, at which the rights of local and indigenous peoples were emphasized, the World Conservation Union revised its protected area categories (IUCN, 1994) to allow indigenous peoples (not just state agencies) to own and manage protected areas (chapter 2). Starting in 1996, the emerging rights of indigenous peoples in international law have been endorsed by such mainstream conservation organizations as the WWF-International (WWF), IUCN, and TNC, with indigenous rights now recognized in conventions that range from desertification to intellectual property, forest conservation to plant genetic resources (Mauro and Hardison, 2000). The challenge now lies in putting these principles into practice (IWGIA, 1998).

Unsurprisingly, as the movement toward respect for local residents' identity and land rights broadened, the label *indigenous* became too restrictive. After all, who counts as indigenous? Except in cases where there is a clear history of colonization, conquest, or genocide, there is no easy answer. Current definitions of indigenous peoples include at least some of the following elements, criteria that are in themselves hard to pin down: descent from original occupants; possession of a common history, language, and culture that is distinct from the national identity; possession of a common land, or simply exclusion from mainstream political decision making. Given all this ambiguity, and the additional question of who legitimately can make such definitions, increasing emphasis is placed simply on self-identification as indigenous. As a result, it has become common practice in some conservation, development, and activist circles to abandon the controversial label *indigenous* altogether, or at least use it interchangeably with such terms as *local, traditional,* and *resident,* to refer to communities with cultures that are different from those of the mainstream of national population (Brechin et al., 1991; Orlove and Brush, 1996); only in certain contexts (e.g., bioprospecting; see below) is the distinction between indigenous and farming/peasant/local populations pronounced, reflecting the specificity of tribal peoples' needs vis-à-vis the national peasantry (Maybury-Lewis, 1997; see also Stevens, 1997b). For the most part, conservationists give equal attention to all local communities affected by conservation actions, irrespective of whether these communities are true first peoples, acculturated to some extent into mainstream culture, or participate in local markets. The range in current estimates of the number of distinct traditional/indigenous societies worldwide (5,000 to 8,000) reflects these definitional discrepancies (Furze et al., 1996; Maybury-Lewis, 1997; Stevens, 1997a).

8.3 Room for Alliance, or Cover for Dalliance?

Given the emerging politicization of local populations, the recognition that most remaining areas of high biodiversity lie in their ancestral homes, new calls for social justice in conservation policy, and the shift from preservation to sustainable use, it is hardly surprising that conservation agencies have scrambled to work with, for, or at least in the name of local communities (see chapter 2). Conservation organizations heavily promote the concept of alliance (or partnership) with local, indigenous, native, resident, or other populations in or adjacent to areas subject to protected area management (McNeely, 1995; Redford and Mansour, 1996; see also Schwartzman et al., 2000a,b). The substantive component of these partnerships varies dramatically from case to case, from little more than the use of an indigenous name for a reserve to ambitious attempts of NGOs, such as the TNC, WWF, or the Rainforest Action Network (RAN), to join forces with neighboring communities in securing conservation objectives.

Do these alliances work? In a summary sense, yes. In a recent quantitative comparative analysis of Asia and the Pacific region, the BSP (Margoluis et al., 2000) found considerable evidence of conservation successes (measured as threat reduction to natural resources), and that this was best where partnerships were small, where project goals and partner responsibilities were clearly specified in advance, and where local bodies had the authority to make decisions. More broadly, environmentalists can further genuine humanitarian concerns through such alliances (as well as claim ethical high ground in defending the rights of the politically oppressed, not just flora and fauna), and local people on their part can enjoy publicity, global visibility, and measures of success in their struggles with nation-states that they never had before (Gedicks, 1995). For example, Indian land rights were formally recognized in Brazil in the early 1990s, no doubt in part because the debt-ridden government, looking for international acceptance, was embarrassed by all the negative press coverage regarding Indian rights (including international letter-writing campaigns) (Conklin and Graham, 1995; see also Box 4.6 on

Kayapó). The fight of the Dayak peoples for their forests in Sarawak, Malaysia, is a classic example of how an initially weak forest-dwelling population and their environmentalist allies achieved some impressive, and perhaps decisive victories (Box 8.2).

Such alliances do not necessarily offer clear sailing for either party. Indeed, their inherent problems have generated a suite of cautionary tales. Intriguingly, in a corner of the literature so driven by ethical and political biases, there is tremendous overlap between such tales (despite a very distinct terminology), whether they are found in the anthropological critique of modern environmental development encounters (e.g., Fisher, 1994; Conklin and Graham, 1995; Escobar, 1999) or among biologists skeptical of too tight a coupling of conservation and development (e.g., Terborgh, 2000). The issue for the most part is not that environmental NGOs aim to shortchange local communities or steal their land (their objectives are generally admirable); rather, the rub lies in the execution. Do these alliances work for both sides? Let us look at the potential difficulties.

First, all sides warn of the dangers in assuming that the local inhabitants' views of nature are consistent with those of modern conservationists (see chapter 4). Clearly, these partnerships incorporate two partially contradictory agendas, with environmentalists set to promote sustainable natural resource management and local communities committed to self-determination and control over their own resources, positions captured in a series of exchanges stimulated by Redford and Stearman's (1993) question "Interests in Common or Collision?" If the distinction between conservationist and local interests is not acknowledged with an explicit trading of each side's priorities, all parties are vulnerable to severe disappointment or manipulation. A stark example of such misunderstanding can be found in Hill's (1996; see also Pereyra, 1998) account of Aché confusion over the legal framework of a reserve drawn up for them by TNC in association with the Funda-

Box 8.2. Environmentalists Find Common Cause with Rainforest Dayak Peoples

In the mid 1980s deforestation was at its highest global rate in the Malaysian state of Sarawak, on the island of Borneo, with 2,100 acres felled each day. Here in the forests also lived more than 200,000 tribal peoples collectively known as Dayaks. Routinely, logging concessions were granted to timber companies for land held under customary law by the Dayaks, even within traditional village boundaries and without notice or consultation. Native communities' requests to the government for communal forest reserves to protect the land around their villages were ignored, unsurprisingly since the minister for the environment and tourism himself owned a prosperous logging company. Illustrative of this man's attitude is his response to an inquiry about the local climatic effects of deforestation: "We get too much rain in Sarawak. It stops me playing golf" (Hanbury-Tenison, 1990, 29, cited in Gedicks, 1995).

So the Penan decided to erect barricades across logging roads, with men, women, and children standing in the way of the bulldozers. Aided at the outset by the Malaysian chapter of Friends of the Earth (SAM), Penan activists were soon joined by other Dayak groups. Paramilitary police poured into the forests in 1987. The ensuing arrests of Penan and Kayan blockaders, prominent environmentalists, lawyers, and tribal rights' activists provoked worldwide anger. An international campaign was mounted by the Rainforest Action Network (RAN) and SAM to draw attention to the situation. With all the international publicity about how the Sarawak government was treating the Dayak, and high-level interventions from Survival International (SI) and IUCN, the prosecution withdrew its charges against all Penans and Kayans in custody the following year (Gedicks, 1995). Continued blockades led to the shut down of logging in nearly half of Sarawak, and to a European ban on imports from Sarawak (Colchester, 1993).

Human Barricades.
Photo courtesy of World Rainforest Movement.

(continued)

(Box 8.2 continued)

Despite positive developments on the ground, anthropologists viewed these environmental alliances with suspicion, noting that Penan reliance on international advocacy groups was just another form of dependency. Since initial colonial contact Penan have lived first as pirates (while the Portuguese, Spanish, Dutch, and British vied for trade in the South China Seas), then as squatters (in response to plantation land grabs for palm oil, cocoa, rubber, and tea), and most recently as poachers in newly established national parks (Colchester, 1997). With the arrival of environmentalists Penan are increasingly depicted as FernGully icons rather than political actors (Brosius, 1997), an image that obscures existing structures of domination (the real cause of the problem) and that is easily shuffled to the sidelines. After 1990 the Malaysian government did just that (Brosius, 1999). They cleverly deflected discussion from bulldozers threatening Penan barricades to the Uruguay Round of the General Agreements on Tariffs and Trade (GATT), the International Tropical Timber Organization (ITTO), and other bodies restricting trade barriers. Suddenly the Sarawak campaign no longer claimed the moral high ground of human rights protection, and became enmeshed in broader issues relating to free trade, sustainability, and Northern over-consumption, all of which are more complicated. Were the Penan to be orphaned again?

Apparently not. Though the Dayak still face threats from logging and dam projects, they won a key victory in the High Court in May 2001 in which Iban villagers' customary rights to forested (not just agricultural) areas were upheld (Warren, 2001). This is a victory for all Dayak villages and elsewhere, since such legal precedents apply to all other British Commonwealth countries. According to the judge, key to his decision was a locally based mapping project supported by a U.S.-based NGO "Borneo Project." Though the world's eyes are not on Dayak blockades as they were in the late 1980s, environmental alliances with local people can clearly be effective.

ción Moisés Bertoni (FMB), Paraguay's leading environmental NGO (Table 8.1). For the NGO disappointment often lies in wait, as when a local community has achieved its autonomy or land grant, and then decides to manage its resources in ways not anticipated by the conservationists (Terborgh, 2000). For local peoples, manipulation is often their fate, as in Alaska's timber battles, where pulp companies managed to win the hearts of local peoples by claiming to expand native lands (when their real interest was in circumventing federal regulations; Dombrowski, 2002). Anthropologists frequently warn of dangers to indigenous groups of becoming symbols, even puppets, in the crusade for conservation; they point to examples of how indigenous groups are adopted in the drive for raising funds, securing approvals, and achieving currently fashionable development goals, and then abandoned with expediency (Fisher, 1994; McIntosh, 2001). All in all, setting indigenous (or any other local) communities up as nature's guardians when they are not, and failing to distinguish the agendas of the different partners to the alliance, can spell trouble for everyone, as we saw in the case of the Kayapó where, as a result of journalistic fiascoes (discussed in Box 4.6), all parties came out with egg on their faces at the Rio UNCED conference.

TABLE 8.1
Aché Questions on a Paraguayan Law Creating the Mbaracayú Reserve

The text of Law 112/91, Article 13 (from Hill, 1996). "In recognition of the prior use of the forest by the Aché indigenous community, these groups will be permitted to continue subsistence hunting and gathering in the area of The Nature Reserve, as long as they employ *traditional methods*, and according to that which is allowed in the administrative plan of the reserve. This use right pertains only to members of the Ache community and cannot be sold, authorized, or ceded to third parties. The *members of the local Aché community* will be allowed to harvest species of wild animals and plants that are *not threatened nor in danger*, under the regulations established for the conservation of the nature reserve. This *use will be regulated by the Honorary Council* and based on technical studies and the management plan of the reserve. The *participation of the local Aché community in the protection and administration of the reserve will be encouraged*, and they (the Aché) will be *offered permanent employment* that comes about as a result of the development of scientific studies, recreation, and tourism in the reserve and the protected areas around it."

Phrases from the Law	Questions from the Aché in Conservation with an Anthropologist
"traditional methods" (of hunting and gathering)	Our methods have changed radically over the twentieth century. Which of these are traditional, and who decides? Why are only traditional methods allowed? Isn't it important how many animals we kill, not how we kill them?
"members of the local Aché community"	Who is a member of the "local" Aché community and who decides this? Can visiting Aché from other communities hunt in the reserve?
"not threatened nor in danger"	What does it mean to say an animal or plant species is "threatened" or "in danger"? Who decides this?
"use will be regulated by the Honorary Council"	What guarantee do we have of any use rights? Will these change from year to year according to whim of the Council? What guarantee do we have that our grandchildren will enjoy use rights?
"participation of the local Aché community in the protection and administration of the reserve will be encouraged"	What does "encouraged" mean? Will Aché have a formal say in decisions about reserve management?
"offered permanent employment"	How often will the Aché be offered employment? What fraction of the employment that comes up will be offered to the Aché?

Second, there is a problem if authentic indigenous status is defined in ways that contradict the realities of most native peoples' lives. Romanticizing traditionalism might bring in money, but it pushes families and communities into selling exotic images of themselves that may be false. Most communities worldwide are abandoning elements of their traditional lifestyles as they struggle to participate in the market economy. Poignantly, as in Redford's (1990, 27) apt rhetorical question, "Why shouldn't Indians have the same right to dispose of the timber on their lands as the international timber companies have to sell theirs?" This all speaks to the debate surrounding "enforced primitivism," so central to the management challenges of the so-called anthropological reserves. A common theme in such reserves is that local communities' continued use of the reserve is contingent on their maintaining a "traditional lifestyle"; though many funding agencies (such as the World Bank as early as 1982) have rejected policies of forced primitivism, the practice does not go away (Box 8.3). Native activists consequently confront a quandary: they can forge alliances with outsiders only by framing their cause in terms that appeal to Western values, but these conditions do not necessarily encompass their own contemporary worldviews and priorities. Alternatively, they can choose to abandon elements of their traditional ways, for example, to ignore the spiritual limits to hunting offtake, but then they risk tainting their image and losing the support of their more media-focused partners (Conklin and Graham, 1995). The problem is only exacerbated by population growth: most anthropological or indigenous reserves lie in frontier zones where medicine is routinely introduced before birth control, a point commonly noted by the more skeptical biologists (Terborgh, 2000; see also Barrett and Arcese, 1995). As the population grows, inevitable cultural transformations occur (as we see in the Kuna later this chapter), with the younger generation developing very different attitudes and aspirations for material wealth. In the longer timeframe enforced primitivism will become not only ethically questionable but completely impractical.

Third, there is the very immediate problem of communication. There are many reasons why complex objectives can be misunderstood, as we saw with the Aché's questions concerning the management of their reserve (Table 8.1). Canny Kuna leaders smoothed over the local people's fear of hiring foreign technicians to help with the setting up of a forest conservation project (see below) by drawing a parallel with the Kuna cultural hero, Dad Ibe (the Sun), who in the distant past, had to befriend the dangerous spirits of the Earth to learn their secrets and thus control them. Accordingly, Kuna leaders portrayed outside conservationists as dangerous spirits invested with knowledge. But often there is no neat cultural analog. An intriguing anecdote from an anthro-

pologist working in West Africa points to how easily communication with conservationists can collapse, even in the first steps of translation (Box 8.4). Associated with these problems of communication is the question of who should speak for the local community, and indeed for the outside agency. As regards the former, in societies with no leadership institutions there may be no individuals with whom negotiations can be broached or binding agreements reached; and even where leaders do exist there lies a danger in investing undue authority in these people, exaggerating potential conflicts of interest within the community, or destabilizing a village political system already out of equilibrium because of the presence of loggers and miners (as so clearly analyzed on the Amazonian frontier, Fisher, 2000). As regards representation of the external agency, outsiders are ill equipped to coordinate with East African societies organized into age sets, with Amazonian natives whose leaders exhort rather than command, or with Andean peasants who demand village by village assemblies for even minor decisions (Orlove and Brush, 1996). Increasingly, international conservation agencies work through local NGO partners to forge communication with local actors and stakeholders (Brandon et al., 1998).

Fourth, the internationalization of indigenous struggles can feed nasty nationalist backlashes (Maybury-Lewis, 1991). Temporary local victories exacerbate violence and repression, as in fights with biotechnology companies in the Ecuadorian Amazon (Conklin, 2002) and campaigns against dams in Quebec (Gedicks, 1995). The situation in Brazil throughout the 1990s was typical. Mainstream Brazilian media portrayed Indians as pawns of foreign economic imperialists (and this included conservation organizations) seeking to interfere in national affairs so as to connive control of the country's natural resources, particularly in the already politically insecure Amazon. In this journalistic climate international conservation alliances unwittingly introduced a dou-

Box 8.3. Enforced Primitivism and the "Bushman Problem"

Are anthropological reserves an appropriate alternative to relocation? The San make up less than 4 percent of the population of Botswana and are heavily discriminated against by the dominant Tswana. In conjunction with other strategies for incorporating San into mainstream Botswanan society through the Remote Area Development Programme (RADP), the government manages those San who still engage in hunting and gathering within the Central Kalahari Game Reserve (CKGR). In this reserve, established in 1961, controlled hunting is permitted, but only if the San use traditional means (spears, bows, poisoned arrows, clubs, and snares) and do not sell any meat, skins, feathers, or other products. In 1963 larger commercial interests were removed from the reserve with the banning of livestock and firearms, and since 1979 San have had to identify themselves as remote area dwellers (RADs) to acquire hunting licenses (Hitchcock, 1999; 2000). Similar policies now apply to the wildlife management areas that cover 17 percent of Botswana's land area.

The reserve brought some benefits to the San, in particular, powerful European and North American environmental groups as allies, protection from adverse "modern" encroachments with exclusion of pastoralists and commercial hunters from the reserve, as well as nutritional benefits contingent on hunting. But "enforced primitivism" has a downside, and jars with the RADP mission to integrate San into Botswanan society.

First, there is an inherent racism in advocating the treatment of San as if they were just another curiosity for the visitor to a game reserve, particularly insofar as their status is conditional on abstaining from the use of modern technology, on staying "backward." The San are caught in a trap—they must either eschew modernization or lose their home. The same dynamics occur in neighboring South Africa, where one of the last groups of Bushmen living adjacent to the Gemsbok National Park were expected to survive merely on government handouts and "traditional" hunting. Unsurprisingly, as they began to intermarry with other local Africans, they soon demanded clothes and decent housing. At this juncture a park warden noted with disgust that "their desirability as a tourist attraction is under serious doubt, as is the desirability of letting them stay for an indefinite period in the park. They have disqualified themselves" (Gordon, 1985, cited in Colchester, 1997, 106). The same complaints were subsequently heard in Botswana, where San hunt very effectively with horses (which they have done since the early nineteenth century; Hitchcock, 2000). Currently, regulations regarding what counts as traditional vary between districts, and even wardens, leading to confusion and abuse; in fact, Hitchcock mentions one warden who arrests any RAD hunting in trousers.

Second, there is jealousy. The Tswana see parks and reserves as wasteful concessions to foreign tourists and hunters, and they are resentful of the San and their special treatment. Consequently, whenever foreign environmentalists claim that San are causing ecological damage, Tswana officials leap at the opportunity of San eviction with a vengeance. Though threats of such enforced resettlements elicit international outcry by human rights groups such as SI (Hitchcock and Holm, 1993), they continue to this day.

(continued)

(Box 8.3 continued)

Third, the current arrangements foster utter dependency, opening loopholes for mistreatment. With foreign aid organizations and top-ranking civil servants determining the rate of social change among the Botswanan San, their fate is solidly in the hands of jockeying conservation groups and their academic advisors, some supporting a San presence in the reserve and some objecting. When the views of the objectors prevail, basic services like borehole maintenance are not attended to (Hitchcock, 1999). Although traditional hunting is still technically allowed in CKGR, San are periodically arrested for doing just that; worse, these arrests increasingly entail intimidation and torture at the hands of game scouts (Hitchcock, 2000; see also Gall, 2001), with some scouts taking the law into their own hands (see photo).

Finally, enforced primitivism retards political mobilization among the San. Treated as curiosities in the landscape, they remain as such. If the San are allowed to stay in the reserve and retain their foraging lifestyle for a few decades in the face of radical economic change engulfing the rest of Botswana, is it a refuge against adversity or an obstacle to change? Perhaps the newly founded grassroots organization (First People of the Kalahari) will change this. In the meantime, it seems enforced primitivism favors neither conservation nor human interests.

Hunting by Traditional Methods is Legal. This sign prohibiting hunting was put up by a wildlife officer from Ghanzi in 1988 of his own will (Robert Hitchcock, pers. comm., October 2001). Photo courtesy of Cultural Survival.

ble-edged sword, easily turned against Indians to undermine their credibility in domestic politics (Colchester, 1997). Even more problematic, according to observers such as Conklin and Graham (1995), the Pan Indian tribal identity politics that cropped up all over Amazonia in the early 1990s derived from Westerners' ideas about Indian identity and not from any genuine grassroots movement against power or patronage. The fear then is that Westerners' dalliance in indigenous affairs may ultimately undercut the power of these communities within the national fora in which ultimately they must participate, and possibly even thwart their political growth.

Last, there are some inherent philosophic contradictions within these alliances (Bonner, 1993). International conservation organizations and environmental campaigners need to use emotionally charged symbols to raise money, such as sensational posters of clubbed seals, harpooned whales, or butchered rhinos; these evoke a sense of outrage that is quite

---

Box 8.4. "Forget about Gola Forest!"

Sitting round the campfire one evening, Paul Richards, an anthropologist who has worked with the Mende of Sierra Leone for many years, proposed a small competition for some visiting student environmentalists from Njala University College, asking them to translate various conservation slogans used on posters in Sierra Leone into either Mende or Krio (the lingua franca). "Save Gola Forest" was the one that caused most difficulty.

The Mende word for save is generally understood as "take care of" or "protect," much as you might care for an absent friend's sheep or chickens. But since for the Mende the forest is not personal property but a common-pool resource (see chapter 6), a word with such private custodial implications was inappropriate. The translation that won most approval was, to Richards's surprise, one that translates literally as "Get away from behind (stop living under the protection of) Gola Forest," or "Forget about Gola Forest!" The idea behind this translation is that the forest is like a patron with too many clients. A patron in Sierra Leone supports his or her clients through life's varied crises—in hunger, bereavement, or court cases—in return for loyalty and labor. But big patrons become tired, they fade and lose their following. The client then must come out from "behind" the patron in question and look elsewhere for support. "Perhaps freed from social burdens the patron's stamina and fortune will recover? So with Gola Forest," Richards speculates (1992, 152).

The winning translation precisely captures the cognitive dissonance between conservationists and local villagers. For the former the forest is client, for the latter it is patron. The current slogan "Save Gola Forest!" advises the former clients of Gola to become its protectors, and is surely meaningless to them if not offensive. A more appropriate slogan then would be "Forget about Gola Forest," and look for another patron.

---

specific to Western sensibilities. In pitching their appeals on the intrinsic value of nature (see chapter 1), and in highlighting the unattractive aspects of utilitarian value, these campaigns inevitably foster a mandate for protecting nature by putting it "off limits" to human use. Local people suffer badly from these kinds of campaigns, as in the arctic, where a ban on trade in fur animals has undermined the livelihood of many communities, with people now living on welfare dependency and besieged by alcoholism, divorce, truancy, and emigration (Lynge, 1992). Indeed, these kinds of campaign can turn local communities against the whole notion of Western environmentalism (see below).

In conclusion, one could mount a pessimistic litany of why alliances between local communities and conservationists will have little to no impact on environmental outcomes. First, local indigenous peoples are not necessarily conservationists but opportunistic; second, traditional lifestyles cannot be retained indefinitely, particularly given the relentless conversion of forest peoples to wage laborers (with more interest in consumerism than in the environment); third, there are deep problems with communication; fourth, the opposition of world capitalism (which often includes local elites) simply dwarfs alliance initiatives, particularly as states seek to secure their control over natural resources, both vis-à-vis other states and their own people (as in the discussion of bioprospecting below). But is it true that with environmental alliances tribal peoples have simply substituted one form of political dependency for another? In our view such a conclusion is unnecessarily negative. These alliances are primarily strategic, and should be recognized as such. They have created opportunities for opposing environmental destruction and for keeping land in native hands, opportunities that would have been inconceivable thirty years ago. Though victories will remain fragile until institutionalized within the broader national framework (perhaps as co-management arrangements; see chapters 7 and 11), these are still victories. And there have been some institutional shifts too. With international recognition of the role of local communities as partners in conservation, the focus of most protectionist debates has now shifted from an exclusive concern with how a project will be developed to who will be involved in the decision-making process. This is at once an ethical and a pragmatic gain. Environmental NGOs can play a major role in helping local communities in negotiations with national government representatives (Kaus, 1993), and scientists can provide invaluable technical support (Getz et al., 1999; Lewis, 1995), both issues we explore further in chapters 10 and 11. Furthermore, as we see in chapter 11, such alliances can successfully transmute into co-management.

8.4 Bioprospecting or Biopiracy?

Pharmaceutical prospecting provides the classic, though problematic, model of incorporating local communities as partners in conservation, and is based on the idea that lucrative drug development can be used as a catalyst for the protection of undiscovered and unstudied biodiversity. The development of medicines and drugs based on traditional remedies has a long history (Figure 8.2), and was described as a utilitarian service of nature (option value) in chapter 1. Though relatively few tropical species have been studied for their pharmaceutical potential, tropical forests have yielded some classic drugs: the anticancer agent vincristine from *Catharanthus roseus*, the muscle relaxant d-tubocurarine from *Chondodendron* and *Strychnos* species (originally used in the Amazon for arrow poisons), and steroids from *Dioscorea* species. These discoveries grew from research programs, initiated in the early or middle parts of the twentieth century, but largely abandoned by the late 1970s (Balick et al., 1996). For the most part the people or communities who originally identified these medicinal properties were not rewarded for their knowledge.

Absence of compensation reflects entrenched practice set in place many years ago, when the European colonial empires began in the early seventeenth century to move plants from one side of the globe to the other, to establish botanical gardens, and to exploit the economic significance of plantation culture in the service of enriching the empire and curing newly encountered tropical diseases (Merson, 2000); hence the very term *bioprospecting,* with its implications of appropriation, and its more rhetorical analogues—*biopiracy* or even *bioimperialism* (Moran et al., 2001). The ethical dimensions of these appropriations were of course not missed at the time, but the process was unstoppable. For example, in the nineteenth century several South American states tried to prevent the export of cinchona bark or seedlings, the fever tree (*Cinchona officinalis*) from which quinine is derived, but with little success given European disregard for the intellectual property rights of anyone other than their own industrialists (Juma, 1989). In the late twentieth century there was a revitalization of ethnomedical and ethnobotanical research, in line with moves to make biodiversity pay for itself (chapter 9). In this context the CBD (article 3—Principle) signed in 1992 has proved a

Figure 8.2. The Many Uses of Neem.
What value do we assign to neem (*Azadirachta indica*), a tree that for over 2,000 years has provided an insecticide, a fungicide, a contraceptive, and an antibacterial agent, as well as toothbrush, sacrament, and spiritual food, to many of India's 900 million people? The estimate that puts India's medicinal plant-based exports at $50 million does not even include this domestic value of the neem (Duke, 1996). Since 1993 the U.S.-based W.R. Grace and Co. has been producing a patented bioinsecticide developed from neem, a hugely effective product used in greenhouses from Karnataka to California. Despite claims of biopiracy (Shiva, 1990), the collective intellectual property rights of traditional users are still not recognized. Reprinted with permission from *Neem: A Tree for Solving Global Problems*. Copyright 1992 by the National Academy of Sciences. Courtesy of the National Academies Press. Washington, D.C.

landmark achievement in redressing the legacy of such global inequalities, essentially by overriding the view of nature as a common heritage and defining biological resources as the sovereign property of nation-states. At the same time this agreement opened a Pandora's box of controversy and complexity that we examine here in relation to pharmaceutical prospecting and in its wider context in chapter 9.

Let us look first at the principle of intellectual property rights (IPR). Typically IPR permit a person to exclude others from using his or her ideas, plants, genes (or any other products of nature), except under license or royalty (Brush, 1996a); as such, granting IPR converts public goods into private ones (Demsetz, 1967). Defining biological resources (such as unimproved genetic material as encountered in nature) as eligible for intellectual property protection has a certain appeal, since herbalists or farmers can exclude others from using their genetic resources just as pharmaceutical firms can ensure the inviolability of their patents. Accordingly, NGOs and scholars felt it was in the best interest of beleaguered indigenous and local communities to act quickly to claim individual or collective property rights to their forests, the crop races in their fields, and their traditional ecological knowledge, and then bargain for deals with commercial bioprospectors seeking genetic information on behalf of transnational pharmaceutical companies and agrochemical firms. In the context of shrinking foreign aid, stagnant prices for other tropical resource exports, and continued pressure to meet daunting debt payment and structural adjustment targets, selling genes to foreign firms looked like one of the few new options for earning internationally negotiable currency.

The notion of IPR in nature was strongly endorsed by the CBD, not only because it redressed the inequities of colonialism but in anticipation that it would provide incentives for local communities to create, identify, and preserve valuable biological resources, and thereby indirectly support *in situ* conservation of biological resources, particularly highly threatened tropical forests. Furthermore, the raw numbers involved in this "grand bargain" (Ten Kate and Laird, 2000) seemed to offer real promise for conservation. It had been calculated that the forgone sales in drugs expected under current rates of plant extinction would amount to $3.5 billion, and therefore that an annual expenditure of that very amount was justified to preserve genetic diversity (Principe, 1996).

So was genetic gold to be the next tropical miracle crop? There was a rider to this conservation realpolitik. From the point of view of the pharmaceutical companies, the profits from discovering and developing biomedical agents appear to be much more modest than originally anticipated (as reviewed in Barrett and Lybbert, 2000), and in fact far too small to induce companies to underwrite broader conservation. From the point of view of indigenous peoples, it also soon became clear that the granting of IPR brought their communities right into the ring of global capitalism and free trade. Their biological resources and local knowledge became subject to the 1994 World Trade Organization's (WTO) multilateral Agreement on Trade-Related Aspects of Intellectual Property Rights (TRIPS), which requires universal patent protection for any invention in any field of technology, including pharmaceuticals. Accordingly, they became prey to eager bioprospectors, organizations that link developing country governments with private-sector pharmaceutical companies. In some senses this was not a bad thing. Deals were stuck that required bioprospectors to share benefits with source populations and to support technological growth in the source country (under CBD, article 16—Access to and Transfer of Technology). Lucrative patents have also been won, as with the recent ruling over the spice tumeric, which sets a precedent for how existing patents on natural derivates that are not truly novel can be overthrown and returned to the source populations (Moran

et al., 2001). But we need to look more carefully.

The huge dangers attendant on turning nature and knowledge into a commodity that can be owned privately are familiar to resource economists who study how and why no country in history has ever advanced up the international economic ladder by exporting primary commodities on "free market" terms (McAfee, 1999). All the cards are in the hands of the purchaser, who can substitute products when this is cheaper, play source countries off against each other, and accordingly reap the lion's share of the profits. For these reasons many fear that bioprospecting will go through a cycle of commodity boom, market saturation, and then bust—just as what happened in the case of indigo, rubber, sugar, and so many other once-touted tropic miracle crops, leaving the exporting countries poorer and their ecosystems degraded (Hobbelink, 1989). As Browder (1992) so eloquently details for Amazonian forests, after the rubber fiasco there has been a succession of forest products—sarsaparilla, spotted cat skins, peccary hides, cacao, Brazil nuts, mahogany—all leaving corporate bosses richer and producers poorer than before the boom/bust cycle. Why will this differ for the next forest product, genetic gold?

And there was a further worry. Already in the 1980s the world had witnessed the bitter "seed wars," in which Northern rich countries with seed industries fought against an international initiative to maintain seeds as a common heritage of humanity. Now the same thing was happening to nature more generally. As a result of the Earth Summit deliberations, biodiversity was relegated from its status as "common heritage," which implies common ownership and might exclude future patent rights, to being simply a matter of "common concern" of humanity (Reid et al., 1996). Indeed, under the CBD biodiversity was expressly identified as a sovereign national resource. Social scientists and activists were uncomfortable with these changes, recognizing that the privatization of cultural knowledge and living nature inherent in the IPR system would bring unacceptable damage to community social cohesion, spiritual well-being, economic livelihoods, and ultimately the environment (e.g., Dove, 1996).

Attention has accordingly turned from patenting and IPR per se to contractual agreements (Moran et al., 2001). The parties to the agreements (on the supply side) may be individual shamans, communities, tribes, or nations, and on the demand end companies, government agencies, or individual scientists. In some cases well-balanced contracts were struck: for instance, the California-based Shaman Pharmaceuticals company agreed to study the knowledge of traditional healers in remote tribal communities in return for providing primary health care (King, 1996). Contracts by no means sidestep the problems outlined above (since pharmaceutical firms are after all in the business of making money), but to the extent they are handled by nonprofit intermediaries and tailored to local conditions, source communities are buffered from market forces. Furthermore, unlike the pure market IPR system, contracts do not establish a global monopoly over any specific invention or resource (Brush, 1996a) and, in contrast to patent law, they can be designed to fit differing circumstances (Moran et al., 2001). As such, contracts have the potential to steer the evolution of new biosprospecting institutions.

A classic and now "model" contract of this type was announced in 1991 by Costa Rica's private nonprofit National Biodiversity Institute (INBio, founded in 1989 as a biodiversity-prospecting intermediary) and the U.S. pharmaceutical firm Merck. Under this agreement INBio provided chemical extracts from wild plants, insects, and microorganisms from Costa Rica's conserved wildlands for Merck's drug-screening program in return for a two-year research and sampling budget of $1,135,000 and royalties on any resulting commercial products. INBio agreed to contribute 10 percent of the budget and 50 percent of the royalties to the National Park Fund, and Merck to offer assaying technology and training for the development of Costa Rica's drug research capacity (Reid et al., 1996). INBio is currently responsible for one of the most comprehensive biodiversity inventories ever conducted at a national level, primarily using parataxonomists, lay people of rural origin who are offered basic training. The Merck–INBio agreement is just one of a number of such biodiversity-prospecting ventures. These kinds of arrangements are facilitating systematic screening for compounds active against HIV and cancer, and are often brokered by nonprofit intermediaries, such as NGOs or activist advocacy groups. Beyond the United States, Japan has launched a major research program in Micronesia, and Indonesia, Mexico, and Kenya have initiated national inventory programs similar to that of INBio; in these cases public organizations often act as intermediaries. The extent to which royalties are shared with host nations and communities is nevertheless still highly variable (Reid et al., 1996).

The Merck–INBio model seems to be working well, with INBio itself now in a position to develop products with other international groups from a position of relative strength, as in its development of a bionematicide from a tropical legume in collaboration with the British Technology Group (Merson, 2000; Moran et al., 2001). It clearly demonstrates what can be achieved with genuine North-South collaboration within the dictates of the CBD. But its success may in part reflect Costa Rica's democratic system, educated populace, and lack of an indigenous sector (Edelman, 2000). Reaching such agreements in less developed nations is more complex, raising the following kinds of questions (Blum, 1993; Merson, 2000). Can the local community (or broader legal apparatus) effectively limit the collection and shipment of genetic resources to prevent overexploitation? When a resource is reaped from an area controlled by many villages or districts, can a cartel of resource-producing communities form to receive the compensation? How can compensation be maintained once the compound can be chemically synthesized, or even grown from *ex situ* sources set up in botanical gardens (Ten Kate and Laird, 2000)? How can spending on conservation be enforced when, even in the model Merck–INBio agreement, only half the proceeds actually go toward conservation (Dove, 1996). And finally, with the long time lag of research and development between collection and marketing, can companies pay compensation *prior* to reaping the benefits, the very compensation that is needed to ensure conservation at the local level?

But perhaps the biggest problem for these contracts stems from the ambiguities concerning the role of the state within the CBD. First, the CBD (article 3—Principle) specifies all states have the "sovereign right to exploit their own resources pursuant to their own environmental policies." First, "sovereign right" is interpreted by some governments as a carte blanche for appropriating biodiversity's newly discovered "value" at a national level, with little regard for the communities in the areas where the medicinal product was found (Brush, 1996b). Second, how can European property rights laws, never designed to represent knowledge held collectively, be modified to deal with cases like *neem* (Figure 8.2)? Finally, how can compensation for IPR be made when few states have incorporated indigenous interests and rights into state legislation (see the "soft" laws and unratified conventions above; Dove, 1996)? Unsurprisingly, the provisions of the CBD for bioprospecting work best in nations where collective and/or indigenous rights are already recognized by law, as in India.

Given all these ambiguities, and a vigorous critique by advocates for indigenous communities in the developing world (Mer-

son, 2000), the contractual model is starting to change. Most notably, academics and activists are becoming more involved. Indeed, contracts with universities are proving a fruitful way of preserving and overseeing the marketing of indigenous knowledge and products in a nonprofit context (Bannister and Barrett, 2001), and of establishing trust funds for the return of monetary benefits to source communities. An interesting example is the International Cooperative Biodiversity Group (ICBG), a multicountry bioprospecting program working in twelve developing countries under the coordination of a multinational group of academic investigators in anthropology, ethnobotany, and drug development. Dedicated to a research paradigm of ethical drug discovery, ICBG has developed a remarkable cooperative arrangement with the Ministry of Health in Lao (Riley, 2001), as well as running a particularly innovative program in Nigeria and Cameroon (Box 8.5). This program seems to do well because so many local scientists are involved and because so many of the medicinal benefits are for treating local diseases (Moran, 2001). Good fit to local political circumstances is key to its success, since a similarly structured ICBG project in Mexico was recently abandoned in a morass of conflict among its various partners (Nigh, 2002).

It is hard to reach any general consensus on the efficacy of the current system. To date few of the signatory nations to the CBD have introduced legislation requiring benefit-sharing for access to their bioresources; indeed, most CBD signatories have only a patchwork of legal provisions, such as collecting or export permits (Moran et al., 2001). Also, since the CBD was introduced, no pharmaceutical bioprospecting product developed by using traditional knowledge has as yet been commercialized, so there is still no acid test of whether an indigenous population can benefit from a commercially exploited drug, and if so how; however, as of December 2002 prostratin (a possible HIV drug identified from *Homalanthus nu-*

Box 8.5. Drug Development and Conservation in West and Central Africa

The ICBG program offers an integrated approach to the discovery of biologically active plants for drug development and biodiversity conservation, aiming to ensure that local communities and source countries derive maximum benefits. It grew out of collaborative work, predating the CBD, between the Walter Reed Army Institute of Research and Nigerian scientists that focused on parasitic disease drug development and capacity-building in Nigeria. In 1994 it expanded to include the Smithsonian Tropical Research Institute, and subsequently nine other institutions in Cameroon, Nigeria, and the United States (Iwu, 1996; Schuster et al., 1999). Focusing on plant material used in traditional medicines has been highly productive with respect to the number of biochemically active compounds generating further isolated and characterized molecular leads (see table).

TABLE 8.5 BOX
Success in Deriving Active Compounds from Traditional
Medicines for the Treatment of Locally Significant Diseases

Disease	Extracts tested	Number and Percent Showing Activity	Molecular Leads
Malaria	500	343 (69%)	20
Leishmania	130	52 (40%)	6
Cytotoxicity	20	16 (80%)	5
Viral	30	16 (53%)	2
Trypanosomiasis	27	13 (48%)	3
Trichomonas	25	10 (40%)	7

Source: Reprinted with permission from *Pharmaceutical Biology 37*: 84–99. Courtesy of Columbia University Press, copyright 1999.

The key objective is familiar—to increase the net worth of the tropical forest, and to show the feasibility of using drug development as a catalyst for biodiversity conservation. However, this project is unusual in many respects, and serves as a model of interdisciplinary collaboration among multiple partners in the developed and developing worlds. These are some of its special features:

- The group does its own research before calling in commercial pharmaceutical companies for marketing.
- Emphasis is on deriving benefits from the process of drug discovery rather than relying solely on the promise of future royalties that may never materialize.
- An entirely independent trust fund in Nigeria was established to ensure equitable distribution of benefits; compensation decisions are made by town associations, village heads, and a professional guild of healers.
- The main target therapeutic categories are under-researched tropical diseases of primary concern to the source countries (see table), and particularly the rural areas, not necessarily the areas of drug development with highest financial returns.

(continued)

(Box 8.5 continued)

- The focus is on the development of phytomedicines, particularly as low-cost medicines for local use, but also for the international market for processed herbal products.
- Feedback is provided to local healers from the study of active constituents and toxicity in traditional medicines for more appropriate administration.
- There is long-term assessment of rainforest ecological dynamics and sustainable harvesting in 1 ha plots using locally trained staff.
- Training (both in-country and abroad) of ecologists, biologists, chemists, pharmacologists, ethnobiologists, and field taxonomists and development of laboratories in Cameroon and Nigeria are part of the program.
- Source country scientists benefit in same ways as foreign scientists; they are not treated simply as middlemen or brokers with traditional healers.

tans by Samoan healers) was patented, apparently under a controversial agreement that 20 percent of the revenue go to the Samoan government; this is a case to watch if the compound proves effective. It is unsurprising, therefore, that access to genetic resources and traditional knowledge remains one of the hottest areas of controversy in the annual Conference of the Parties that meets to follow up on the CBD (e.g., Entwistle, 2000).

If there is any consensus at all, it is on four issues. First, the argument for conserving forests for their pharmaceutical products is strongest not where we are considering undiscovered drugs for first world ailments, but where the forest produces medicinals that are popular locally. This is the case in the area around the Brazilian Amazonian town of Belém, where both rural and urban populations depend almost entirely on remedies derived from native species, where few substitutes are available, and where Western pharmaceuticals do not yet exist for some of the principal diseases so treated (Shanley and Luz, 2003). Second, as regards the role of Northern pharmaceutical companies, there is a gradual but palpable trend toward more creative benefit-sharing programs with source populations, particularly as academic institutions become more

involved (Ten Kate and Laird, 2000). Third, the most effective benefit-sharing schemes include technology development and capacity-building in the source countries (often referred to as technology transfer), as seen in INBio. Fourth, multilateral international agreements are required to regulate the profit motivations of pharmaceutical and other companies, even when under contract, for example, to make sure that these companies are not polluting the nonmarketable commodities of nature (e.g., air and water), and that source communities are not forgotten about once their products are synthesized in the lab (Reid et al., 1996). In other words, regulations are needed to blunt the lingering colonial and neocolonial legacies of global inequality. In chapter 9 we turn to such international agreements and regulations.

8.5 Green Consumerism

Other incentives can be offered to local communities to conserve their tropical forests, notably the harvesting of fruits, nuts, and oils for an international market. The rationale here is when local communities see the profits to be reaped from sustainably harvesting valuable forest products, the alternatives (handing land over to logging concerns or expanding marginal agriculture) will seem less attractive.

Of all planet-friendly advertising slogans that have emerged in recent years, none has been more powerful than "rainforest harvest," evoking exotic luxuriance and abundance with the impeccable environmental credentials of tropical forests. These initiatives, tagged in the early 1990s as *green capitalism* and *green consumerism*, caught on fast, with articles in such high-profile spots as *The Economist, Scientific American*, and *Time* (Clay, 1992). They had wide appeal because of the benefits they could potentially deliver directly to communities in need and the apparent contributions to forest conservation. The stumbling block lay in identifying and securing markets, particularly international markets, for forest goods. To meet this challenge, the indigenous activist group Cultural Survival (CS), in conjunction with Cultural Survival Enterprises (CSE), worked with politically progressive companies, such as the Body Shop, encouraging them to purchase and market rainforest products. The most successful scheme was Rainforest Crunch, a nutty mix that Community Products and Ben and Jerry's Homemade Ice Cream Company distributed in the United States (Figure 8.3). CSE also began nonprofit trading of forest commodities. Appalled by the fact that in the 1989 Brazil nut harvest, remote collectors were paid 3 to 4 cents per pound, whereas the same weight of shelled whole nuts in New York fetched $1.70 to $2.20, CSE's goal was twofold: to eliminate middlemen and to help forest dwellers enter national and international economic systems on their own terms. During the first six months of trading enough income was generated to finance the first harvester-owned nut-shelling factory in the Brazilian Amazon.

CS was acutely aware of the many pitfalls associated with marketing products from the forest. They recognized that if they were too successful, soaring demand might well drive villagers to destructive harvesting, or even to monoculture plantations of the most prized commodities. Consequently, CS determined to avoid products from areas deforested specifically for the market, and to purchase from deforested/degraded areas only if they were replanted with two or more tree crops, criteria albeit very difficult to police (Clay, 1992). Additionally, CS planned to monitor the health and nutrition of local rural and urban populations, to evaluate negative or positive outcomes from the marketing, and to incorporate these enterprises within the broader humanitarian objectives of CS (e.g., Clay, 1991).

Their critics, however, were unimpressed. They felt that CSE was only intensifying the capitalist relations of production that had caused the problem (the erosion of biological and cultural diversity) in the first place, by replicating the old patron-client relationships that had characterized the production of rubber and sugar in previous decades. Critics then turned to ferreting out problems. Body Shop trade apparently only favored a very few individuals in each community (Corry, 1993), and rainforest crunch schemes were failing to establish local institutions whereby local populations could avoid abuse and exploitation (Dove, 1993). A publication of the Organization of Indian Nations of Colombia noted that with these marketing schemes "our communities' independence is weakened as our well-being is made dependent on Western markets" (cited in Corry, 1993, 12); whether this was a general claim, or a complaint specifically about NGO-coordinated green capitalism, was unclear. Critics speculated on the consequences of this dependency, expecting that once the rainforest fad folded, producers would find their communities wrought by alcohol, firearms, and slavery, as they had so many times before at the tail end of commercial booms. Though there was much rhetoric and posturing on both sides, the overriding message was clear: local people must have ultimate control over their products (and their transportation) before

Figure 8.3. Rainforest Crunch Product Lines.
Community Products ordered Brazil and cashew nuts for Rainforest Crunch, a nut brittle Community Products manufactured and sold as a candy and as a flavoring for ice cream. Community Products gave 40 percent of all profits from the sale of the brittle for rainforest conservation programs. In addition, high school and college groups sold the candy door to door to raise money for projects in Brazil (Clay, 1992). Photo courtesy of Kenneth Graham.

any marketing system can be fair. Since forest dwellers are so often marginal to economic and political activity, green capitalism is on these grounds essentially a nonstarter. In the view of its critics it is a mere commercial sideshow to the destruction of the forests and local livelihoods (Gray, 1990).

By the time the Rainforest Harvest Conference was held in May 1990 to determine whether rainforest marketing was feasible, there was already mounting and coordinated opposition to green capitalism (Hyndman, 1994), both within and between indigenous activist groups. Furthermore, technical problems continued to plague the actual enterprises, with an organization not designed as a commodities-trading firm burdened with containers of rancid brazil nuts that the buyers would not accept and for which the local producers in the forest had not yet been

paid. Ultimately, the day was won by the critics, though as much through polemic as sound reasoning. CS was accused of trying to solve all indigenous problems with rainforest marketing, when this was clearly not the case (Clay, 1991), and their critics were portrayed as romantics promoting the preservation of indigenous peoples in some kind of time-warped protected reserve under the supervision of paternalistic do-gooders, also not a fair characterization.

Two more recent initiatives that embody similar ideals to the rainforest crunch marketing schemes of the late 1980s—"fair trade" and "green-certified" products (see also Box 9.2)—are plagued by the same kinds of problems. These movements, though distinct, share a concern with developing markets in which the consumer recognizes the importance of equity and respect, both for the producers and by extension for their environments (Richards, 2000). While the objectives are to ensure that workers receive a fair price and natural resources are not depleted, these are often hampered by issues relating to certification, sustainability (social and ecological), and vulnerability to international markets. As regards certification there is always the nagging question of consumer trust (McNeely and Weatherly, 1996). How can I know this product was harvested in a sustainable way or that the harvester received fair compensation? As regards sustainability, consider "green-certified," biodiversity-friendly, or shade coffee initiatives (Johns, 1999; Gobbi, 2000). These schemes aim to promote agroforestry techniques that combine commercial production of coffee (or cocoa) with (often traditional) methods of crop production, most notably, retaining the forest cover. Shade production is generally a good thing for biodiversity in that it protects forest ecosystem functions, in particular, the microclimate and soil fertility; but because of its lower yield such programs often require extensive outreach work and financial assistance to ensure social sustainability, initiatives that can be expensive and hard to coordinate (Rice and Greenberg, 2000). Ecological sustainability can become even more problematic: as the international price for coffee rises, there is huge temptation to bring remaining tracts of primary forest into commercial production. Recent analyses show that natural forests are being converted at an alarming rate to cocoa and coffee, that occupy 11 and 8 million ha, respectively; for example, in Ivory Coast their cultivation has replaced 80 percent of the original forest (Hardner and Rice, 2002); new work in Sumatra shows a linear relationship between the price of robusta coffee on the international market and the rate of deforestation (O'Brien and Kinnaird, 2003). Finally, there is the volatility of the international commodity market. When prices plummet, millions of farmers are displaced as coffee farms are converted to other crops or pasture, causing human hardship and further

ecological degradation. Clearly, these green-certified markets have their limits.

In a recent review Hardner and Rice (2002) summarize the limitations of green consumerism in its many forms as a conservation strategy. Only a tiny slice of the international market is occupied by green goods, consisting of consumers with money to purchase such products. For example, almost all the timber certified by the Forest Stewardship Council from Asia, Africa, and Latin America goes to Europe and the United States, which together import less than 6.5 percent of the 228 million m^3 of all timber, green or otherwise, produced in the Tropics each year. The majority of forest destruction is by forces not susceptible to green marketing—local timber trade, livestock raising (Brazil), oil palm production (Indonesia), and firewood (Madagascar). In the case of Madagascar, almost all the timber harvested will become charcoal for local peoples' cooking fires. In these massive rural sectors, beset with poverty, there is no room for green consumerism. With these considerations Hardner and Rice propose a very different kind of green consumerism, the direct purchase or lease not of biodiversity's products but of biodiversity itself (chapter 11).

8.6 Conservation through Self-Determination

The new model to replace green consumerism is conservation through self-determination, a strategy with which indigenous populations worldwide (e.g. Aborigines, Maoris, Mayans, Sherpas, Maasai, and Inuit) are now experimenting. Self-determination entails indigenous groups redefining their identity in the newly globalized context of interethnic politics. Particularly remarkable has been the mobilization of native Amazonian shamans against the threats of bioprospecting, with healers from over forty different native groups in Brazil denouncing the activities of pharmaceutical companies. With a message that resonates across many sectors of Brazilian society, shamans have presented themselves as true citizens and patriotic guardians of national patrimony; even the Brazilian military, suffering an identity crisis since the fall of the dictatorship in 1985, now embraces biopiracy as a focus of national security anxiety, and hence is sympathetic to the shamans. These developments demonstrate how creative indigenous identity can be reformulated to negotiate novel tensions and opportunities (Conklin, 2002).

Self-determination fits closely with the interests of political ecologists, particularly its strands of environmental justice (see chapter 7). Indeed, many social scientists are primarily concerned with issues of cultural and economic self-determination, protection of legal and human rights, and implementation of reparations for past damages, stressing that no just environmental outcome is possible unless truly democratic decision-making mechanisms or procedures are in place (Warren, 1999; Schroeder, 2000). Accordingly, observers unimpressed with existing CBC developments, particularly their susceptibility to being appropriated by state or outside interests behind the smokescreen of community rhetoric, call for democratic decentralizing reforms, reforms that guarantee participation in decision making about resource extraction, utilization, and preservation (Ribot, 2002). Only with such legal changes are local voices more likely to be heard. While such reforms lie beyond the direct ambit of most environmental NGOs, their significance highlights the need for conservationists to collaborate with local forces working toward greater democracy. At the same time we should not forget that decentralization, democratic or otherwise, can be risky in both ecological and social respects (chapter 7).

Self-determination inspires territorial claims. This development is epitomized in the Miskito Indians' establishment of an autonomous homeland, consisting of coastal lagoons and mangroves, offshore coral reefs, and seagrass pastures off the northwest coast of Nicaragua (Nietschmann, 1991a). Buffeted by back-to-back occupations by right-wing and left-wing dictators who promised development and brought none, and highly destabilized by the arrival (on the heels of the 1980s civil war) of illegal commercial fishing operations for lobsters, turtles, and shrimp (as well as of heavily armed drug traffickers), the Miskito decided to pursue their 500-year struggle for autonomy by asking WWF and other agencies for help in establishing a reserve, the Miskito Coastal Protected Area. Nietschmann (1991b) argues that the truly self-deterministic aspect of this conservation effort might be why a project like this succeeds whereas so many other conservation-development projects fail. Indeed, the Miskito and adjacent Sumu Indians are active in national-level protest campaigns, demarcation, and mapping projects among coastal communities. And even inland, where one of the largest intact areas of tropical forest in Central America is under siege from loggers and settlers, locally organized Miskito and Sumu communities are demarcating their territorial limits in the face of considerable violence (Chapin, 2001). Similar mapping projects in neighboring Honduras stimulated the formation of seven indigenous federations in the Miskito.

Mapping, often known as counter-mapping (Peluso, 1995; Poole, 1995) or mapping against power, is a popular initiative in this area, with local communities taking an active role in demarcating their resources and mobilizing networks in the service of conflict resolution with outsiders. Clearly linked to progressive political goals of self-determination, this method can be effective, as demonstrated by Miskito initiatives, but it also has drawbacks. In a detailed study of counter-mapping in northern Tanzania, Hodgson and Schroeder (2002) show how the very process of drawing a map cements a notion of fixity in land claims that may not actually exist.

In fact, it is in this same area, northern Tanzania, where we can see troubled and conflicted local movements for self-determination (Hodgson, 2002). Since 1990 over one hundred indigenous NGOs have emerged in the predominantly Maasai areas of this country, called "indigenous" even though Maasai are relative newcomers to the area. These organizations are structured around diverse claims of identity based on ethnicity (Maasai), mode of production (pastoralism), or a long history of political and economic disenfranchisement by first the colonial and now the postcolonial state. While democratization has provided new space for such organizations, the wealth of opportunities (development funds, workshops, overseas tours, etc.) showered by international donors has led to bitter competition and splintering among these groups. Issues at stake include whether women should have equal rights to men, the right of young educated men to have greater decision-making power than traditional elders, and of course the legitimacy of identity, for there are other (non-Maasai) pastoralists in northern Tanzania, other Maa-speaking nonpastoral communities, and many other disenfranchised populations. As one participant observed at a workshop in Arusha in 2000, "Some people just put on red cloths and call themselves pastoralists!", referring to the Maa-speaking Arusha people who have recently reclaimed their Maasai heritage in light of donor preferences to help "the Maasai" (Hodgson, 2002, 1090). Accordingly, the indigenous rights movement in Maasai areas of Tanzania is unable to form a coalition, or to advocate very effectively at the national level. As a result, they continue to make deals with different donor organizations, which themselves represent no monolithic entity; for example, while one NGO is funded to create community wildlife management areas, another lobbies against further encroachment of reserves onto pastoralists' lands. Part of the problem here is that the so-called indigenous rights movement in northern Tanzania came into being directly in response to donor aid (Saugestad, 2001, cited in Hodgson, 2002). Unlike in Canada,

Scandinavia, Australia, and New Zealand, where local organizations consolidated and mobilized long before they engaged in dialogue or litigation with their respective governments, Tanzanian NGOs had to launch straight into the toughest issues of all, rights to water and land, with no time to build unified local support.

Perhaps the most instructive and encouraging example of self-determination comes from Panama's Kuna Yala, the very first internationally recognized forest park created by an indigenous group (Chapin, 1991), that became an overnight cause célèbre among environmentalists and indigenous peoples' rights advocates and then disappeared from view (Box 8.6). The checkered history of the Research Project for the Management of Wilderness Areas in Kuna Yala (PEMASKY), the instrument whereby Kuna designed and ran their reserve, speaks clearly to issues of indigenous environmentalism, alliances with conservationists, and the new model of self-determination. As regards Kuna's concern with biodiversity protection, it is quite clear that, as in the Chipko case, the primary goal of PEMASKY was territorial defense, not conservation; indeed, protection of the forest was primarily strategic. As Arias and Chapin emphasize (1998), the Western idea of a "reserve" came from expatriate scientists but fell on fertile ground among people with a strong belief that areas of forests constitute spiritual domains (Chapin, 1991) and a deep fear of losing their land to settlers. As regards the alliance with conservationists, this played a key catalytic role in the early 1980s but soon outlived its usefulness.

So what was the importance of self-determination? The transformation of PEMASKY into a suite of local NGOs integrated closely with other Kuna and Panamanian organizations, the involvement of ex-PEMASKY trained staff in institutions all over Panama, and the heightened awareness among the Kuna that biodiversity is their key resource reveal how the Kuna themselves seem to have woven PEMASKY objectives deeply

---

Box 8.6. Whatever Happened to PEMASKY?

The 50,000 Panamanian Kuna Indians of the Comarca de San Blas set up the Research Project for the Management of Wilderness Areas in Kuna Yala (PEMASKY) in 1983 to defend their territory. Although they had been granted the Comarca as an autonomous tribal homeland by the Panamanian government in 1938, the 1970s saw non-Kuna farmers and cattle ranchers settling along the newly constructed Pan American Highway inside the Comarca. Kuna first experimented with a small agricultural community to defend the entry to their homeland, an initiative that failed. They then focused on protecting the forest (Breslin and Chapin, 1984; Chapin, 1994).

The Kuna homeland (now known as Kuna Yala) extends from the San Blas islands on the Caribbean coast to the top of the mainland Cordillera, and the reserve incorporates a 60,000 ha^2 rainforested agroecosystems. PEMASKY was set up by an enterprising group of Kuna within the Kuna Workers Union after listening to the recommendations of the Research and Training Center for Tropical Agriculture (CATIE). Its brief was to regulate natural resources and protect Kuna historic culture by promoting scientific investigation, environmental studies, and ecotourism. It received ample international financial support for the protection of the relatively undisturbed rainforest, from the Inter-American Foundation, the Smithsonian Tropical Research Institute (STRI), WWF/US, and USAID. In 1987 the Kuna General Congress approved a PEMASKY request that the park be considered as biosphere reserve (though the paperwork never went through). With control over the title to their homeland, the Kuna designed and managed the reserve from the outset (Arias and Chapin, 1998; Clay, 1991).

PEMASKY became an overnight success, touted as a model alliance between conservationists and indigenous peoples. Throughout the mid 1980s it successfully defended its borders, halted a mining initiative, stopped development of a U.S. naval base for drug control, controlled scientific research in its territories, reaped benefits from ecotourism, and conducted extensive education programs in the San Blas communities. But as Clay (1991) notes, Kuna Yala was unusual in two respects. First, the San Blas Kuna never relied heavily on the forest to provide signifi-

(continued)

---

into their broader agendas. Had PEMASKY been imposed and run by outsiders, the foreigners would have gone home when the money ran out and that would have been the end of it. Perhaps PEMASKY's enduring if transmuted success lies in its self-determination. It is precisely because of the education and political savvy of ex-PEMASKY personnel that PEMASKY's agenda lives on. The Kuna have not "lost the way of their ancestors" (the worry at the end of the 1980s that education and emigration was

undermining PEMASKY, see Chapin, 1991), but found a new way. Kuna Yala provides a good antidote to enforced primitivism (Box 8.3), and shows that when an organization is truly self-determined, it will reemerge in other forms. In sum, while it is obvious that self-determination provides no quick solution to the problems facing environmentalists and indigenous peoples, it undoubtedly plays an indispensable part in the development of any long-term solution. Enduring success, if it can ever be achieved, will de-

(Box 8.6 continued)

cant income or even for subsistence agriculture; they live on a chain of 375 tiny coral islands 12 miles off the Atlantic coast of Panama, and depend primarily on fishing, trade, near-shore agriculture, and employment in the Canal Zone; use of the forest is mainly for medicines, fruits, and handicraft materials, and tree-felling is prohibited for spiritual reasons. Thus, Kuna objectives diverge little from those of Western conservationists. Second, unlike most indigenous peoples, the Kuna already held title to their land as a consequence of a series of armed uprisings in the 1920s and a deal with the U.S. government over the cutting of the canal through their land.

Seemingly at the apex of its fame, PEMASKY's image began to fade. By 1987 PEMASKY was running out of money, and most of its staff were dismissed in 1988. In Arias and Chapin's (1998) analysis, the principal problem was that PEMASKY never developed a long-term management or fundraising plan, nor properly integrated the scientific investigations that were ongoing into its own plans. Furthermore, it was split between its headquarters in Panama City, its field station on the Continental Divide, and the islands where lived its most important constituency—the Kuna people. In addition, the Kuna themselves, particularly the youth, were losing interest in the forest as a spiritual homeland (Chapin, 1991). Perhaps worst of all, PEMASKY was burdened by its global but somewhat mythological reputation as a success story. It was collapsing, victim of its own success.

But PEMASKY may have sown enviromentalism more broadly. Numerous young Kuna inspired by PEMASKY now study biology, geography, economics, agronomy, and anthropology at the National University, and enjoy an impressive library put together by PEMASKY. Kuna can negotiate with outside scientists more as equals, as in the Kuna/STRI initiative on creating a marine protected area in the Cartí area of the Comarca (Ventocilla et al., 1997). They are also working on enlarging the autonomous region to include other Kuna communities (Arias and Chapin, 1998). All in all, Kuna steered PEMASKY along their own agenda, and though the internationally hailed project did not survive (to the chagrin of outside observers) it has amply succeeded in training Kuna to confront their own environmental problems, as well as in defending the forest.

---

pend heavily on the extent to which solutions are sought by insiders rather than outsiders.

8.7 The View from the Other Side

While Westerners were tempering their approaches to traditional and local communities in order to craft humane solutions to the conservation crisis, non-Westerners' views of the North were undergoing their own evolu-

tion. Perhaps the best articulated indigenous responses lie in the writings of Indian scholars. For instance, in the view of Guha (1997), Western environmentalism, and particularly the movement for preserving wildlife in India, is fueled by different interest groups—city dwellers and foreign tourists who value recreation time in the wild; ruling elites who view the conservation of flagship species, such as tigers, as important for national prestige; international conservation organizations concerned with educating rural people about the value of conservation; and state functionaries who work in the forest and are motivated by trips abroad funded by conservation organizations. These observations have of course been well rehearsed over the last twenty-five years, as we saw, for example, in the critique of protectionism in chapter 2, and have been taken seriously by many conservation biologists (e.g., Robinson and Redford, 1991; Redford and Padoch, 1992).

What is remarkable, then, is that Guha's lament is still pertinent today, despite huge swings of policy in the wake of the Brundtland Report (chapter 2). This is true not only of India, where areas set aside for wildlife conservation have displaced over 600,000 tribal and forest people (see Box 2.3), but also in many other parts of the world where the principle of working with local people has still not really transformed into practice (Colchester and Gray, 1998). Clearly, rhetoric is far ahead of the reality. Indeed, despite the worry of some conservationists that conservation organizations now pay more attention to the rights of human communities than to wildlife, non-Westerners still feel alienated from the very Western blend of environmentalism that emphasizes intrinsic value so heavily (see chapter 1). Non-Westerners' mistrust was spiked by the environmental movement's abandoning of its anti-industrial stance, which to third world observers looked suspiciously like Westerners being ultimately unwilling to give up their own standard of living to obtain conservationist ends (Sarkar, 1990). Insofar as the West, for better or worse, sets some sort of model for the aspirations of people in developing countries, this is a lamentable outcome.

There are, however, important areas of convergence between the thinking of conservation biologists and indigenous activist writers. Let us consider, for example, the "flagship approach" to conservation. For many years conservationists have relied on attracting the attention and donations of a potentially fickle public by publicizing threats to high-profile large mammals and birds, whether at home or abroad. Local communities and their spokespersons have been alert to the injustices of such campaigns, both with respect to how these ignore issues of local livelihood and how they bolster the interests of national states in ways that have little or nothing to do with conservation. The WWF's flagship conservation program to protect the panda in China is a good example (Colchester, 1997). Almost all the early WWF funds went into a hugely expensive captive breeding program carried out in the Wolong Nature Reserve (albeit at the demand of the Chinese government), while hardly any money was spent either on relationships with the Han and other ethnic minorities who lived in the area, or on combating the main threat to panda habitat, logging (Schaller, 1993). Furthermore, there was a sting in the tail. The campaign rendered the panda a potent symbol of both conservation and Chinese nationalism. Accordingly, pandas became prestige gifts to visiting dignitaries, they were coveted for zoo collections, and their skins became increasingly valuable in the illegal fur trade, all putting more pressure on their wild populations (see Figure 3.5). Very similar critiques have been made of another of WWF's high-profile campaigns, the Save the Tiger program in India (Gadgil and Guha, 1992). Though some tiger populations have stabilized as a result of this project (see Box 2.3), manslaughter by tigers reached serious proportions in areas such as the Sunderbans Delta, where, with an increase of the tiger population from around 130 to 205 in the 1980s, a toll of 1,000 human lives was reported. In a similar vein, Bonner (1993) based much of his critique of the Western conservation movement in Africa on its obsession with elephants, animals that can cause major damage to subsistence crops.

The interesting point, however, is that, albeit for different reason, biologists (Caro and O'Doherty, 1999; Leader-Williams and Dublin, 2000), conservationists working on the ground (Zhi et al., 2000), and indigenous peoples (Lynge, 1992) are now questioning the validity of the flagship approach, and indeed more broadly single-species conservation techniques (see chapters 3 and 11). Flagship species are often selected in a highly ethnocentric way, to appeal to Western or international donor sensitivities, and these choices can easily alienate local populations, as when dangerous carnivores or large mammals are used.

8.8 Conclusion

Local communities and conservationists are both taking initiatives to work together toward protection of the environment. This approach, based on alliance rather than conflict, has become prominent in international meetings, declarations, and conservation policies. It would, however, be premature to believe that this constitutes a "new model" of conservation, though it is certainly a new vision (IWGIA, 1998, 300). In this chapter we have looked at both the opportunities and the pitfalls. Much of the old mistrust persists, with local communities suspicious that old-style conservationist interests are seeking to displace them, and conservation biologists nervous that these communities will be tempted by market forces to degrade their lands. However, a common concern with protecting indigenous and other traditionally occupied lands from unrestrained infusions of the market and immigrant settlers provides the basis of what can be a productive alliance. Furthermore, increased communication within and between such alliances, as demonstrated in conferences held by groups such as TNC (Redford and Mansour, 1996), nudge participants toward better understanding and more effective practice. And in the end, it may be that when these alliances break down, a more locally sustainable form of biodiversity conservation is in the wings, as

we anticipate (albeit somewhat optimistically) for Kuna Yala.

Clear positive outcomes have emerged. On the ground there have been major successes in slowing biodiversity loss, as we saw with the Chipko movement, the Dayaks, and the Kuna. In policy, too, there have been breakthroughs. For example, the IUCN revised its protected areas categorization to recognize that the owners and managers of these areas could now be NGOs, such as individuals, local communities, and indigenous peoples, and both IUCN and WWF explicitly endorse the principle of free and informed consent in all interactions between indigenous peoples and conservation organizations, in line with the current draft of the United Nations Declaration on the Rights of Indigenous Peoples (Arias and Chapin, 1998). How these changes will shake out in the real world of politics, particularly given the sovereign power of nation-states over the biodiversity that lies within their borders (see chapter 9), remains problematic. Certainly conservationists will need to place the recognition of indigenous rights by nation-states firmly on their agendas (IWGIA, 1998). Since conservationists rightly dread such cluttering of their agendas (Brandon, 1998; see chapter 2), particularly with objectives they are not qualified to deliver on, this is where conservation efforts coordinated by multiple NGOs that specialize in different areas of policy could be so valuable. It is also clear that alliances between conservationists and local communities can run into trouble when economic payoffs are presented as the principal incentive for conservation, throwing into relief the whole question of the marketing of biodiversity (chapter 9). It is to these issues—the marketing of biodiversity, the role of the private sector, the international rules and regulations regarding the stewardship of nature, and financial responsibilities for global conservation—that we now turn.

CHAPTER 9

Global Issues, Economics, and Policy

9.1 Introduction

Over seventy years ago the forester and conservationist Aldo Leopold realized that a set of incentives was needed to induce private landowners to manage their land in ways that would serve public interest and future generations. This principle he termed "conservation economics." But he was not naïve, and recognized that sustainable management of land would need to be overseen by something beyond the pale of economics—ethical principles. With his famous "land ethic" he claimed that the "extension of ethics to . . . human environment is . . . an ecological necessity" (Leopold, 1970, 239). In short, Leopold wanted to bring two disciplines, economics and ethics, into the service of conservation. In our present era conservation economics, under the aegis of the discipline of ecological economics, is blossoming and lies at the heart of a huge raft of international economic initiatives (many still under negotiation) geared to fund conservation worldwide. Conservation ethics, by contrast, despite the environmental movements' avowed multicultural and multireligious roots (Callicott, 1994; and see Box 1.3), has not yet percolated very far beyond Western liberalism; business and politics worldwide is largely unaffected, and Leopold's latter wish remains unfulfilled. On account of the dearth of widely held ethical principles relating to conservation, conservationists must turn increasingly to policy,

international agreements, or simply nonlegally binding multilateral protocols for guidance and control, if conservation economics is to be held in check.

In this chapter we look at conservation from a worldwide perspective. We start at the heart of the dominant contemporary conservation strategy—selling nature to save it, in fact, precisely at the point to which bioprospecting (chapter 8) brought us. Looking in particular at ecological economists' practice of environmental valuation, we assess from a more political vantage than in chapter 1 the strengths and weaknesses of this salesman-like approach to biodiversity conservation (9.2). Thinking about nature in terms of money and markets inevitably raises the possibility of handing over its management to business and private enterprise. We consider this idea, particularly in connection with the "use it or lose it" strategy (9.3), before examining a new strategy emerging in the face of imminent species extinction and habitat loss, the purchase by both private individuals and nonprofit organizations of protected areas in the developing world (9.4). Given how poorly business is equipped to provide public goods and how tenuous property rights in foreign countries are, we end the chapter with a consideration of international policy initiatives regarding who pays for the protection of biodiversity (9.5) and under what kinds of international agreements or protocols (9.6). The conclusion is that whatever strategies are devised to protect global biodiversity, international regulation is critical, fraught though it be.

9.2 Ecological Economics and Environmental Valuation

We have already seen in chapter 8 how a monetary value can be put on the pharmaceutical products of nature in order to conserve them. Now we apply this idea to all nature's goods and services, returning to the discipline of ecological economics introduced briefly in chapter 1. The central rationale of ecological economics is that by defining all environmental goods and services as commodities, in other words, giving them a price tag, nature can no longer be regarded as a free resource. Nature is now thought of as capital, natural capital (Hawken et al., 1999), whose price varies with scarcity and demand like any other economic resource. With the true costs and benefits of conservation thereby internalized in the price of products, for example, when the price of hydroelectricity includes the cost of controlling erosion and silting caused by the dam, a market model will ideally set both economically and ecologically efficient levels of resource use and pollution.

How do we put a price on nature? Environmental valuation was first implemented in 1936 with the passage of the U.S. Flood Control Act (Gatto and De Leo, 2000), but the last ten years has seen a mushrooming of analyses pricing biodiversity and ecosystem services. There are now numerous empirical applications of different techniques for valuing not just tropical forests (see Box 1.2), but wildlife, coral reefs, mangroves, rural water supplies, air purity (Costanza and King, 1999; Rietbergen-McCracken and Abaza, 2000), and indeed the whole of nature (Costanza et al., 1997; Pimentel et al., 1997; and chapter 1). Economic valuation appeals to diverse constituencies, ranging from free-market advocates who believe it will im-

TABLE 9.1
Hypothetical Australian Farm Business in Twenty Years

Commodity	Share of Farm Business (%)
Wheat	40
Wool	15
Water filtration	15
Timber	10
Carob sequestration	7.5
Salinity control	7.5
Biodiversity	5

Note: In this model traditional agricultural commodities account for 55 percent of revenues, as opposed to 100 percent today. Other income derives from a mature market for ecosystem goods and services.

Source: Reprinted with permission from Daily et al., Volume 289:395–396. Copyright 2000 American Association for the Advancement of Science.

prove economic efficiency (as outlined above), to managers in search of an integrative currency with which to guide decision making, to environmentalists who believe that the standing of neglected natural services will be enhanced by recognition of their true economic value (Carpenter and Turner, 2000). Accordingly, valuation methods percolate beyond the academic community to policy makers and the business world. For instance, the Kenya Wildlife Service used data on how much foreign visitors value wildlife (based on contingent valuation and travel cost methods, see below) to raise the national park entrance fees for non-Kenyans by 310 percent. And in another example Daily et al. (2000) report how the world's first conservation company to go public (Earth Sanctuaries, Ltd.) won a change in accounting law to have rare native animals and ecosystem services counted as assets (Table 9.1). In this section we turn to the strengths and limitations of these controversial new developments. We focus primarily on the socio-political considerations associated with these methods of evaluation, though these inevitably raise ethical and philosophical issues already touched on in chapter 1, where we discussed the dangers of putting nature on sale to save it.

Ecological economists view nature's goods and services as resources, and quantify them as such in their account keeping (see again Table 9.1). Species contributing to the provision of ecosystem services, such as pollinators, can therefore be given a price. It is important to realize that this value varies across cultures and over time; thus, peat bogs were once economically important in the British Isles, but no longer since the emergence of labor-saving fuel substitutes; pollinators, by contrast, are gaining value as they become more scarce. The extent to which any item in nature is considered a commodity will depend on many things, like knowledge, technology, access to alternatives and substitutes, and cultural values, and will therefore differ not only between but also within nations, districts, or even communities, and over time. Consequently, in calculations of the true economic value of different natural resources, consideration must be given to broader social, political, and cultural dynamics, as well as to internal differences within communities.

Take, for example, the study of non timber forest products (NTFPs) in the vicinity of the Iquitos market examined in Box 1.2. In the flurry of research determining the economic value of such tropical products (de Beer and McDermott, 1989; Panayotou and Ashton, 1992), Peters et al.'s (1989) evaluation of this particular Peruvian watershed showed that collecting plants and animals from the forest for subsequent marketing was economically more productive than logging over the long term. From this kind of study managers might recommend the area be run as a reserve in which native populations are granted harvesting rights (see chapter 10). Deriving policy recommendations on the basis of market prices nevertheless conceals many assumptions, for example, that forest dwellers are motivated to collect NTFPs for market sales. It actually turns out that in much of the Amazon the

harvesting of wild foods and products is viewed as an inferior occupation (Browder, 1992), and may even be abandoned for marginal value considerations once incomes rise, as discussed in chapter 5. In economists' terms Peters et al. failed to give full consideration to forest dwellers' opportunity costs or lost opportunities. In other words, a more politically and economically grounded assessment of the assumptions underlying calculations of NTFP values challenges Peters et al.'s conclusion that pure economic considerations favor traditional forest harvesting.

Estimating the benefits and opportunity costs associated with conservation gets increasingly tortuous when value must be inferred for goods that, unlike NTFPs, are not currently traded on markets. Ecological economists have developed a number of methods to deal with this challenge. These methods are acknowledged as somewhat experimental (e.g., Rietbergen-McCracken and Abaza, 2000), and include unconventional pricing techniques, such as avoided costs, household production functions, hedonic pricing, and contingent valuations (Table 9.2). The last method is commonly used by conservation biologists examining the economics of protected areas, but is also the most controversial.

Contingent valuations are based on surveys designed to detect how much a person is willing to pay (or accept in compensation) with respect to a particular natural resource or ecological service. The big question, however, is who is surveyed, and with whom their opinions are compared (McAfee, 1999). Consider what happens when we ask foreign tourists how much they would pay to spend a day in Serengeti National Park and compare this to how much a Maasai family would be willing to accept in compensation for establishing a national park (Figure 9.1). Contingent valuation surveys would undoubtedly conclude that the most economically efficient form of land use for the Serengeti would be to continue managing it as a park and to evict pastoralists (insofar as

TABLE 9.2
Methods of Environmental Valuation

Method[a]	Example	Difficulty
Avoided costs pricing	The value of a service is estimated by the cost of replacing it. *Example*: Vegetation clearance for farming and grazing regimes around villages in developing countries lead to shortages of fuelwood. Fuelwood must be replaced by animal dung, with the result that dung, normally used as a fertilizer, must be replaced by chemical fertilizers. By computing the cost of chemical fertilizer the monetary cost of vegetation clearance is determined.	Relatively straightforward (and the most conventional of methods) because chemical fertilizers have a market value. Additional costs (e.g., increased fuelwood collection time costs incurred at the household level) missed, unless a monetary value can be put on fuelwood collectors' time
Household production functions	The value of a nonpriced service can be determined by examining, as a proxy, household expenditures related to this service. *Example*: Harold Hotelling, in a letter to the U.S. National Park Service in 1947, proposed that the actual traveling expenses incurred by visitors could be used as a measure of the recreational value of the sites visited (the travel cost method).	Relatively straightforward, until expenditures of visitors are weighed against evaluations made through contingent valuation (see text)
Hedonic (or indirect revealed preference) pricing	A value for goods not conventionally traded is derived from prices of similar goods that are traded. *Example*: Clean air can be valued by comparing land rents in clean versus polluted neighborhoods.	Specific to the value of a commodity at a particular site; e.g., land rent differences may not reflect preferences at another site, and therefore cannot be generalized
Experimental methods (contingent valuation and existence value surveys)	A direct estimate of people's preferences or willingness to pay for a service through explicitly designed surveys. *Example*: Offer respondents hypothetical choices, often phrased as how much would you be willing to pay for X, how much would you be willing to receive in compensation for Y, or how much would you pay simply to know Z exists?	Unreliable when applied to unfamiliar and/or technical issues Nonquantitative answers are disregarded Difficulties with aggregating preferences

[a] Basic categories follow Pearce (1993; see also Armsworth and Roughgarden, 2001).

they deplete the "wildlife experience"). Efficient, yes, but is this a fair comparison? The structural inequalities and disarticulations between Northern, rich, and industrialized economies and Southern, poor, and primarily commodity-based economies means that the "value" of any given resource, as assessed by a Northerner, will always outweigh the uses of the same resource by most residents of Southern countries. As such,

Figure 9.1. Mismatched Values: Tourists and Maasai.
A Maasai herdsman in conversation with a guide in a tourist vehicle on the edge of the Serengeti plains. The Serengeti plains were first gazetted as a national park in 1940. It is highly unlikely that most Maasai, then or now, would reject as compensation the amount that tourists to Tanzania are willing to pay to visit the Serengeti. But is this the right justification for maintaining the park? Photo courtesy of Tim Caro.

the wishes of communities in the developing world can never compete with those of the rich. Because of this distributional issue (the economists' term for difference in wealth), it is simply not meaningful to weigh the amount a professional earning US$50,000 a year is "willing to pay" for the continued existence of a tropical ecosystem against the "willingness to accept compensation" for the loss of an ancestral homeland of a resident of that same ecosystem, particularly when he or she has little or no cash income and a vastly different worldview. In a similar vein, it was only because of the high earnings of U.S. citizens in Alaska that Exxon paid the huge US$5 billion to compensate for the Valdez oil spill of 1989; if the accident had happened in Siberia, where income is lower, compensation would have been minimal.

The Serengeti example shows the problems that can arise from managing nature in relation to its value on the international market (counting biodiversity in dollars, eu-

ros, or yen), with international conservation strategies simply reinforcing the claims of global elites to the greatest share of the Earth's value. When only money talks, other inequities (North—South, rural—urban, landed—landless) become irrelevant. In the view of political ecologists, environmental valuations merely validate shifting the costs of the ecological crisis onto the inhabitants of the developing world, onto countries that are cash poor but also, according to the logic of ecological economics, both "under-polluted" and "over-endowed" with natural resources (lacking the capital, technology, and expertise to "develop" that wealth; e.g., Guha, 1989b; Arizpe and Velazquez, 1994; Dove, 1993; Goodland et al., 1993; Heang, 1997; McAfee, 1999). In many respects this critique parallels that of bioprospecting outlined in the previous chapter. Through the sale of nature's products, be these shamans' recipes, timber, or access to prime research or ecotourist opportunities, the grip of international capitalism merely tightens on communities blessed with these natural resources, particularly if they lie in the developing world. In short, the thrust of this critique is that by framing global ecological dilemmas in terms of dollar values and efficiency we are obscuring more fundamental issues of comparability and environmental justice (see chapter 7), thereby promoting a bias toward technical solutions and away from policies that require innovative social change.

The problems inherent in environmental valuation methods go beyond the contentious question of whose pocket book expenditures are scrutinized, and at least with respect to the contingent value method raise somewhat intractable philosophical matters. First, there is the value of knowledge itself. How can people rank ecological goods and services when they do not fully understand them (Diamond and Hausman, 1994; Ludwig, 2000)? For example, most people would place low in the hierarchy a species (small insect) or natural area (mosquito-infested swamp) with little emotional or aesthetic appeal, simply because they do not understand its ecological significance. The question boils down to this. In what areas of human action should the knowledge of experts be substituted by the preferences of the people? Or, as Diamond and Hausman (1994) so tellingly pose, why do we use contingent valuation methods in decisions regarding the environment but not regarding surgery or bridge construction? A second tough ethical and philosophical dilemma is how to incorporate discount rates, the extent to which future costs and benefits are devalued (chapter 6). The notion of equal treatment of generations, captured in the famous quote "We have not inherited the earth from our fathers, we are borrowing it from our children" (Lester Brown cited in Paehlke, 1989, 140), would have future generations treated the same as current ones. But who speaks for the unborn, and with what right? Third, beyond all these difficulties, hovers the most impenetrable of all, the logic of assembling into a single preference

ranking criteria that are utterly distinct—economic, social, philosophical, and aesthetic. Is it legitimate to roll them into a single economic index? Many have proposed that it is not (e.g., Sen, 1970); for example, Ludwig (2000) asserts categorically that personal and social values are of a higher order and generally incompatible with economic values. Many agree with him, and suggest that incommensurate goals should be explicitly recognized and analyzed as trade-offs between, for example, economy, equity, aesthetics, and social values (Gatto and De Leo, 2000). For these reasons critics of environmental valuations maintain that clever technical measures can never substitute for wise judgment.

There are nevertheless two strong points to be made in ecological economics' favor. First, although economics cannot be used to determine whether a conservation goal is appropriate, it can usefully indicate the most cost-effective way of achieving that goal; conservation is, after all, all about politics, and projects with a positive bottom line will always be the most attractive. Looking, for example, at the case of Kenya (Box 9.1) we can see how keenly environmental valuation pinpoints the particular challenge faced by that nation, how critical international financial support is to future conservation efforts, and incidentally how sensitive the conclusions of such valuation studies are to the methods used.

Second, environmental valuations are key to examining opportunities and constraints for conservation, although their utility differs by scale. Local studies can be useful and quite accurate. For example, research that measures the marginal value of a wetland area could compare the forgone opportunities for agriculture and development to the measured benefits of nitrate filtration and water purification and, at least for a relatively homogeneous group of stakeholders, arrive at useful results regarding optimal land use. Broader-scale studies cannot claim as much accuracy, but can be arresting with respect to the light they shed on obstacles

Box 9.1. Does Kenya Profit from Protectionism? Ecological Economic Calculations

The desire by overseas visitors to see wildlife in its natural habitat is an important reason for tourism in many African countries. Overseas visitors are willing to pay substantial sums of money to view African wildlife, and place a high premium on visiting unspoiled natural habitats. For example, in Kenya the viewing value of a lion in Amboseli National Park exceeds US$0.5 million and the annual tourist value of the flamingoes on Lake Nakuru National Park lies between US$8 and US$15 million (studies cited in Earnshaw and Emerton, 2000). A contingent valuation survey for the nation of Kenya found that foreigners value visits to protected areas at US$450 million (Moran, 1994). It looks as if wildlife policies cannot fail to pay. However, when such contingent valuation studies are compared to more conventional methods, and the opportunity costs of conservation explored, a new picture emerges.

Still focusing on Kenya Norton-Griffiths and Southey (1995) estimate the conservation value of lands set aside in parks, reserves, and forests, and compare this with the opportunity costs of such set-asides. Using methods in Table 9.2 to calculate the net benefits to Kenya of setting aside lands for parks, reserves, and forests, they then sum the net benefits from direct uses (US$27 million from tourism) and the net benefits from indirect uses and nonuses (US$15 million from watershed and erosion protection, pharmaceuticals, and carbon sequestration), to come up with a total net benefit from these set-asides of US$42 million. Turning now to the opportunity costs, these are calculated as the potential net returns from agricultural and livestock activities in these areas (were such activities not banned), and are based on productivity figures from regions of similar land quality that are not under conservation. The results (see table, using 1989 as a baseline value) show that agricultural and livestock production in the parks, reserves, and forests of Kenya would generate gross annual revenues of US$565 million and net returns of US$203 million. Potentially the parks of Kenya could support 4.2 million Kenyans, 5.8 million livestock, and 0.8 million hectares of cultivable land.

TABLE 9.1 BOX
Potential Population, Gross Revenues and Net Returns from Parks,
Reserves and Forests if Converted to Agricultural and Livestock Production

	Forests	Parks and Reserves	Total
Area (km²)	19,200	41,420	60,600
Population (million)	2.1	2.1	4.2
Livestock (million)	1.9	3.9	5.8
Cultivated Hectares (million)	0.4	0.4	0.8
Gross Revenues (US$ million)	280	285	565
Net Returns (US$ million)	104	99	203

Source: From Norton-Griffiths and Southey, 1995.

(continued)

(Box 9.1 continued)

We can conclude from Norton-Griffiths and Southey's study that the net benefits to Kenyans from wildlife tourism and forestry within the parks, reserves, and forests of Kenya (US$42 million) are completely inadequate to offset the opportunity costs of leaving these lands underdeveloped (US$203 million). This conclusion is very different from that of Moran, for two reasons. First, Norton-Griffiths and Southey calculate the benefits of tourism in terms of what Kenyans actually accrue from wildlife set-asides (US$27 million), whereas Moran measures this in terms of what foreigners would be willing to pay just to know Kenyan wildlife exists (US$450 million). Second, Moran ignores opportunity costs. Taking the Norton-Griffiths and Southey study as more thorough, we can see that despite the popular belief that Kenya epitomizes the economic benefits enjoyed from biodiversity conservation, the reality is quite different. For conservation ends Kenya is forgoing a huge sum of money (2.2 percent of its 1989 GDP), and denying land for settlement to over 4 million Kenyans. Given the inappropriateness of cash-strapped developing nations like Kenya bearing this awesome burden, we can anticipate enormous pressure for development on parks, reserves, and forests in Kenya within the next twenty-five years. To avert this outcome, international funds would be needed to subsidize the government's conservation strategy.

As it turns out, in the intervening years since this study was published international funds have been inadequate. Already by 2001 the government proposed to clear 170,000 acres of forested land in the East and West Mau, Mount Kenya, and Londiani areas to meet both the economic and political exigencies of settling dissatisfied families from overpopulated areas. Solutions lie both in capturing more of the profits of the tourist sector for Kenyans (as in a recent initiative in the Maasai Mara National Reserve; Walpole and Leader-Williams, 2001), and in improving the management of wildlife outside protected areas (Norton-Griffiths, 1998). As regards tourism, either the number of tourists who come to Kenya should be expanded or the costs of park entry increased, although there are risks associated with both strategies, especially since the elasticity of tourist demand to price changes is not well understood (see chapter 10). As regards the management of wildlife outside protected areas, there lie possibilities in diversifying the wildlife sector, especially through legalizing sport hunting again after the 1977 ban or loosening regulations on game ranching (see later in this chapter).

and opportunities for policy. Though every step of these ambitious broader evaluation studies will be contested by each stakeholder concerned, insights emerge as gems (Table 9.3). Thus, from James et al. (1999) we see how comparatively cheap the conservation of biodiversity in protected areas is (at least when compared with the cost of conserving biodiversity beyond protected areas), and how easily this could be funded if perverse agricultural subsidies were to be scrapped; from De Leo et al.'s (2001) study we see how Italy could actually save money by conforming to the United Nations Framework Convention on Climate Change (the Kyoto Protocol); and from Kremen et al. (2000) we observe the pressing need for supporting countries (like Madagascar) that set conservation priorities, and for doing this at the national level. Environmental valua-

TABLE 9.3
Environmental Valuations Highlight Opportunities and Obstacles

Question	*Answer*
How expensive would it be to maintain and expand protected areas (in line with IUCN's objective of 10 percent of land area in each region) compared to, first, the greening of agriculture, and second the maintenance of perverse agricultural subsidies (James et al., 1999)?	Cheap! *$27.5 billion per annum.* Protected areas maintenance and expansion (based on the cost of conserving an ecologically representative global network). *$290 billion per annum.* Greening agriculture (based on the cost of protecting biodiversity in human-dominated landscapes). *$950–$1,450 billion per annum.* Perverse subsidies (based on current payments)
Is there any net saving in reducing greenhouse gas emissions to 1990 values by 2008–2012 (the Kyoto Protocol) for a Western industrial nation such as Italy (De Leo et al., 2001)?	Yes! *1.5 to 2.0 billion euros net savings per annum* could be achieved by bringing the Italian energy-producing sector into compliance with the Kyoto Protocol.
What are the overall benefits and costs of Madagascar's Masoala National Park Integrated Conservation and Development Program? This is a 23,000 km² primary rainforest buffered by sustainable community forestry project (1,000 km²), threatened by slash-and-burn rice farming and commercial forestry, and supported by donor investments (Kremen et al., 2000).	Positive at the local and global levels, negative at the national level! *$92,000–$527,000.* Local: net benefit (due primarily to NTFPs) *$27–$264 million.* National: net deficit (due to opportunity costs of commercial forestry) *$68–$645 million.* Global: net benefit (due mainly to avoidance of damage caused by greenhouse gas emissions from deforestation) Ranges represent net present values, calculated under different time discount functions and over ten and thirty years

tions are more usefully considered as offering a nexus of opportunity rather than a statement of truth.

In short, ecological economics offers useful tools. However, its evaluation methods cannot get around an old issue, one we introduced in chapter 5, explored conceptually in chapter 6 as the public goods problem, and applied to conservation policy in a coarse geographical sense (see Table 7.1): how to distribute the costs and benefits of conservation in a world that is already riven with inequity. Before turning to this tough issue, we take a look at what happens if conservation is handed over to private concerns.

9.3 Business—Dancing with the Devil?

Once we start thinking about nature in terms of money and markets, the corollary of handing its management over to business needs to be taken seriously. Prescriptions for the future inevitably are divided. While some emphasize the need to strengthen and redesign multilateral environmental agreements (see below), others feel that meddling with national and international policies and protocols will achieve nothing. Adherents of this latter group stress the limited abilities of governments to enforce regulations, and argue that the key place to change practice and find funds is in the corporate world

(Daily and Walker, 2000), ideally in partnerships with private businesses (e.g., McNeely, 1995; McNeely and Weatherly, 1996).

In this view corporate enterprise is powerful, innovative, adaptable, and efficient; once nature is thought of as a critical limiting resource, natural capital (as discussed in the previous section), then these corporate qualities can be exploited for environmental ends. Some corporations are clearly beginning to recognize the market edge associated with environmental sustainability, and operate on the assumption that natural capital, like any other form of capital, must be safeguarded from liquidation (Hawken et al., 1999). Encouraging transformations have occurred (Daily and Walker, 2000). For example, Shell and British Petroleum, and more recently Dow and Ford, have resigned from the Global Climate Coalition, a Washington-based group opposed to the Kyoto Protocol (a protocol attempting to slow global warming; see below). Certified green investment portfolios are increasingly available, and there is lobbying for "green" business schemes within frameworks like the North American Free Trade Association, though as yet (as we saw with the marketing of environmentally certified products in chapter 8) these occupy a miniscule slice of the market and are difficult to regulate.

Since businesses face high initial costs in moving toward sustainable practice (particularly loss of competitiveness), companies cannot act unilaterally. In this context, organizations like the World Business Council for Sustainable Development and the International Chamber of Commerce are major achievements, as is Keidanren, a nationwide business organization that states as an objective the need to find an ecologically sound base for Japan's industry (Nagami, 1995). Though voluntary, these mark an important first step. They are stimulating many private-sector investors looking for image, dedicated staff, and loyalty into the direct funding of biodiversity and its promotion. Such companies extend far beyond the pharmaceutical sector discussed in chapter 8,

and include companies like Exxon, which made a US$5 million grant to support conservation of the tiger in Asia, its advertising symbol, and other large recent contributions (Balmford and Whitten, 2003). In short, there is considerable mileage for industries, primarily in the developed world, to look green.

By inviting into the environmental arena big business, notorious for the most part for externalizing the social and environmental costs of their operations, are we simply dancing with the devil? International capitalism is after all one of the principal causes of habitat degradation (chapter 7). For some, struck by the fact that of the 100 biggest economic entities in the world half are corporations (not nations), this is mere academic speculation, and there is indeed no choice. As Daily and Walker (2000) presage, the conservation movement's failure to engage industry and private interests will guarantee that an environmentally sustainable future remains nothing more than a fanciful whim; nonprofit organizations and charities simply do not have enough money or clout to halt global environmental degradation (McNeely and Weatherly, 1996).

There are plenty of constructive views for how to harness these market powers, to both fix and fund environmental problems and biodiversity protection (McNeely and Weatherly, 1996; Richards, 2000). We can think of these strategies as economic instruments or mechanisms that will put in place a global market in biodiversity, a market that in the parlance of economists is currently "missing" (Box 9.2). Most of these international incentive programs are still in their infancy, and many remain mired in deep politics (like the United Nations Framework Convention on Climate Change, the Kyoto Protocol; see below). They are, however, important, insofar as ultimately both the power and the excesses of business will have to be harnessed to global environmental policy. In this context there are also promising developments within the World Trade Organization (WTO). Currently free trade

Box 9.2. Harnessing the International Market: Innovative Incentive Mechanisms

1. Charging for the use of the global commons. The global commons (technically, common-pool resources as defined in chapter 6, insofar as they are depletable and nonexcludable) include the oceans, outer space, and the electromagnetic spectrum, and have enormous economic value; for example, the oceans provide over a billion tons of fish per year, serve as a means of transport, help regulate the climate, and act as a major carbon sink, yet are grossly overused. Users should be charged for access to these commons and the revenue allocated for conservation purposes (Bezanson and Mendez, 1995). Raising revenues in this way, which would require establishment of a system of user rights, regulations, and rents, would be an enormous political challenge. Maybe this is why the idea has not gone very far.

2. Joint Implementation. Carbon offset trading is an example of joint implementation, and is much in the policy limelight. In the context of the United Nations Framework Convention on Climate Change, joint implementation entails voluntary cooperation between two or more countries to reduce production of greenhouse gases. Effectively, an industrial nation, rather than cutting its domestic carbon dioxide (CO_2) emissions (a very costly course of action), will invest funds in some form of forestry in order to offset its carbon production. It turns out that tropical forest protection represents the cheapest way of reducing CO_2 emissions, because of low land and labor values in most tropical countries. Carbon offsets have brought some Northern electricity companies and Latin American reforestation schemes together, often brokered by NGOs. Currently such offsets are being experimented with by the World Bank's Prototype Carbon Fund (Kiss et al., 2002). Ultimately they will occur under the Clean Development Mechanism (CDM), within the as yet unratified Kyoto Protocol). The CDM is a kind of carbon banking system that has been in place since 1995 but is plagued by difficulties concerning both compliance and fair accounting for how different programs ensure carbon abatement (Parson and Fisher-Vanden, 1999; Richards, 2000); for instance, as yet the CDM allows carbon credits to be claimed only for afforestation projects, and not for avoiding deforestation (Niles et al., 2002). Other problems with joint implementation schemes include the risk that existing natural forests will get replaced by fast-growing non-native species such as pine and eucalyptus, and all the ecocolonial issues prompted by the fact that while Northern countries deforested without penalty, they are now pushing for economic slowdowns abroad rather than at home (Spergel, 2002).

3. International taxes. Direct taxation of international trade in products like tropical timber hardwoods could be used to raise funds from Northern governments to support conservation (Spergel, 2002). This idea has several attractions but faces political opposition from importing countries; it might also prove difficult to ensure the money is well spent (Barbier et al., 1994; Richards, 2000). Like carbon taxes, without binding international legislation, there is insufficient political will to implement them. One exciting related possibility is the proposed Tobin tax on the estimated US$2 trillion traded by currency speculators which, at a 0.1 percent rate, would raise US$100 billion annually for conservation and poverty reduction (Balmford and Whitten, 2003).

(continued)

(Box 9.2 continued)

4. Certification. To ensure the sustainable management practices of products like tropical timber, only certified products can be traded. The rationale is that an environmentally discriminating market will force those pursuing unsustainable practices to improve their forest management in order to sell their products on the world market. Though these ideas have drawn strong donor support, problems have arisen with low demand for certified products, with infractions to WTO rules (but see text), and with monitoring the honesty of labeling (Richards, 2000; and see the discussion of green consumerism in chapter 8). Similar "fair trade" initiatives have not yet been much used for conservation purposes, and also run up against conventional WTO regulations.

among nations signatory to the General Agreements on Tariffs and Trade (GATT) exacts an enormous environmental toll, with nations able to produce a natural resource (say timber) more cheaply than others simply dominating the market. What this means is that nations with more regulation are simply exporting their environmental damage to countries with less regulation, rendering regulation ecologically irrelevant. A recent move within the WTO to curb these apparent price distortions with respect to embargoing imports of tuna and shrimp that are fished in such a way as to minimize damage to other species of concern (particularly dolphins and sea turtles) marks a potential legal milestone in reconciling free trade with the environment (Yu et al., 2002). Perhaps GATT can be used to get rid of environmentally perverse strategies.

At the national level, the level charged with implementing CBD objectives, there are more advanced, and often more practical, economic instruments for valuing nature. These include green taxes, as collected on the use of fossil fuels in Scandinavia and the Netherlands; tax deductions and easements for landowners who manage their lands in ways to conserve biodiversity, as in Australia, the United States, Canada, and several African countries (see Table 9.1); the removal of perverse economic incentives for habitat destruction (see chapter 7); and legal

changes that allow the nonprofit sector to be as dynamic and innovative as the for-profit sector. Innovative ideas for putting business profits to conservation ends are emerging: for example, hydroelectric companies in Colombia must by law pay a percentage of their revenues toward the management and conservation of upstream water sources, and in El Salvador municipalities downstream of El Impossible National Park contribute to park management in return for watershed services (Kiss et al., 2002; and see Box 11.1). All these reforms move toward putting a proper price on biodiversity, since without true pricing the signals of overexploitation will be lost in the noise of marketing, subsidies, and global trade, and an undervalued biodiversity will continue to suffer.

If the links between business and environmental goals at national and international levels are still quite ephemeral, there are encouraging developments at a somewhat more modest scale that are gaining prominence and that speak directly to the potential of sustainable utilization by profit-minded individuals and organizations. The "sustainable use" movement is based on the view that the value of biodiversity can itself be turned into an instrument for long-term, broad-based biodiversity conservation. Often dubbed as the "use it or lose it" school, we explored its philosophical grounding in a utilitarian view of nature in chapter 1; we have also seen how these ideas are put to work with such (albeit troubled) initiatives as bioprospecting and green consumerism (chapter 8). Recent developments in southern Africa demonstrate how utilization can provide strong economic incentives for biodiversity conservation.

Game ranching, in conjunction with sport hunting, live animal sales, and even wild animal farming, has become the fastest-growing form of land use in southern Africa and plays a major role in that region's conservation strategy (Box 9.3). Comparisons between southern and eastern African countries are useful in that they highlight the factors that make game ranching profitable—supportive state policies, available marketing, slaughterhouse and veterinary facilities, a good network of roads to encourage marketing and tourism, and strong legal systems that recognize secure rights in land and the wildlife found therein (Hearne and McKenzie, 2000). Critical too is the phasing out of subsidies supporting unviable agriculture; even where wildlife utilization has a higher economic return than farming, agricultural subsidies and land tenure laws that encourage settled agriculture will distort market prices and force landholders to persist with farming. Abolishing subsidies is difficult since, for a combination of food security, economic, and political reasons, the agricultural sector has been protected in most African countries for many decades (Earnshaw and Emerton, 2000). Equally important are land tenure systems that keep land physically undivided; government support for the subdivision of group ranches into little privately owned plots in Kenya is creating huge obstacles for successful wildlife management schemes (Homewood et al., 2001). Finally, also critical are institutional factors that allow an enterprising landowner to incur the high startup costs of moving into this new and very extensive form of farming, such as fencing and the purchase of endemic species that have

long been absent in the area, as we see in Box 9.3. Accordingly, in East Africa, where many of these conditions are missing and trade in wildlife products is closely regulated, only game ranches that are well established and have easy access to large markets such as Nairobi are working well (Boss et al., 2000). Elsewhere the fate of wildlife living on unprotected land in Kenya is dire (Norton-Griffiths, 1998); many see a pressing need for policy changes that allow landowners to profit from wildlife, and for the ban on all hunting to be lifted.

In sum, despite the problems identified in the previous section with putting a price on nature, game ranching schemes demonstrate that there are ecological and sociopolitical conditions in which capitalizing on the value of nature seems to work as a conservation strategy, albeit a broad-based program focused mainly on large mammals. Despite the potential dangers of bringing business in as a partner for conservation, practicable schemes have appeared. Indeed, the promoters of the idea of managing wildlife consumption as a business claim that what a traditional conservationist approach has failed to achieve over the past forty years, an economic approach has finessed through pricing and policy reform within a few years. Since wildlife utilization purports to offer real alternatives to purely nonconsumptive utilization such as tourism in some instances, in chapter 10 we examine further issues associated with ecotourism and extractive reserves.

There remains nevertheless a worrying rider. Given the enormity of the challenge to integrate business more profitably to the cause of conservation, it is disappointing that the corporate community (with the exception of biotechnology) is so notably absent from the regular meetings designed to put the CBD into practice, the Conference of the Parties (CoP). For example, at the CoP held in Nairobi in 2000, there was virtually no effective dialogue with business or industry (with the exception of biotechnol-

Box 9.3. The Business of Game Ranching

Though many large mammal species have always been important sources of food in Africa, until recently they were not farmed. This is largely because most are difficult to manage. Some, like buffalo, are overly aggressive, and others, like many of the ungulates, are almost impossible to fence because of their skills in leaping. With the advent of domestic livestock (cattle and sheep), Africans had even less reason to domesticate indigenous species. Over the last twenty years, however, largely in response to declining profits in the cattle business, game ranching has become an integral part of South African land use. There are now 4,000 privately owned ranches keeping game, covering more than 80,000 km^2, a lot of land in comparison with the 28,000 km^2 under the control of the National Parks Board. For the most part these ranches are economically successful, providing animal products in the form of meat and hides as well as recreational game-viewing, photographic safaris, and tourist hunting for both meat and trophies (Hearne and McKenzie, 2000).

Often game ranching starts out as a hobby for the struggling cattle entrepreneur, who from year to year replaces his dwindling cattle numbers with game holdings. If neighbors engage in the same strategy, internal fences can be removed, effectively enlarging range size for large-bodied species. Increasingly in the last decade ecological consultants were hired to make recommendations (often derived from computer simulations and models) regarding the ideal mix of game species to stock on the property. Though in some areas economic analyses indicate that cattle-raising yields more meat per hectare than game ranching, because of the higher efficiency of domestic stock that have been subject to thousands of years of artificial selection, in other areas the indigenous arid-adapted species are more productive. In these areas further economic modeling is used to determine whether tourists should be part of the plan, and whether they should engage in consumptive (hunting) or nonconsumptive (photographic safaris) activities; and, if hunting is to be allowed, whether this should be for trophies or meat. These decisions are based on such factors as location, access to markets, and tourist appeal. Inevitably the values entered into computer-based calculations are to some extent based on educated guesses, the "rule of thumb" method (Hearne and McKenzie, 2000). Nevertheless, the models provide useful guidelines for managing populations toward a desired goal over a specified period, against which the economic and ecological health of the ranch can be assessed at the end of the period.

A major hurdle to setting up a game ranch is the initial capital required for fencing, wildlife purchases, tourist huts, and other infrastructure and equipment. Purchasing a suite of game to attract tourists (elephants, lion, cheetah, white rhino, buffalo, and other species) from surplus stocks auctioned by the National Parks Board can cost over US$300,000. Increasingly, the South African government is developing programs to facilitate the transition between cattle and game, such as the share-blocking scheme, whereby a number of people can buy shares in a company that has recently acquired a farm, thereby spreading risk and attracting capital. Share-block owners get exclusive rights to a cottage on the farm, thereby increasing the attractiveness of the investment.

Game ranching in South Africa is driven by the economics of private enterprise. Land increasingly reverts to game without the need for campaigns for conservation, and the area under game ranching is still growing (Grootenhuis and Prins, 2000). The economic success of game ranching can be attributed primarily to political factors, specifically the demise of apartheid and the phasing out of subsidies to farmers. Insofar as habitat destruction accounts for biodiversity loss, the developments in South Africa are promising.

ogy), not even in key areas of tourism and finance (Entwistle, 2000).

9.4 Buying a Nature Reserve

Big business does not have to be invited into the environmental arena to engage in consumptive wildlife use. There is another kind of private-sector deal that is becoming increasingly prominent as a solution to species extinction and habitat protection—the purchase of areas to be protected.

Consider the sorry state of mahogany. New World mahogany (*Swietenia* spp.) provides a telling example of difficulties in conserving commercially important species. Traded for more than 500 years, most species are now commercially extinct, strong evidence incidentally of why big business is not a good dance partner. Timber companies do not adopt sustainable practices because, with limited financial incentives and government oversight, conventional logging is more attractive than any other harvesting regime. While the Convention on International Trade in Endangered Wild Fauna and Flora (CITES) (see below) has been effective for large, charismatic species like elephants and whales, it has done disappointingly little to help conservation of mahogany and other timber species. Efforts to list mahogany as a protected species on the CITES Appendix II (for which the ban on trade is not total as in Appendix I but left to the discretion of the exporting state) have failed three times, basically because of the pressure domestic logging concerns place on national governments (Gullison et al., 2000). When CITES failed, boycotts were used; though effective in the United Kingdom, these were entirely offset by increased consumption in the United States. Defenders of mahogany, utterly frustrated with the ineffectiveness of international agreements and boycotts, proposed private purchase as a safety net for hardwood in Bolivia. This they saw as the only remaining feasible and affordable way

of protecting mahogany. In one case The Nature Conservancy (TNC) purchased for $US2.50 per hectare the logging rights to 630,000 ha of timber concessions adjacent to Noel Kempff Mercado National Park; in another, Conservation International (CI) bought at $2.22 per ha logging rights to a 45,000 ha concession that was later added to Madidi National Park. We look at these conservation concessions or leases more closely in chapter 11.

Private purchase of nature reserves has been used very successfully in the United States over the last forty years by TNC, a private nonprofit conservation organization. In fact, TNC now owns and manages over 14 million acres in the United States and 83 million acres overseas, often managed through cooperative agreements with academic institutions, government agencies, private individuals, businesses, and other conservation groups (Murray, 1995; Swanson, 1999a). Key to this NGO's success in its land acquisition policy has been its emphasis on local involvement with the purchase, its skill in raising money, and the selection of sites not subject to intense conflict. Though TNC increasingly purchases reserves in Latin America, the Caribbean, and the Pacific (see Redford and Mansour, 1996; Brandon et al., 1998), the feasibility of such enterprises overseas needs careful consideration, as we examine below.

Money is certainly not the problem. Land prices are low in much of the developing world (see the Bolivian prices above), and environmental philanthropists of the Northern conservation movement have deep pockets (Spergel, 2002). The bequests of high-profile business leaders, such as the Turner Endangered Species Fund that owns over 2 million acres, serve as a model for others contemplating such ventures. Since money for foreign purchases is probably forthcoming, many conservation organizations see purchase (by individuals or institutions) as a watertight solution, and are prepared to invest accordingly.

Politics, however, are another matter, especially insofar as overseas purchases raise questions of national sovereignty and neo-colonialism. Essentially, when foreigners (or foreign interests) propose to buy land over-sees for conservation purposes their at-tempts to determine local land uses must be seen as a short-term strategy (a safety net) at best. This is because property titles are simply a state's promise to enforce the own-er's right to the use of the indicated re-sources to the exclusion of all nonowners. As with all entitlements of this sort, if the owner's interest in the use of land comes into conflict with that of the host govern-ment, the state has the right to reclaim the property. Hence, insofar as the conservation of biodiversity is not a national priority, or is in conflict with state interests, the tension between domestic and foreign interests ren-ders outright purchase somewhat tenuous as a conservation solution, despite its imme-diate appeal. In short, the notion of property titles, so successfully used by TNC within the United States, may not transfer easily across national boundaries (Swanson, 1999a).

A good example of this difficulty comes from Chile. In the mid 1990s a multimillion-aire clothing magnate of U.S. citizenship tried to buy a thousand square miles of Chile, a slice of land that includes glaciers, whitewater rivers, an active volcano, and a good part of the world's reserve of alerce (*Fitzroya cuppressoides*). His deal ran into trouble (Ryle, 1998). Effectively, the pro-posed Pumalín Park, potentially the world's biggest private nature reserve, would have split Chile in two, awakening the fears of the Chilean military and others concerned with national sovereignty and security. Lo-cal Chilean conservation organizations, too, were divided over the rights and wrongs of a foreigner taking control of so much of Chile's biodiversity. As it turns out, a smaller set of purchases was approved, and by sidestepping government regulations a private Chilean foundation was established to protect the area. Pumalín is currently the world's largest private reserve (Langholz, 2002). But will it be respected? The project

is tainted with suspicion and mistrust, with Chilean officials and visitors to the resort viewing the protected land as a forest fief-dom, the playground of a rich Californian, and home to a cult (deep ecology; see chap-ter 1) which, in the eyes of some nervous observers, is dangerously linked to Nazism. It is also the case that in these kinds of deals the lines between private business and non-profit ownership become blurred, with own-ers being directly accused of running the reserve for personal gain.

But does privatization necessarily cause such problems? Private reserves are in fact proliferating throughout the world, particu-larly in Latin America, as holdings of TNC, in southern Africa as game ranches (see Box 9.3), and in Costa Rica as family-owned recreational areas. The problem is that as yet rather little is known of their success. In the case of Costa Rica private reserves add the equivalent of thirty-six national wildlife refuges to the protected area system, thereby providing an enormous boost to na-tionally held conservation lands, especially since these areas often provide buffer zones to existing protected areas (Langholz, 2002). On the other hand, privately owned areas, in Costa Rica and elsewhere, are not subject to monitoring, are not selected according to habitat prioritization schemas (see Box 3.6), are vulnerable to excessive ecotourism (see chapter 10), and are viewed as islands for the elite (as we saw in Pumalín) since they are often ostensibly held primarily for fam-ily pleasure or corporate entertainment. Pri-vatization may nevertheless provide a key preemptive step against biodiversity loss be-fore a better solution can be engineered. For example, conservation biologist Daniel Janzen (1987, 1037) began raising money in the mid 1980s to purchase an old hacienda-holding of dry forest that was being annu-ally burned. His plan was to "grow" a na-tional park at Guanacaste. Initially, this was seen as a hostile "takeover" by critics of the protected area movement, who quote Janzen as having made the inflammatory claim "we have the seed and the biological expertise: we lack control of the terrain" (Guha, 1997, 16). However, over the years, as a result of numerous conflicts and compromises, something very different than a classical protectionist national park has emerged, under the label of the Area de Conservacion Guanacaste. It has built and managed an endowment that has put an end to burning in the area, continues to purchase land, teaches biology to all school children in a 20–30 km radius, and attempts to decentralize its management (Janzen, 2000; Allen, 2001).

One thing that is clear from these private purchase deals is that the more the foreign interest group works with and is represented by domestic institutions, both governmental and nongovernmental, the more successful such initiatives are likely to be. Increasingly, as the new park is seen as a domestic interest and not a foreign imposition, conflicts will be dealt with as a matter of national politics, thereby avoiding opportunities for neocolonial conspiracy theories. A good example of this is Sierra de Las Minas Biosphere Reserve, established at the end of Guatemala's long civil war by a private NGO, Defensores de la Naturaleza; when the legitimacy of Defensores was challenged by local landowning and timber interests, its rights to manage the reserve for conservation were upheld by the Constitutional Court. This was clearly a great victory for conservation interests. In a sense, then, we should think of foreign purchases, like that of the Guanacaste hacienda holding, as merely the first step in a process whereby outsiders attempt to influence local choices by creating incentives that are consistent with conservation objectives, objectives that may subsequently be backed by national policy decisions, as in the Sierra de Las Minas ruling.

Though private purchase can work in the right kind of political climate, it is critical for such projects to watch carefully developments beyond the borders of the protected area, given the lessons of chapters 2 and 6. Privatization and enclosure, by placing property rights in the hands of a few, inevitably increase the pressure outside of the reserve, and areas traditionally managed as a commons are at greater risk of degradation; this has been called the "tragedy of the non-commons" (May, 1992, 361; see also McNeely and Weatherly, 1996). In the case of privately purchased reserves, this could lead to recurrent problems in the future. As such, the simple solution of privatization, so alluring to conservation biologists, should probably only be considered in a narrow set of politically appropriate circumstances, in cases of extreme urgency, or as a stopgap measure.

Where solutions such as Guanacaste or Sierra de Las Minas are not feasible, or at least unleash concerns about sovereignty and protectionism, other less radical but closely related options are available. One that can work, again in developed countries such as the United States, New Zealand, and Australia, is emphasizing the protection of endemic biodiversity in the hands of private owners (Norton, 2000). Here the objective is not to buy up land for conservation, but rather for the government to work with already existing landowners, providing fiscal policies such as taxes or subsidies that promote conservation. Generally, in these countries government incentives (compensations, tax breaks, etc.) work better than regulations (fines and permits), since private landowners will try to escape regulation by seeking to hide information from the government, thereby eliciting extra enforcement costs (Shogren et al., 1999). Thus, in Costa Rica the government has since 1997 been paying landowners for providing ecosystem services, such as carbon sequestration and the protection of watersheds, biodiversity, and scenic beauty. The payments, about US$50 per ha per year, are financed in part by a tax on fossil fuels and are resulting in significant forest conservation and restoration (Daily et al., 2000). Another example of successful conservation on private lands is the growth of commercial ranching in southern and eastern Africa (see Box 9.3 again). With appropriate international policies concerning trade, sustainable wildlife utilization can provide strong economic incentives for conservation initiatives. One clear advantage of this strategy is that a large proportion of threatened biodiversity is found on private land, often in heavily populated areas, precisely because private land is where most people live and where nature is most at risk (e.g., Schwartz et al., 2002). To date, however, there has been little evaluation of the effectiveness of government environmental subsidies, at least in the farming sector (Oldfield et al., 2003).

All in all, the recent interest among individuals, private-sector institutions, and non-profit organizations in buying land for conserving purposes is encouraging, as are the incentives for conservation on already privately owned land. Though initially one might assume that private-sector purchases of land overseas for conservation purposes are unethical (or politically unfeasible) this is not necessarily the case, particularly where the purchasing entity is linked to respected non-profit organizations. Similarly, the assumption that the private sector is not well equipped to provide public goods related to the global environment is also not necessarily true, though as we will see in chapter 10 conflicts emerge in ecotourism when the management of biodiversity dictates different strategies than does the management of a successful tourist business. There are, however, some caveats to make. First, though purchases by individuals and nonprofit organizations are appealing, the real money lies in the for-profit private sector (McNeely and Weatherly, 1996); as such, the developments documented here do not substitute for the need to make partnerships with corporate business (as discussed above). Second, despite the key role played by private-sector institutions and nonprofit organizations in biodiversity conservation, government spending remains axiomatic (James et al., 1999). Private and corporate interests in conservation cannot substitute for government policies, nor can they be used as an excuse for releasing governments from their environ-

mental responsibilities; this would surely be the ultimate "selling off of nature." Third, where privatization is inappropriate the purchase of leases (or concessions) in biodiversity may be an attractive option, particularly in developing nations (see chapter 11 on direct payments). Finally, corporate and private investments in biodiversity cannot succeed without government interventions, such as expanding legal definitions of property rights, offering fiscal incentives for the provision of public goods, and establishing enforceable regulations, given how poorly business is equipped to provide public goods. As Hawken et al. (1999, 261) aptly put it: "for all their power and vitality, markets are only tools. They make a good servant but a bad master and a worse religion," adding that it is particularly dangerous "when they threaten to replace ethics or politics." So this chapter ends with a consideration of policy initiatives and international treaties.

9.5 International Policy Initiatives: Who Pays?

We now turn to the key question of how to distribute the costs of environmental depletion and toxic pollution at the international scale. Conventional social science wisdom tells us that such collective action problems can be overcome by either selective incentives or hierarchical force, that is, by either market forces or the state (Olson, 1965). However, since global environmental problems emerge in the first place precisely because of market failure and the absence of a world state, international regulatory organizations and agreements become a critical third axis (Sandbrook, 1997). Indeed, without the conservation ethic for which Aldo Leopold yearned, there is nowhere to turn for guidance, control, and justice but to the international regulatory sphere, much maligned as it is. Two questions relate to such regulation. First, who pays for the inevitable opportunity costs entailed in slowing (or stopping) development so that undeveloped areas can continue to offer safe havens for biodiversity? Second, how can we can agree on the international rules concerning biodiversity conservation, and specifically on how governments can work with and control the hidden hand of economics that, as we saw in previous sections, infiltrates every aspect of environmental change (for worse or better)? So who pays and what are the rules?

Who pays for remedying global environmental problems? Here the agreement reached at the Earth Summit in Rio in 1992, the CBD, is clear. Costs for implementation of global biodiversity conservation are to be borne by Northern industrial nations (article 20—Financial Resources). There are several reasons why all ratifying parties to the Convention agreed to this (McNeely, 2000). First, though many of the benefits of biodiversity flow to all citizens of the world, the costs tend to fall on the world's poorer countries because of the current patterning of surviving biodiversity (chapter 1), and indeed on the poor within these countries (chapter 7). Second, most developing countries have insufficient resources (because of their weak tax base) to cover conservation costs, and they have limited cash flows to finance the upfront costs of conservation (with benefits often enjoyed only after many years). Third, external financing is often required to cushion the short-term effects of policy reforms that demand sustainable use, either to compensate those adversely affected by the policies or to build consensus for the reforms, or both. Fourth, it is possible that the early stages of development for underdeveloped nations are more environmentally destructive than later phases (Tisdell, 1994); if this is the case then there is an additional onus on wealthier nations to help less developed nations with environmental challenges. Finally, developing nations need special encouragement (undoubtedly financial) to step outside of "development via imitation" (Swanson, 1999b, 29), and to craft national plans that recognize the importance of building on existing natural resources rather than simply replicating industries of Northern nations (in which they are unlikely to be truly competitive anyway). For some combination of these possible reasons, then, developed industrial nations that were signatory to the CBD agreed to article 20, an ultimate rationale no doubt being that the developed world benefits from the international political security contingent on environmental and economic stability. In this way the CBD recognized the already existing role of international donor agencies such as the World Bank, USAID, the European Union (EU), various programs within the UN, and other very active governmental programs in Germany, Holland, Britain, Denmark, Norway, Canada, and Japan.

There may be agreement in principle over who bears the burden, but there is far less consensus over the mechanisms whereby this burden can be shifted. What is needed are funding mechanisms as opposed to funds (Swanson, 1999a). While the latter consist of charitable handouts, the former offer a permanent dynamic incentive structure whereby funds continue to flow, providing (in the case of protected area management) ongoing compensatory payments for the sustained production of a public good, as well as incentives to protect this public good into the future. Devising such mechanisms with respect to conservation is not easy, given biodiversity's relatively unproductive value in the domestic economy; as we saw earlier, global and local interests can value biodiversity in very different ways.

There are, however, some model international treaties and agreements that have successfully devised mechanisms for funding the conservation of environmental resources. The 1971 Convention on Wetlands of International Importance especially as Waterfowl Habitat (known after the town in which it was signed as Ramsar) neatly creates incentives for developing countries to protect areas of land and water. Under Ramsar, incidentally, the first global treaty concerning conservation, each country designates (or "lists") certain wetlands as protected for waterfowl that migrate across international boundaries; in return this nation receives the right to expect designations of protected wetlands by other neighboring countries. The listing mechanism operates as a form of barter, with each state compensating others "in kind" for their respective listings (Swanson, 1999a). This approach to natural habitat conservation is very effective for shared regional resources such as lakes, rivers, or wetlands, and the species that migrate between them. The so-called listing approach has also worked well for The World Heritage Convention (WHC), adopted by UNESCO in 1972, with respect to the internationally financed management of national cultural resources. UNESCO recognized that if global treasures, like the Pyramids or the Taj Mahal, were to be conserved, international cooperation was needed. Under the WHC domestic governments provide the site and services, and an international trust fund supported by wealthier nations covers the costs; locals do the work, outsiders pay. Nations have a strong incentive to list their buildings since listing entitles them to apply to the trust fund, promotes tourism, and enhances national prestige. The success of the listing mechanism derives in part from the interdependencies among nations within a region that share common interests in rivers and wildfowl migrations (Ramsar), and in part because of the existence of direct benefits from tourism to the host country (WHC).

Another success story in the pantheon of international treaties on the environment is the Montreal Protocol, that successfully reduced chlorofluorcarbon (CFC) and halon use, thereby protecting the ozone layer (Grundmann, 1998). The Montreal Protocol's success reflects the immediate health risks associated with ozone depletion, the availability of cheap substitutes, and the fact that the zones associated with the causes and consequences of ozone depletion overlap. This third point may account for why

this protocol has fared so much better than the Convention on Climate Change (the Kyoto Protocol that attempts to reduce global CO_2 and other greenhouse gas emissions but stumbles under U.S. and third world resistance; Sandalow and Bowles, 2001). Though both Montreal and Kyoto deal with atmospheric processes (ozone depletion and CO_2 buildup) that are global in scale and lie beyond the confines of any individual country, they differ in the extent the negative environmental effects are exclusively felt by the source nations. In the case of CFCs, these were produced largely by Northern nations; it was also these temperate zone nations that were the most likely to suffer directly from the holes in the ozone layer that were appearing first in polar regions. Thus, developed temperate zone nations were motivated to sit down at the negotiating table and propose restraint, even self-restraint. For the Kyoto Protocol the situation is quite different. The benefits of halting climate change are truly global (sea-level changes contingent on global warming threaten many different nations, rich and poor), whereas the cost of one principal avenue for reducing greenhouse gases (putting an end to logging) is borne most heavily by underdeveloped tropical forested countries, making a treaty more difficult to reach (the resistance of the United States reflects other dynamics). In short, consideration of where the costs and benefits of various actions are felt, and the extent to which they overlap, lies at the heart of designing international agreements (see Figure 5.5).

From this perspective we can highlight as the particular challenge in conserving biodiversity, the target of the CBD, the fact that biologically rich terrestrial and marine areas lie firmly within the boundaries of individual countries. Worse still, for international treaty brokering, biodiversity is concentrated primarily in the poorest countries of the world and the poorest regions of these countries. Accordingly, to the extent that the CBD aims at the global management of biodiversity and encourages the setting up of parks and protected areas, it impinges on land use planning, a matter that has traditionally been considered wholly national policy.

So how can the CBD develop institutions for intervening within national development choices regarding land use, and provide an efficient funding mechanism that entails the incentives for making these changes? It is in this context we assess the Global Environmental Facility (GEF). The GEF was set up, operated by the World Bank, various UN agencies, and a host of bilateral aid agencies, with the mandate to provide several hundred million U.S. dollars per annum for biodiversity protection according to the priorities identified in the CBD (articles 20 and 21). Specifically, GEF is expected to cover incremental costs developing countries incur while achieving global environmental benefits, incremental here referring to the benefits enjoyed beyond the national level. (Thus, if a forest yields US$1 million regional watershed benefit, and US$10 million global carbon storage, GEF would finance the US$9 million). With this novel feature GEF can compensate developing countries for projects that would not be warranted in the domestic economy. Since its inception in 1991, GEF has invested over $1 billion in biodiversity projects, most of which are related to protected areas, offering a mechanism for the transfer of subsidies from rich to poor countries. It supports biodiversity conservation trust funds in nations like Brazil, Bhutan, Uganda, and Thailand (McNeely and Weatherly, 1996), and promotes a hitherto unseen level of global cooperation between states, intergovernmental organizations, and the nongovernmental sector in supporting actions to conserve biological diversity.

On the other hand, GEF is poorly funded. The CBD failed to establish a mandatory level of funding from developed nations (McNeely, 1999); in fact, donor governments have not increased their GEF commitments since its conception (Balmford and Whitten, 2003). Consequently, GEF is woefully under-funded given the magnitude of the task at hand (James et al., 1999), and will have to turn to the global marketplace, in other words, back to big business, to fill its yawning coffers (McNeely and Weatherly, 1996). The GEF also fails to provide a permanent dynamic incentive structure to its funding activities (Swanson, 1999a; Richards, 2000), with developed nations seeing their contributions as charitable handouts. Finally, the operation of the fund shows political bias toward the interests of Northern industrialized countries (Edwards and Kumar, 1998). From this critical perspective it is clear that GEF cannot operate alone, nor can it justly substitute for necessary policy reforms and international controls.

The GEF is not, however, the only international funding instrument available. An initiative for funding that preceded GEF by several years was the debt-for-nature swap scheme (Box 9.4). This idea was first suggested by Thomas Lovejoy in the pages of a 1984 *New York Times* editorial, but did not catch on until the international debt crisis intensified in the late 1980s (Thapa, 1998). The plan in a nutshell was to retire some proportion of a developing country's foreign debt in return for commitment to host, not necessarily fund, an environmental project of some sort. As an example, in 1993 the WWF (with money from USAID) bought US$19 million worth of Philippine government debt from international commercial banks (actually for the discounted price of $US13 million—the banks were so keen to get rid of these debts), and in exchange the Philippine government agreed to establish an environmental trust fund to the value of US$17 million, in other words, 90 percent of the debt's face value. Though these clever schemes have helped developing countries initiate environmental trust funds (as in the Philippines; Spergel, 2002), support national park and reserve systems (e.g., Costa Rica; Resor, 1997), and bolster local environmental NGOs (e.g., Dominican Republic; Visser and Mendoza, 1994), they are

Box 9.4. Debt-for-Nature Swaps

Debt-for-nature swaps entail the exchange of a certain proportion of a nation's foreign debt for national commitments to conservation. Only indebted countries that are unlikely (usually unable) to pay are eligible. The debt (heavily devalued on a secondary debt market that emerged in 1982 to try to clear these obligations) is purchased by an NGO from a developed country, and then "swapped" for an environmental program valued at a higher price than the debt's secondary market value, such that the impact of every dollar invested by the NGO is increased by a multiplier effect. At all stages the foreign NGO works with a domestic counterpart, and of course with the national government.

The first debt-for-nature scheme was between CI and the government of Bolivia, and involved canceling a foreign debt of $650,000 in exchange for local currency to be used toward the Beni Biosphere Reserve. In the same year WWF purchased a $1 million debt from Ecuador for $355,000, and converted this into a $1 million bond for conservation. Since then other NGOs, including TNC, several international governmental consortiums (like the Paris Club), and European banks have become involved, working with a host of nonprofit fundraising organizations. By 1993 over thirty debt-for-nature swaps in more than twenty-four different countries resulted in retired foreign debts that contributed more than US$200 million to conservation; the most heavily involved countries are Costa Rica, the Philippines, Ecuador, and Madagascar, with Poland and Bulgaria as European cases (Deacon and Murphy, 1997). Costa Rica stands out for its strong system of parks and reserves, and is often used as example of the success of the scheme. However, as with the INBio project (chapter 8), this may in part reflect its unusual political culture (Edelman, 2000). Indeed, debt-for-nature swaps are most commonly adopted by countries with democratic institutions, as well as those with heavy foreign debt and high levels of endangered species (Deacon and Murphy, 1997).

Areas of concern with this funding instrument have included inflationary effects, sovereignty, inefficiency, and inequity between debtor and donor nations. As regards the potential inflationary effects on the local economy from a sudden influx of cash, these were mitigated by the issue of long-term bonds. Sovereignty was an issue in the late 1980s; it held up the first deal in Bolivia for twenty-one months, and kept Brazil out of the scheme until 1991 (Mahony, 1992; Visser and Mendoza, 1994). However, as more projects were put into place and domestic NGOs got increasingly involved, it became clear that foreigners had no direct claims of ownership to the protected area. Similar fears of ecoimperialism have been countered by the observation that, since there are no enforcements written into these deals, the debtor countries clearly have the upper hand. Broader equity issues have invited even more critique. Some suggest that it is the commercial banks and not developing countries that benefit most from debt-for-nature swaps (Mahony, 1992). It appears that because of the volatility of the secondary debt market, in some cases debtor nations ended up owing just as much as they did before the swap, while creditors (the banks) managed to redeem losses they had almost given up on with the hard cash of conservation organizations. It is also possible that debt-for-nature swaps legitimized foreign debt at a time when many indebted countries were campaigning that these very debts were incurred illegally.

Debt-for-nature swaps are limited in being one-time payments. Increasingly, they are being replaced by (or evolving into) independent environmental trust funds, the income of which is used for conservation purposes. These offer the critical permanence to the funding, absent in the debt-for-nature scheme. All the same, many of the issues raised by debt-for-nature initiatives are not easily circumvented, and this scheme teaches many lessons.

not the magic bullet once hoped for, and with the exception of a recent purchase of threatened forests in Belize, have become less popular.

We can see why by looking at the original rationale for the swaps. The debt-for-nature scheme was based on the assumption that foreign debt promotes environmental degradation, specifically deforestation. With the austerity programs of the International Monetary Fund and World Bank in the 1980s, debt-burdened nations were believed to be forced into the production of timber, minerals, meat, and monocrops to generate hard foreign currency through export. Governments caught in the trap of international debt would accordingly subsidize these activities through environmentally perverse policies (incentives to clear forest, tax credits, agricultural pricing policies, etc., Repetto and Gillis, 1988; see chapter 7). Plausible as this scenario sounds, empirical support for these dynamics is mixed. Kahn and McDonald (1995) use multiple regression analyses on country-by-country data to show that international debt promotes deforestation, whereas Gullison and Losos (1993) find that different nations' rates of deforestation are more closely tied to population size than international debt. In fact, the effect of debt on deforestation varies between countries, and undoubtedly depends on the scale of the analysis (see chapter 3), insofar as so many interacting factors (in addition to foreign debt) stimulate deforestation (Geist and Lambin, 2002), such as land availability, inequitable land ownership, and urban migration (see Figure 7.2). As Gullison and Losos (1993; see also Barraclough et al., 2000) conclude, even the elimination of all Latin American debt would not halt the degradation of marginal lands or the clearing of primary forests, because so often there is simply no other land available for landless peasants to colonize under current land tenure arrangements. Furthermore, freeing governments of their debt could (perversely) stimulate the construction of more highways and dams, thereby further opening up

remote areas. The general conclusion, then, is that alleviating even a small proportion of foreign debt may slow environmental destruction in some contexts, or even remove obstacles to the funding of new domestic environmental programs, but it cannot be regarded as a panacea for debt and deforestation.

Debt-for-nature swaps have in recent years fallen out of favor for two reasons. First, because they entail an international transfer of property rights, in the sense of the rights to forestall development, they can run up against national political priorities (as discussed with the buying of reserves earlier in this chapter). Second, with no dynamic incentive structure they essentially consist of one-time payments to countries in exchange for a promise not to develop a specified habitat, and as such are not economically sustainable. To counter the political insecurities problem, debt-for-nature swaps were increasingly placed in the hands of local conservation organizations throughout the 1980s, effectively to establish a local political power base with interests consonant with those of the global conservation movement, in other words, an internal pressure group empowered with hard currency. To counter the fiscal difficulties associated with one-time payments, there have been some clever moves; for example, the Foundation for Philippine Environment chose to swap their debt not with a private commercial bank but rather with the Philippine Long Distance Telephone Company, whose stock was appreciating four-fold (McNeely and Weatherly, 1996). More generally, to deal with both the political and fiscal issues, conservation trust funds are now replacing debt-for-nature swaps, emerging as the funding mechanism of choice.

Conservation trust funds offer a new mechanism for conserving individual parks, a country's entire protected area system, or even a particular species, and have in recent years been established in more than forty countries (Kramer and Sharma, 1997; Oates, 1999; Spergel, 2002). These trusts or endowments are based on capital funds from which, in most cases, only income can be used. In this sense they offer the dynamism and permanence required of a funding mechanism, and are ideal for covering long-term and recurrent costs that are hard to raise from other sources. The boards of these trusts consist typically of government representatives, NGOs, the donor agency involved in the establishment of the trust, and scientists. If well designed, trust funds may represent the best hope for managing wildlife and habitats in places like the central African forests, where the opportunities for other conservation strategies, such as ecotourism, are small (Wilkie et al., 2001; and chapter 11). Though conservation trust funds can face difficulties in attracting donors (since large capital funds are required at the outset, to yield the interest), they have been successfully established in Uganda, Malawi, Bhutan, Bolivia, Brazil, Peru, and Mexico (Balmford and Whitten, 2003). In this category of funding mechanism we can also consider National Environmental Funds. Since the early 1990s these funds have been set up in about twenty tropical countries, and almost all the transitional economies of eastern Europe, mainly on the basis of debt swaps and GEF contributions (Richards, 2000).

Trust funds can certainly bring in global money to support conservation actions. Furthermore, insofar as they are established under national rather than international law they have the potential of affecting the course of national investment paths (indeed, often the trust fund requires the government to make financial commitments), while at the same time allowing some input from outside conservation interests. Ultimately, however, their influence over key domestic resource management decisions is only indirect (Swanson, 1999a). For example, even with a fund to assist with the conservation of a unique coastal ecosystem, political pressures to build a port may prevail. Securing such an objective requires international mechanisms that have a more direct bearing on

domestic politics. And so, to explore how funding mechanisms can be tied in with further policy reform, both national and international, we end this chapter by returning to international institutions and regulations.

9.6 What Are the Rules?

Here the CBD, a distillation of policy decisions and discussions that had evolved over the thirty years prior to the Earth Summit, is technically the centerpiece achievement. The potential of such international agreements can nevertheless be discerned from earlier initiatives. One example is CITES. This treaty, signed in 1972, defines endangered species and, recognizing that consumption of such products as ivory or medicinal plants in one country may affect biodiversity profoundly in another, bans trade in them. For a few lucky species, such as elephants and rhinos, this has worked well (Swanson, 1999a). Powerful as CITES has been in controlling trade, it does have problems. First, it has proved toothless in the control of such commercially valuable species as mahogany in Bolivia (Gullison et al., 2000) or Hawksbill turtle in Japan; indeed, charismatic megavertebrates, such as whales and elephants, continue to absorb its attention to the exclusion of other equally threatened species (Micklebugh, 2000). Second, its strength is also its weakness. By placing blanket and therefore effective bans on trade, say in ivory, which successfully reduced poaching rates worldwide (see Box 2.6), these restrictions stifle what might be construed as constructive trade; southern African countries are arguing that if trade in ivory were permitted for some nations (under a CITES exemption), this would generate funds that could subsidize local conservation efforts (as suggested in Box 9.3). The general problem lies in the fact that global bans (the only bans likely to be really effective) by definition treat all countries equally, whether or not any specific country might actually be able to draw conservation benefits from controlled trade. Despite these limitations, CITES remains a success story.

Not all international agreements have fared as well, as noted above. Often they become mired in fundamental differences among richer and poorer nations or hamstrung by strong domestic interest groups. Thus, the Kyoto Protocol, aimed at reducing greenhouse gases, as well as many attempts to regulate harvests from the world's oceans, remains largely stalled (Sandalow and Bowles, 2001). We take carbon emissions as an example of the difficulties in reaching an agreement that will slow the buildup of greenhouse gases and ultimately global warming. Consider Agarwal and Narain's (1991) critique of the carbon emissions quota system designed by the World Resources Institute (WRI). The WRI had proposed that to reduce the emissions of carbon dioxide and methane, both gases that lead to the greenhouse effect, each country should take responsibility in proportion to its net production of greenhouse gas (the amount accumulating in the atmosphere minus that absorbed by natural sinks). Concerned with issues of equity, Agarwal and Narain (1991) recalculated the WRI figures, arguing that a country's share of the world's greenhouse gas emission quota should incorporate consideration of how many people live there (Figure 9.2). The recalculation shows that although India and China are among the world's five top emitters (from the WRI figures, Panel a), on a per capita basis their emissions are low (Panel b); indeed, most developing countries (with the exception of Brazil) have a long way to go before they exceed their per capita share of the carbon sinks of Earth. It is precisely these kinds of still unresolved issues pertaining to the fairness of the ground rules that are holding up international conventions like the Kyoto Protocol (Heil and Wodon, 2000).

So what about the Rio agreement, the CBD, and its various articles that we have referred to at many junctures in this book? The CBD represented a confluence of international dialogues that had existed for a number of decades. It grew out of the recognition that in addition to controlling trade in endangered species (CITES's mission), problems of habitat loss, pollution, introduction of exotics, and domestic overconsumption also needed to be addressed if biodiversity was to persist, hence the lead up to the Earth Summit at Rio. The CBD's central objective is "the conservation of biological diversity, the sustainable use of its components and the fair and equitable sharing of the benefits arising out of the utilization of genetic resources" (article 1—Objectives), with as stated means to this end the need to "develop national strategies, plans or programmes . . . that integrate, as far as possible and as appropriate, the conservation and sustainable use of biological diversity" (article 6—General Measures for Conservation and Sustainable Use) and to "establish a system of protected areas" (article 8—*In-situ* Conservation). The CBD had no entirely novel points of departure, nor many wholly new obligations distinct from those inhering under already existing programs and agreements discussed in chapter 2. Nevertheless, it marks a milestone along the route toward increasingly active intervention in the process of national development planning and decision making (Swanson, 1999a), and as such embodies the existing state of play among international conservation organizations, businesses, and governments. Though these objectives are still far from implemented, the Conference of the Parties to the CBD meets regularly, and continues to grapple with issues of implementation and interpretation.

The CBD has certainly produced some positive outcomes. The establishment of the GEF represents a significant avenue for international cooperation, as we saw above. Other productive agreements are over sharing the benefits of biotechnology between provider and producer nations (article 16—Access to and Transfer of Technology). Although implementation has proved complicated (as seen with bioprospecting in chapter 8), it has certainly stimulated huge

Top 10 Emitting Nations (WRI Calculations)

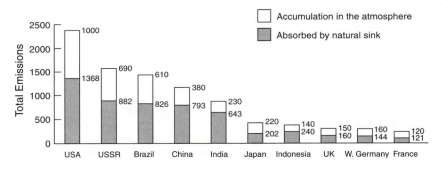

WRI'S Top 10 Emitting Nations (CSE Calculations)

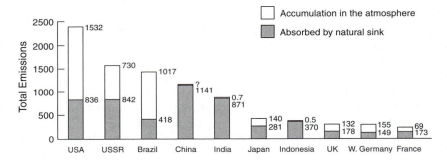

Figure 9.2. Emissions and Equity.
Comparative figures of total emissions of greenhouse gases of WRI's top ten emitting nations (in million tones of carbon equivalent) as calculated by (A) the World Resources Institute and the (B) New Delhi's Centre for Science and the Environment (from Agarwal and Narain, 1991). The CSE estimates differ from the WRI estimates in that they are calculated on a per capita basis. Using the CSE estimates four of the WRI's top ten emitting nations (China, India, Indonesia, and France) no longer fall into this category; though not shown in the figure, their place is taken by Canada (ranked fifth worst in the world), Australia (eighth), Colombia (ninth), and Saudi Arabia. Reprinted with permission from *Global Warming in an Unequal World: A Case of Environmental Colonialism.* Copyright 1991 by the Centre For Science and Environment.

activity in this area. Similarly productive are the proposals for cooperation among neighboring countries (article 5—Cooperation), for example, in situations where species migrate across international borders; these principles have been formally recognized in new programs at IUCN and WWF, as well as in transboundary biosphere reserves. And finally, as McNeely (1999) notes, some governments have taken significant steps toward achieving the Conven-

tion objectives, such as commissioning national biodiversity strategies and action plans, and in some cases passing new legislation.

The Convention also has some sticking points. Constructed in committee rooms, several passages across the different articles of the CBD contradict each other. Had this not been the case, fewer states would have ratified. The downside, however, is that the remit is not always very clear. This is the case in several areas, notably the rights of local communities with respect to their sovereign states (discussed in chapter 8), the balance between development and conservation, and the extent to which international interests can interfere in domestic issues. As regards the balance between conservation and development, there is considerable disagreement over whether or not the CBD is hampered by the legacy of the Brundtland Report, the intellectual shift toward linking development and conservation discussed in chapter 2. Thus, Guruswamy (1999) believes the CBD emphasizes economic growth (particularly poverty eradication) as the first and overriding priority of developing countries, citing both the Preamble and critical articles dealing with financing, and McNeely (1999) can counter this accusation, also with reference to CBD articles. In essence both are right. It depends which parts of the Convention you read most closely, and which contradictory statements you choose to ignore.

Perhaps the boldest move at Rio, and the one that has caused the most controversy, is the dismissal of the view that nature is a common heritage of humankind. In fact, in complete reversal of this view, the CBD proclaims that states have the "sovereign right to exploit their own resources pursuant to their own environmental policies" (article 3—Principle). We encountered the force of this position in the context of bioprospecting (chapter 8), with nations now being directly compensated for their genetic resources and traditional knowledge. However, when we consider this principle in relation to the totality of biodiversity, not just those tantalizing option values in medicinal plants, or to such global problems as CO_2 buildup, the picture gets murkier. This is because national governments may not be as interested in biodiversity writ large as in its potentially lucrative medicinal compounds. Thus, in denying biological diversity the status of common heritage of all humanity, the CBD essentially rules out a common obligation to preserve and protect

this heritage. In the CBD responsibility lies ultimately with the nation-state, generating in turn the need for only more international regulation; in this respect, the fragility of article 3, without any enforcement, is striking. It mandates that states have "the responsibility to ensure that activities within their jurisdiction or control do not cause damage to the environment of other states or areas beyond the limits of national jurisdiction" (in other words, the global commons). Apparently, earlier drafts proposed much stronger obligations on national states, but these were overruled. As it turned out, rejecting the concept of common heritage, the convention settled for the effete and rather weak claim that biological diversity is a common "concern" of humankind, a notion that was ultimately relegated to the Preamble (even though it had ranked as a fundamental principle throughout the drafting process; Guruswamy, 1999).

All in all, despite the euphoric welcome to the CBD, the passing of time and the struggles over implementation necessarily unmask its weaknesses. To what extent these derive from fundamentally flawed foundations, or simply the inevitable strains of conservation realpolitik, is hard to determine. It cannot be denied that the Rio Convention is weakest in its recommendations for implementation; indeed, relatively few countries have as yet developed legislation to support its proposals. As such, it remains a convention without legal clout. The CBD does not yet fall into the category of successful international protocols reviewed earlier in this chapter, such as Ramsar, WHC, CITES, and the Montreal Agreement, perhaps in part because of its ambitious scope and in part because of the specific features of biodiversity (its low immediate value, its largely nonshared nature, and its inequitable distribution among nations along lines already heavily drawn in history and ideology). The convention nevertheless stands as a landmark of some level of international commitment, and has put in place useful ideas and institutions that will, one would hope, stimulate further and more precise international agreements, treaties, and conventions in the future.

9.7 Conclusion

Because of the intrusion of global capitalism into almost every corner of the world (the focus of political ecology; see chapter 7), conservation practice and protected area management can no longer be thought of in isolation from these forces. Furthermore, to the extent that modern conservationists actively embrace the global economic system as an ally in support of conservation efforts, and strive to work with business, these forces are exacerbated. This chapter addresses the challenge of international capitalism by looking at global developments in the management of biodiversity.

There are many inherent problems that arise from thinking and acting at this scale, addressed in this and chapter 8. We dealt, for example, with the pitfalls associated with indigenous-environmental alliances, the difficulties with marketing tropical forest products sustainably, the dangers of biopiracy, the biases inherent in some popular environmental valuation techniques, the potential instability of private reserves, and the strain between domestic economic interests and global environmental concerns. In each case, global inequity lies at the root of the problem. Our conclusions have been boringly repetitive: maybe market forces can be bridled to the cause of conservation, whether at the individual, corporate, or national level, but watch out! Putting nature

on the market is dangerous, both for people and for the rest of nature. Remember that global markets are driven primarily by Western preferences, whether these be for a holiday in the village of exotic "ecosystems people," for Brazil nut hair conditioner, or simply for the unsupportably high standard of living to which so many of us are addicted. How can these preferences and political weights substitute for informed policy development? Guidelines and regulations at the international level are essential, as are coordinated attempts to educate Western consumers about more environmentally responsible uses of their disposable income.

So finally we turned at the end of this chapter to international financing and policy initiatives. As regards the establishment of efficient funding mechanisms for the support of remaining areas of biodiversity, we have seen that their low immediate value, their largely nonshared nature, and their inequitable distribution across nations present the GEF with a huge uphill task. As regards regulations, it is clear that the remit of the CBD entails a massive tension between national and global interests, in addition to the tensions we have already looked at between individual and community (chapters 5 and 6), and community and nation (chapters 7 and 8). Overall we have seen that conserving biodiversity requires a combination of policy reform and appropriate economic instruments. Policy reforms remove the underlying causes of the loss of biodiversity and create incentives for efficient use, and economic instruments generate the additional financial resources required to fund investments in biodiversity.

CHAPTER 10

From How to Think to How to Act

10.1 Introduction

Now we bring all the previous material on how to think about conservation to a focus on action. We outline the instruments available to conservationists, and review them in terms of what we have learned. The strategies are ordered loosely in terms of their emphasis on protectionism and utilization, concepts first introduced in chapter 1 in the context of different rationales for conserving biodiversity; as such, they range from protected area management in conjunction with outreach activities entailing education and nonconsumptive tourism to broader regional programs integrating conservation with extensive utilization. Clearly, conservation initiatives can be usefully ordered along other axes, such as the extent to which local communities participate in the conservation initiative (IIED, 1994). Thus, Adams and Hulme (2001) propose a continuum ranging from centralized or top-down protected area management in association with outreach programs to the complete devolution of resource management to local authorities. As we point out in chapter 11, recognition of the distinctiveness (and orthogonal nature) of these two axes, utilization and devolution, is key to the creation of novel solutions to persistent problems associated with protected area management.

The principal strategies we deal with in this chapter are pure protectionism (10.2), protected area management in association with outreach (10.3) and conservation education (10.4), ecotourism (10.5), ICD projects (10.6), and various forms of utilization (10.7). In reality there is a lot of variation within each instrument, with different projects combining several instruments, such that it can be quite difficult to pigeonhole contemporary conservation projects according to their dominant strategy, especially since the rhetoric surrounding the project implies everything is involved. Our objective is nevertheless to present an overview of the strengths and weaknesses of each method in relation to arguments and concepts introduced in earlier chapters, and to give some examples. In the penultimate section we discuss some of the problems faced by evaluators of these projects, both with respect to biodiversity persistence and the monitoring of economic and social achievements (10.8) before concluding that each of these conservation strategies has value in specific contexts, particularly when used in conjunction with one another (10.9), and that the challenge now lies in combining these strategies with appropriate institutional-level structures, the focus of the final chapter.

10.2 Protectionism in the Name of Science

With the exception of a caucus of social scientists who see only a very limited role for protectionist strategies in the modern world (e.g., Pimbert and Pretty, 1997), protected areas remain a key component of modern conservation thinking; indeed, in many parts of the world protectionism is gaining, not losing, momentum (see chapter 2), and is augmented further by the rush to purchase private reserves (see chapter 9). While the rest of the strategies outlined in this chapter concern programs designed to mitigate the biological, logistic, fiscal, social, and ethical difficulties with protectionism introduced in chapter 2, this section evaluates the tenet of some conservationists that the best form of conservation lies in putting land or water under legal protection and enforcing restrictions on entry (e.g., Oates, 1999; Spinage, 1998; Terborgh, 1999). Inherent in this position is the view that CBC initiatives and compensation schemes should be limited: in Soulé's (2000, 60) words, "if there is too much money the community can (1) become a magnet for entrepreneurs and settlement by outsiders, (2) be infected by corruption and graft from inside, and (3) be subject to egregious 'taxation' from more powerful government entities." Linked to the protectionist position is an emphasis on research. One argument here is that humanity has an obligation to focus, first and foremost, on conducting species inventories of the world's remaining areas of pristine biodiversity before they disappear (Wilson, 1992). Another is that the presence of a researcher in a protected area, even though this person cannot offer full protection to any area (however small), signals its importance and, in some cases, can alert outside agencies to its encroachment. Approaches such as these wave red flags in the face of social scientists, who see protectionism as anathema for community welfare (see chapter 2), and narrowly defined scientific objectives as both a foil for "people-free" parks and a façade for proffering unmerited privilege to a chosen few conservation biologists. The position of Terborgh and others is thus much lambasted in the literature, but it is good to remember that

the natural world has no other voice (Soulé, 2000). Accordingly, the somewhat extreme views of pure protectionists should not be belittled, just as these ardent conservation biologists themselves would better serve their position by not ridiculing what they see as the "excesses" of CBC, since there is clearly a place for both strategies.

There is obvious merit to the clear, uncluttered scientific focus embodied in the position of pure protectionists. After all, despite the rise of protected areas (chapter 2), only tiny fractions of the Earth are actually protected: according to various estimates, terrestrial and marine reserves cover 7.9 and 0.5 percent, or less than 5 and 0.01 percent, respectively (Roberts and Hawkins, 2000; Balmford et al., 2002; depending on different definitions of closure and of land and sea coverage), and these figures shrink even further when one considers that 70 percent of large protected areas (>5000 km^2) are inhabited by people (Bruner et al., 2001). Nevertheless, protectionism raises at least three related questions. First, when is protectionism most likely to work? Second, can protectionism succeed in ensuring biodiversity conservation? And third, could these outcomes be achieved in more socially sustainable ways?

There are clearly instances where pure protectionism, as classed under IUCN (see Table 2.1), is essential to achieving conservation ends, for example, where only a few fragments of intact habitat remain, or where rapid growth in population, market access, and material aspiration puts an area of high biodiversity at significant risk. The success of protectionism may depend strongly on how and why the park was gazetted. As Brandon (1998) notes, parks that were set up in remote areas, with little productive potential, low population pressure, and minimal political representation are the most likely to prosper, precisely because their establishment entailed little social cost. Rio Bravo Conservation and Management Area, tucked away in the northwest corner of Belize in an area of naturally low population density, sealed by a ring of Mennonite communes, and secured through a private land deal, is an example of such a success story (Wallace and Naughton-Treves, 1998). Similarly, in the United States parks with sparse local human populations at their borders experience lower rates of large mammal extinctions than parks in densely settled areas (Parks et al., 2002). Given these observations, it seems logical that parks set up in densely populated areas, with the remit of controlling regional economic or demographic pressures on a threatened area, face massive problems (Brashares et al., 2001) and require major political effort to be effective. These latter parks must not only prohibit illegal utilization within the protected area but also stabilize threats from outside that creep into the park. Costa Rica's Corvocado National Park exemplifies how difficult management of such parks can be, particularly when national conservation policy fails to resolve critical logging and tenure issues in adjacent lands. For instance, in an attempt to compensate displaced local residents the Costa Rican government set up a buffer zone next to the park (Gulfo Dulce Forest Reserve); within this area hunting, mining, and timber extraction were allowed, but the lack of political will and national funding to monitor extraction has led to a spillover of these activities into the park (Cuello et al., 1998). It is in these kinds of circumstances where biosphere reserves (with zoned areas allowing different levels of utilization; see Box 2.4) or ICD projects (see chapter 2 and below) may be more suitable than parks, which can simply be written off as "paper parks" when violated.

The answer to the question of when and where protectionism works is still largely unclear, and systematic comparisons between protected and nonprotected areas are not as helpful as they might at first appear. Consider again Bruner et al.'s (2001) demonstration that areas under protection show less habitat destruction than areas not under protection (see Figure 2.5). There are two problems with a study like this. First, the difference could reflect any combination of the factors identified above, such as the distance of the protected area from centers of population, or the suitability of areas originally set aside for protection for more intensive forms of land use. Second, proper comparison of strategies for conservation lies not in contrasting protected and unprotected areas, but rather in comparing the results of protectionism with those of other conservation strategies (see below). Only in this way can we determine where strict protection is most effective and where other IUCN categories (allowing utilization; see Table 2.1) would work better. In this respect analyses of the effects of different levels of protection on biodiversity within a particular country are useful, such as Carillo et al.'s (2000) study determining the importance of hunting restrictions on maintaining species abundance in Costa Rica. Similarly, in Tanzania measures of biodiversity have been linked directly to different levels of protection. At the national level heavily protected areas with and without utilization show greater large mammal abundances (Caro et al., 1998) than lightly protected areas, and at the local level the same effect holds (Caro, 1999).

As regards the second question, are parks and protected areas effective in conserving biodiversity, in the full sense of genes, species, and ecosystem processes described in chapter 1? To date the answers to this question are mixed. Social scientists for the most part (e.g., Ghimire and Pimbert, 1997; Orlove and Brush, 1996) assemble cases from which to argue that parks fail. Sometimes the criterion of success, for example, poverty elimination, falls outside of the protected area's original mandate, and the critique of protectionism is misplaced (see chapter 11 for the need to determine clear objectives). Often, however, social scientists rightly point to environmental degradation over a broad regional scale, noting that even if the protected area remains intact, the wider ecosystem loses its functionality. The cause of this degradation outside (and some-times inside) the protected area may be re-sentment among resettled local populations, a lack of fiscal and political commitment to conservation from the central government (both discussed in chapter 2), regional-level cycles of poverty and landlessness (Figure 7.2), and loss of community management practices as common-pool land comes un-der state ownership (chapter 6). Despite all these very real problems, rigorous protec-tionism can be effective in slowing biodiver-sity decline. For example, in Kenya, even though 44 percent of wildlife was lost be-tween 1977 and 1995, losses were much greater in unprotected (48 percent) than in protected areas (31 percent; Norton-Grif-fiths, 1998). Similarly a review of selected parks across Latin America and the Carib-bean suggests that protectionism, particu-larly in conjunction with local NGOs, can work very well (Brandon et al., 1998). Fur-thermore, although the effectiveness of pro-tectionism depends heavily on government expenditure (Leader-Williams and Albon, 1988), harsh penalties (the traditional mili-tary-style anti-poaching strategies endors-ing such policies as shoot-to-kill) are proba-bly much less effective in reducing poaching than directing funds toward increased de-tection and patrol (Leader-Williams and Milner-Gulland, 1993). In short, the claim that conservation initiatives that focus on protectionism are failing is certainly not true in all cases. Establishing protected areas re-mains the front line of the "battle to con-serve biodiversity" (in McNeely's terminol-ogy; see chapter 2).

Turning to the final question, we will see in chapter 11 that a protectionist agenda can be achieved in novel and imaginative ways that may prove more socially, politi-cally, and hence ecologically sustainable (Box 2.5). For example, protected areas can be identified and policed by local communi-ties themselves, as with the no-take areas or sanctuaries within the marine fisheries in Thailand and the Philippines (Johnson, 2001; Parras, 2001) or the indigenously es-tablished no-entry zones of the Amazon (see chapter 11); the case of Kuna Yala, despite its false starts (Box 8.6), and the Miskito Coastal Protected Area (see chapter 9) are early examples of such local initiative. Fur-thermore, local people can be trained in var-ious fields of ecology to catalogue species diversity and analyze ecological processes in their protected area, thereby playing an active role in the necessary scientific research see the innovation of "citizen science" in chapter 11).

In conclusion, strict protectionism re-mains a critical conservation strategy in some cases. Furthermore, as we will see in chapter 11, protectionism has the potential to evolve into a more participatory process. This is very important, given the potential dangers associated with private reserves, their vulnerability to local resentment, and their fragile legal and political position (chapter 9). Nevertheless, while extreme threat and limited local resistance may jus-tify protectionism (and even private pur-chase) in some circumstances, it is best com-bined with outreach and education (see below).

10.3 Protected Area Outreach

Even where protected areas, often as fully protected national parks, are well estab-lished and resource extraction is prohibited, outreach programs are needed to offer tan-gible benefits and compensations to whom-soever is negatively affected. Outreach ac-tivities are extremely broad (some of their elements are addressed in later sections, for example, on ecotourism and integrated conservation development), and appropri-ate outreach activity is clearly situation-specific. Here we discuss benefit-sharing; conservation education, another kind of protected area outreach, is addressed in the following section. A rather different ap-proach, direct payments and concessions, is examined in chapter 11.

Benefit-sharing entails the provision of development-orientated services such as school, health, roads, and water services. These projects can look impressive, although decades of rural development have shown that small-scale interventions are only useful if followed up with operational expenses, such as road rehabilitation, supplies of pharmaceutical and school materials, teachers' and health clinicians' salaries, and technical training (Barrett and Arcese, 1995). A very early benefits-sharing program linked to the establishment of a national park in southern Kenya demonstrates the importance of such follow-up. The Development Plan for Amboseli (proposed in 1973; see Western, 1994) was designed to compensate the local Maasai, already embittered by a long history of alienation and land grabbing in their area, for the imminent loss of the valuable Ol Tukai swamps that were to be rendered off limits with the new national park. Specifically, the Maasai were to receive piped water outside of the park in return for their former free access to this year-round source of water. Without training in pump maintenance, or any long-term commitment to the purchase of diesel, this high-profile benefit-sharing program failed (Lindsay, 1987), although a broad suite of subsequent development initiatives, relating to tourism, revenue sharing, and agricultural and social services have subsequently helped avert at least some of the tensions that reigned in the 1970s and extend wildlife habitat.

Outreach work of this ilk can nevertheless be successful if well managed, and has drawn the attention of conservationists and policy makers alike, as with Tanzania's Cullman and Hurt Community Wildlife Project (Box 10.1). An important objective of the CHCWP benefit-sharing program is to make villagers aware that the assets they gain are a direct result of their restraint in hunting, in other words, their wildlife stewardship; additionally, villagers are made responsible for the actions of all. Accordingly, if poachers are found to come from villages partici-

pating in the project, benefits are withheld until a concerted effort is made by the village to control the rule-breakers; to date, only one village has been dropped from the benefits program due to excessive poaching. A scheme such as this is still not without complications. Though the bounty scheme has successfully turned some poachers into anti-poachers (and meets the "who shall watch the watchers?" concern raised in chapter 6), the project effectively institutionalizes a vigilante system in which certain community members are hired to monitor the behavior of others; since the incentives for catching poachers far outrank the per capita income of the average Tanzanian, some game scouts become overzealous in their pursuit of poachers (Bonner, 1993; Schroeder, 2002). Another problem, also emerging from the discussion in chapter 6, is that villagers need to form a user group, otherwise the benefits they receive will be whittled away by immigrants. Finally, although benefit-sharing schemes such as this assure that humanitarian goals of conservation are met, the mere hints of vigilantism unnerve political ecologists, who maintain that environmental justice cannot be bought at a human price (chapter 7). In the eyes of many social scientists schemes such as these entailing only minimalist dispensations of justice (such as building a school or dispensary) are more an exercise in public relations than an attempt to improve rural livelihoods; more blatantly they buy off dissent (Schroeder, 2002). In the eyes of others they are a good start in the right direction.

Benefit-sharing programs can be difficult to implement, precisely because they are small and not necessarily consistent with broader regional policy. Returning to the Costa Rican example discussed above, when Corvocado National Park attempted to recompense the displaced residents affected by the establishment of the park, park authorities unwittingly attracted new immigrants to the area looking to benefit from the scheme. Furthermore, since evidence of prior residence entailed demonstrating "im-

Box 10.1. Outreach in Tanzania

The Cullman and Hunt Community Wildlife Project (CHCWP) was started by a safari hunting company operating on the borders of Serengeti National park in 1990 (Bonner, 1993). CHCWP is founded on the conviction that wildlife and associated habitats can only be conserved by sharing the benefits derived from wildlife among the communities where the wildlife live. The project aims to encourage village communities living near conservation areas to protect wildlife and habitat. This is done by establishing a benefit-sharing scheme through which wildlife can be recognized as a renewable and lucrative natural resource. As such, CHCWP aims to instill the idea that wildlife provides a better long-term return through its conservation than its overexploitation.

The strength of CHCWP lies in its three benefit-sharing components. First, revenue generated from the (largely foreign) hunting clients is used to promote economic development in villages surrounding the hunting blocks. Clients contribute voluntarily a community conservation fee over and above the government hunting fees. These donated funds are held by the Project on behalf of the villages, by agreement with each village. At the end of each year the money is totaled. Though the district authorities are advised as to the amounts available, they receive none of the funds. To determine how a village will use its funds, a village meeting is organized. Here with debate, discussion, and voting a project is identified, a budget drawn up, and a project committee of three women and three men is formed. Where building is involved, local builders are contracted and villagers donate their labor. Strong emphasis is put on accountability and openness regarding the utilization of CHCWP village benefit funds; for example, materials are purchased in the presence of the village committee members and transported to the village, and all plans are presented to the district or regional center. Second, quotas of wildebeest, zebra, and other game are regularly hunted for meat distribution within selected villages.

Finally, project villages are encouraged to self-monitor poaching in and around their village through a rewards program: for destroying a poachers' camp (US$6.00), for each wire and steel cable snare graded for size (US$0.30–0.90), for a rifle or shotgun handed over to the Wildlife Department ($US75.00), for an elephant or rhino poacher convicted (US$300). In conjunction with three CHCWP anti-poaching teams operating in all hunting blocks allocated to this safari company, between 1991 and 2000 412 poachers have been caught and convicted, and over 150 firearms have been confiscated. The number of wire snares recovered and destroyed has declined from an average of 3,000 (1990–94) to 1,800 (1995–96) to 1,200 (1997–98). One wire cable snare is estimated to strangle at least five animals before the wire deteriorates. Assuming these numbers measure a reduction in the numbers of snares set, rather than less fastidious detection, the decline in the number of seized wire snares each year suggests that involving and rewarding local communities in the protection of their natural resources has great potential for success. Independent wildlife surveys in areas adjacent to participating villages are nevertheless needed for proper evaluation.

provements" to the land in the form of settlement, agriculture, and cattle grazing, the ostensibly conscientious outreach work only escalated the rate of illegal forest cutting as immigrants hastily cleared plots to prove their eligibility for the scheme (Cuello et al., 1998).

In short, benefit-sharing schemes depend heavily on follow-up. The ethics of buying off communities needs to be addressed, as does the validity of the intervention without taking into consideration the broader sociopolitical context.

10.4 Conservation Education

Another very popular outreach activity is conservation education. Because conservation actions require a change in people's behavior and compliance with new legislation, the success of any conservation program depends upon active public support, participation, and understanding. As such, education programs play a key role in wildlife conservation and resource management. For this reason, conservation education has become an increasingly common form of outreach. There are five common objectives to education programs: to encourage general interest in nature, to generate more conservation awareness, to bring about a specific change in opinion, to disseminate specific information, and to provide training (Sutherland, 2000). But does conservation education work? Or is it just a supremely manipulative way of heading off dissent through changing people's minds?

We will consider Sutherland's objectives in reverse order. Disseminating specific information and providing training are great ways to impart conservation objectives and involve local people in protected area management. With training they can better participate in ecotourism, experiment with new farming methods or other development initiatives, and find employment as guards, accountants, or managers. In later sections of this chapter we will see that a shortage of adequately skilled local people can impede a project's performance, making locals feel marginalized and disempowered. Furthermore, there is enormous mileage to be achieved by training selective members of the community with a view toward their becoming environmental leaders in their own right (as has happened in Annapurna; Box 2.7). Educated local leaders can play a key role in designing or revitalizing common-pool property regimes, as we saw in chapter 6. Critically important, too, is the education of higher-level officials, who are often responsible for regional policies that render local conservation projects practicable.

What about bringing about a specific change of opinion? Here conservation education takes on the character of a mass publicity campaign. In West Africa, for example, songs about the importance of conservation released by famous singers or respected griots have become top sellers. Such populist approaches can be extremely effective, as with the Saint Lucia Parrot (*Amazona versicolor*), now almost a national mascot and gaining numbers on the island (Box 10.2). Contributing to the Red Data Book, Collar (1992, 413) observes, "the recent history of conservation in Saint Lucia has become a model for other Caribbean countries and reveals an achievement unparalleled in the world." Excitement and national pride in the project serves to overcome the inevitable costs that must be borne by at least some people. A similarly successful conservation story linked to education comes from Canada's Gulf of St. Lawrence (Blanchard, 1995). Members of the local fishing communities used to take the eggs and disturb the nesting grounds of razorbills (*Alca torda*), Atlantic puffins (*Fratercula arctica*), and other seabirds, all of whose populations were declining, but as a result of an education campaign working through schools, a nine-part radio series, and film documentaries, they stopped doing this. A follow-up survey to the campaign revealed dramatic change in residents' knowledge, attitudes, and behavior concerning seabirds.

Box 10.2. The Saint Lucia Parrot's Comeback

The RARE Center is a U.S.-based nonprofit organization whose mission is to protect wildlands of globally significant biological diversity by enabling local people to benefit from their preservation. Since 1973 RARE Center has been working closely with rural communities, NGOs, and local governments in more than thirty countries in Latin America, the Caribbean, and the Pacific. The RARE Center's approach is based on intensive, short-term interventions through local training and technical assistance to meet specific community needs. Their goal is to (1) build local conservation awareness and support, (2) generate sustainable economic opportunities for local communities, and (3) provide resources for parks and protected areas.

One way in which RARE Center meets these objectives is through its conservation and education program. With its emphasis on *Promoting Protection Through Pride*, the program trains local organizations to use social marketing techniques to build grassroots support for conservation. A sterling example of the success of this program is the Saint Lucia Parrot (Butler, 1995; Wells, 2001). *Amazona versicolor* has been placed in increasingly dire straits as the result of hunting pressure, capture for the international pet trade, and increasing habitat destruction. By the mid 1970s the bird was in serious threat of extinction.

Working with the Saint Lucia Forestry Department, a *Promoting Protection Through Pride* campaign was conducted in 1990, building upon work begun in the 1970s. Borrowing from Madison Avenue-style advertising and marketing, the program appeals to people on an emotional level and dramatically influences attitudes and behaviors. The campaign focused public attention on the plight of the Saint Lucia Parrot and what could be done locally to help protect it and its habitat.

Using social marketing techniques such as billboards, posters, songs, music, videos, sermons, comic books, and educational puppet shows (see photo), the campaign broadcasts its message to all segments of the target population during the course of one year. The result was almost instant—local enthusiasm for the parrot, involvement, and tangible support for conservation. In the case of the Saint Lucia Parrot, these efforts resulted in a dramatic increase in environmental awareness, a stronger protected area system, and improved wildlife legislation. The Saint Lucia Parrot, once on the verge of extinction with less than 100 birds, is now coming back, with a current estimate of 450 birds in the wild.

Educational Outreach to All Ages in Saint Lucia.
Photo courtesy of the RARE Center for Tropical Conservation.

Figure 10.1. Gorillas—An Icon of National Pride in Rwanda.
A shop front in the town of Ruhengeri. Picture courtesy of Kelly Stewart.

Yet another example is the promotion of gorillas (*Gorilla gorilla*) to an icon of national pride in Rwanda within a mere ten years as a result of the Mountain Gorilla Project, which combined education with enforcement, ecotourism, and development. Despite the years of civil war (in both Rwanda and neighboring countries), genocide, and mass refugee movement bringing military activity to the conservation area, the Virunga gorilla population has actually increased over the last ten years (Harcourt, 1986; Weber and Vedder, 2001). This speaks loudly to the project's success with linking, in people's minds, gorillas with ecotourism and national prosperity (Figure 10.1). Campaigns such as these are more than mere fads: by working first on local attitudes toward charismatic or economically important species, the groundwork for longer-term environmental initiatives is laid.

And what about the more diffuse programs designed to encourage a general interest in nature and greater conservation awareness? Because it is suspected that most environmental attitudes are formed during childhood, school children are key targets

for education. Hands-on experience, new curriculum development, field trips, tree planting, cleanups, festivals, art, and drama can all bring the natural world alive for young students, as Jacobson and McDuff (1998) show for Malaysia and Brazil. Particularly remarkable have been the extracurricular wildlife clubs that started in a Kenyan high school in 1966 and have now spread to fifteen African countries, anglophone and francophone, with over 4,700 clubs (McDuff and Jacobson, 2000). These clubs rely on traditional means of delivery, like magazines and newsletters, to disseminate information. They also have palpable results: in Kenya during the 1970s clubs organized anti-poaching marches and demonstrations against trade in ivory, and in Ghana they curbed the trapping of shorebirds in some communities. Across Africa wildlife clubs have survived organizational upheavals, complete loss of funding, and even civil war, a testimony to their popularity and grassroots organization. Nevertheless, they are plagued by underfunding, loss of qualified leaders to better-paying NGO work, and poor communication among clubs. In this respect, interschool-based activities are particularly effective. In more than thirty countries teachers affiliated with the Global Rivers Environmental Education Network (GREEN) take their students to explore their local rivers, and then encourage them first to exchange their findings regarding water quality and the socioeconomic determinants of watershed degradation with students from other parts of the world, and later to present their results to their local governments (Stapp et al., 1995). In this way students learn conservation science and action.

Conservation education, of whatever kind, becomes increasingly critical as nature becomes commodified. Think back to ecological economics and the tenet that a project is only justified if it yields economic returns over and above alternative land uses, in other words, if it pays (chapter 9). We raised many problems with an exclusive reliance on this approach, noting how at any time or in any place the goal of biodiversity conservation can be eclipsed by changing prices and economic possibilities; furthermore, it seemed likely that conservation can probably only very rarely satisfy pure utilitarian objectives, at least in the lifetimes of the stakeholders concerned. Under these fickle conditions, then, education emerges as a key component of any conservation strategy. Already there is some evidence that even conventional education pays: each additional year of schooling for a household head in lowland Bolivia is associated with a lower annual rate of old forest clearance by that household (Godoy and Contreras, 2001). But education, and particularly conservation education, may also be important insofar as it instills nonpecuniary values. This possibility has inspired a group of Chileans and foreigners to start a project that provides both formal and informal education at all levels of Chilean society, offering a critical analysis of the narrow economic and utilitarian environmental ethics that prevail in that nation (Rozzi et al., 2000). The project aims to avoid the green colonial-

ism criticized by political ecologists (see chapters 7 and 8), to incorporate indigenous knowledge, and to instill a sense of environmental and ethical responsibility, not just on the island of Chiloé (where the project is based) but across Chilean and broader Latin American society. Its central goal is to galvanize opposition among conservationists, indigenous peoples, developers, and landowners against Chile's much touted economic growth over the last two decades. Ethically committed to instilling nonmaterialist values, this project leads us back to the question raised earlier.

Is conservation education, sometimes referred to as "building constituencies for conservation," merely a strategy whereby outsiders facilitate their own environmental objectives, or more coarsely, is it strategic brainwashing? Though the concept of biodiversity conservation may indeed be a very Western imposition (chapter 8), local communities all over the world have elaborate traditional ecological knowledge (TEK) of their natural environments (chapter 4), supporting practices that can usefully be involved in management. Conservation education programs should concentrate on working with and strengthening TEK. This would serve both to protect TEK against erosion from global and market forces (see chapter 4) and to avoid inciting resentment among the beneficiaries. Simbotwe (1993), for example, delicately intimates how demeaning it can be to have outside "experts" educate Africans about their own environment, particularly when local perceptions of nature are ignored. It does not have to be this way. If conservation education could work toward elaborating and extending indigenous rationales for conservation (or even religious institutions that support conservation; see Box 6.5), in conjunction with instilling broad conservation values, it could be extremely effective for both conservation and political ends. Furthermore, it seems unlikely that conservation education could be effective in the absence of other programs ensuring livelihoods and broader environmental justice (for instance, compensation, direct payments, and other strategies discussed below). As such, the potentially immoral aspects of attempting to instill nonpecuniary values in the minds of others, specifically with the goal of warding off their opposition to Western-imposed conservation projects, seem to be small. Finally, though there seems to be good evidence that conservation education does affect people's knowledge, values, and attitudes (Caro et al., 2003), the question of whether such attitudinal changes actually impinge on behavior remains unaddressed. In a later section of this chapter we examine attitudinal studies.

10.5 Ecotourism

A strategy of growing importance and popularity is ecotourism, potentially the most straightforward and least intrusive way of deriving social and economic benefits from protectionism. Though the world's first national park (Yellowstone) was originally conceptualized as "a pleasuring ground for the benefit and enjoyment of the people" (Chase, 1987, 5) and a "national domain for rest and recreation" (Wearing and Neil, 1999, 40), the term *ecotourism* was only coined in 1981 by Hector Ceballos-Lascursian who, in his capacity as president of PRONATURA (a Mexican NGO), was promoting the conservation of the rainforests of Chiapas through tourism. Ecotourism is loosely defined as nature-based travel (often, but not always, to relatively undisturbed areas of the globe) for activities that range from birdwatching or diving to cleaning up oil spills. Three key elements to this conservation strategy are that ecotouristic enterprises contribute revenue directly to conservation in the destination area, provide funds and motivation for the restoration and monitoring of natural areas, and nurture an appreciation of the natural world in the visitor and host community (Weaver, 1999). As such, there is a great deal of emphasis on local management, on minimizing

the physical (also sometimes the social and cultural) impacts of tourism, and on education. From the viewpoint of the local community, ecotourism is attractive primarily because of the potential it brings for their participation in the economic growth of their area. From the viewpoint of mainstream conservation agencies, an alliance with the tourism industry offers a tantalizing possibility of the much sought after win-win scenario in which nature and local communities thrive (Pleumarom, 1994).

Tourism is currently the world's largest industry and has a growth rate of 3.5 percent per annum (Wearing and Neil, 1999). Ecotourism lies somewhere within this touristic explosion, though its contribution is estimated to vary from as little as 1.5 to 13 percent (or, by other estimates, from a $US10 billion to $US200 billion concern), depending on its exact definition (Weaver, 1999). With such potential growth, ecotouristic ventures are poised to raid the biosphere, demanding continued open access to a renewable supply of raw material—the experience of nature. All the problems inherent in valuing nature for its economic returns, and more specifically, in "dancing with the devil" of commerce (see chapter 9, since most ecotouristic ventures are privately run) are raised, somewhat ironically since the majority of ecotourists probably value nature for inherent rather than utilitarian reasons (chapter 2). The great challenge lies in managing this massive new growth industry (Steele, 1995), particularly in addressing the problems associated with open access (chapter 6) and in ensuring that benefits are captured locally (see below). Until this challenge is met, equating ecotourism with the equally controversial notion of sustainable tourism, or linking it to the growth of an international movement in socially responsible tourism, remains something of a charade.

Despite a commitment to minimizing its ecological footprint, ecotourism, like every other form of tourism, is essentially consumptive, and must be thought of as a form of limited development. Though it may improve local infrastructure, such as roads and electricity, common problems include pollution, natural resource degradation, and more diffuse cultural effects. The difficulties in controlling the development and consequences of ecotourism are demonstrated on the Galápagos islands (Box 10.3). Once a beacon of light to sustainable tourism with its biological research station, naturalist guides, and low-impact floating hotels, the Galápagos Islands now offer a warning light signaling the dangers of ecotourism without sufficient planning, government control, and local involvement in the tourist sector (Honey, 1999). At its heart, the tale of the Galápagos reveals the complex intersecting dynamics between government and conservation authorities (over tourist numbers), between conservation authorities and immigrants from the mainland (over illegal offtake), between ecotourist operators and locals (over tourist revenue), between ecotourist operators and conservation authorities (over appropriate codes of tourism), and between local Galápaguenos and immigrants (over fishing rights).

Potential problems associated with ecotourism are presented in Table 10.1. As regards ecological problems, these are of course associated with all forms of tourism, but ecotourism is perhaps most sensitive to these difficulties since it depends on intact natural areas and is concentrated in ecologically sensitive zones. Classic examples include the disturbance of wildlife as a result of viewing from too close quarters and the driving of vehicles over sensitive grasslands to seek out charismatic megafauna in many of the East Africa's heavily visited parks (Goodwin and Leader-Williams, 2000), and reports that several Himalayan peaks are simply littered with oxygen cylinders, noodle packages, and toilet paper. Educating tourists and zoning their activities to limited areas of the ecosystem is an obvious solution, as happens in the Annapurna Conservation Area in Nepal, where the most popular trekking routes are designated as special

management zones (Wells, 1994; and see Box 2.7).

As regards economic factors, the power of ecotourism as a conservation strategy is the cash it draws. But who gets the money, and what is it used for? A big problem with ecotourism in developing countries is the export (or "leakage") of profits out of the community (even out of the nation). This happens because of a lack of sufficient training, technology, and capital to run the necessary infrastructure. For example, in East Africa most tourists pay U.S. or European-based companies for their airline tickets, lodges, car rental, and package tours. Studies from around the world suggest that some 55 to 90 percent of tourist spending goes back overseas (Koch, 1997). With more coordination at the national level, international leakage can be sealed (see below), but similar problems emerge with respect to assuring revenues stay in the communities and ecosystems most affected by the tourism rather than reverting to the national capital. Since the justification for ecotourism is that visitors bring direct benefits to the area, there must be mechanisms for the money to stay in, or a least trickle down to, the local area. Ecotourist enterprises routinely claim that such economic trickle down occurs through employment opportunities, such as guides, rangers, hotel employees, or in related service industries. But trickle down depends on two factors. First, there must be enough jobs and opportunities to make a difference. Often this is not the case; for example, only 6 percent of households outside Nepal's heavily visited Royal Chitwan National Park received any income from ecotourism at all (Bookbinder et al., 1998). Second, the economic gains must stay in the community; in this context economists refer to a "multiplier" effect, whereby each tourist dollar generates additional benefits for the community rather than, for example, being invested in property in a distant urban center. In many cases, however, the number of jobs is limited and multiplier effects are missing, such that the claim that employ-

---

Box 10.3. Trouble in Paradise? The Galápagos Archipelago

The Galápagos Archipelago lies in the Pacific Ocean, approximately 1,000 km west of continental Ecuador. In part because of its isolation, and in part because of the late (1535 AD) arrival of humans, it has extremely high levels of endemism in its flora and fauna, and its subtle variety of species inspired Charles Darwin's theory of evolution by natural selection. Buccaneers, pirates, and whalers arrived soon after, and with their goats, rats, and dogs decimated many populations of sea turtles, land tortoises, and fur seals. Settlement of four of the islands in 1900 brought a new suite of exotics—pigs, cattle, cats, and plants like papaya and banana. The impact of these invasives on island biodiversity has been enormous, and some of the iguana species are now on the endangered list.

In 1959 over 90 percent of the Galápagos land area was declared a national park, and the Charles Darwin Research Station was established in 1960; in 1978 it became a world heritage site (WHS) and in 1984 a biosphere reserve. Protection of the waters surrounding the islands was upgraded in 1998 with the Galápagos Marine Reserve. The research station, in conjunction with the Ecuadorian National Parks Service, has eradicated introduced species from some of the smaller islands, and reintroduced endemic species onto islands where they had gone extinct. Nevertheless, in 1996 the archipelago's WHS status was put in jeopardy with the threat of a new designation—WHS in danger—which would unravel the islands' very lifeline: ecotourism.

Ironically, one of the main causes of the islands' problems is the growth in tourism. In principle, there are tight rules—a master plan calling first for a maximum of 12,000 per year (1974, increased to 30,000 in 1987), the obligatory escort of a well-trained guide, zoning of tourist use, and strict regulations for conduct on the islands. But already by the early 1980s ecotourism was transforming into mass tourism, both Ecuadorian and foreign. At first locals converted their crafts into small tourist vessels, but increasingly private enterprises from the mainland and abroad bring in much larger boats, even "floating hotels," and now many tourists arrive at one of three island airports. By 1996, 62,000 people were visiting the islands each year (Honey, 1999). Tourism became a milk cow for the Ecuadorian

(continued)

---

ment opportunities enrich the community is more rhetoric than reality (Brechin et al., 1991). There are, however, promising developments, particularly where locals own the facilities hosting tourists and earn money directly, as in the Annapurna Conservation Area; here complaints center more closely on the lodge owners, specifically their tendencies to hire only their own kin in the touristic enterprise, and on the challenge to develop local sources of investment for all this new wealth (Wells, 1994). A clear demonstration of the importance of distributing revenue at the local level comes from a comparison of the five major wildlife tourism districts of Kenya and the severity of their wildlife losses. Losses are related not to visitor numbers, nor to total revenues, but to the distribution of revenues between the central government, the tourism industry, and the local area (landowners and users) (Table 10.2), the implication be-

(Box 10.3 continued)

government, creating an impasse between conservationists, tour operators, and the government (Kenchington, 1989), each of whom had different ideas about how many tourists should be allowed and how the tourist revenue should be divided up. Furthermore, the park guards were too overwhelmed with the traffic to reliably collect revenues. Far too many people walk on any one island at a time (only ninety are permitted), and there is disturbance of animals, damage to trails, frustration, crowding, and breaking of regulations. According to some observers, attempts to control visitor damage merely distract attention from the fundamental problem—visitor numbers (Steele, 1995).

A closely related problem has been the growth in the permanent population of the islands, due primarily to immigration of families from the mainland looking for work (Trillmich, 1992). As tourism boomed, new migrants swarmed to the islands. Those who failed to find work turned to commercial fishing operations, many of which are tied to the Asian market. In addition to overfishing of lobster and sea cucumber, there was a growing black market in seahorses, snails, sea urchins, and black coral taken for sale to the Far East, now effectively banned.

In 1997 the president of Ecuador signed the Galápagos Decree, a consensus document of all stakeholders (tourism, islanders, industrialists, and conservationists) designed to promote conservation by controlling immigration, commercial fishing, and introduced species; this was sanctioned by the Charles Darwin Foundation (Honey, 1999). The decree sparked counter-lobbying by influential politicians, businessmen, and fishing industry magnates who wanted local control, open access to the marine reserve, and more liberal residence permits, leading to an escalation in ecosabotage and violence that had been simmering throughout the 1990s—arson, tortoise massacres, and sieges of the research station. Yet despite this opposition a Special Law for the Conservation of the Galápagos was passed in 1998. As yet the government has eschewed large-scale resort-based tourism (huge cruisers rarely land). Nevertheless, little tourist cash reaches the islanders' pockets, since most foreigners come on specialized tourist carriers, and islanders for the most part profit only from Ecuadorian (low-paying) visitors; furthermore, tourist numbers are still not capped.

ing that receipt of ecotourist dollars inhibits land conversion and poaching.

Regarding social factors, unanticipated social consequences to ecotouristic enterprises have often been observed. First, the trickle down from ecotourism, as from any other development initiative, can exacerbate wealth disparities in a community, as painstakingly explored by those who study the political ecology of tourism (Stonich, 1998; and see chapter 8). As reported for an ecotouristic venture on the Micronesian island of Yap, "the Chief is not sharing the entrance fees to the village . . . and money is making people stingy, therefore harming community spirit" (Stronza, 2001, 269). Increasing wealth differentials arise in part because communities are made up of heterogeneous interest groups, with differential power, influence, and potential to garner new opportunities (chapter 6), in part because only a small subset of households can benefit directly from tourism, and in part because an influx of tourists often inflates

TABLE 10.1
Potential Problems with Ecotourism

Ecological Factors

Disturbance to wildlife
Vegetation damage from means of transport
Litter and human waste
Environmental impact of hotels, camps, etc.

Economic Factors

Leakages
Distribution of benefits (nepotism, failed "mulitiplier effects," limited
    local investment opportunities)

Social Factors

Wealth differentiation—limited opportunities and inflation
New sources of stratification
Social problems
Breakdown of traditional authority
Escalating material desires
Cultural decay

the local price of food, firewood, and transport. Second, and closely related, like other forms of development, ecotourism engenders new sources of stratification. An example comes from the Zambian ADMADE, which generates funds from expatriate hunters for village development (chapter 7). Chiefs have acquired more power than ever before by inserting their own private wishes at the top of the village wish list, while at the same time young men with lucrative jobs as game scouts challenge the traditional village hierarchy, with both dynamics disrupting village institutions (Gibson and Marks, 1995). A third problem is that of escalating wants, attendant on boutique tourism and of course many other kinds of interventions by outsiders. It seems quite probable that

TABLE 10.2
Wildlife Losses and Tourism Benefits in Five Districts of Kenya

	Narok	*Samburu*	*Taita Taveta*	*Kajiado*	*Laikipia*
Loss of wildlife (1977–94)	−65%	−33%	−29%	+2%	+12%
Tourist numbers (1994)	138,000	90,000	238,000	160,000	50,000
Disbursement of funds since 1992	US$0.1m	US$0.2m	US$0.1m	US$1.2m	US$0.2m
Revenue goes to:[a]					
Central government	**	**	***	***	*
City council	***	***	*	*	*
Tourism industry	***	***	***	***	*
Landowners/users	*	*	*	**	***

[a] Minor * through to major *** dispensations.
Source: Reprinted with permission from Norton-Griffiths, M., The economics of wildlife conservation policy in Kenya, eds. Milner-Gulland, EJ and Mace, R. *Conservation of Biological Resources.* Blackwell Science Ltd., Publishers.

the arrival of wealthy tourists in remote areas stimulates new expectations among residents exposed to Western dress and gadgets. If people measure their satisfaction (and happiness) in comparison to the community of people to whom they are regularly exposed (the notion of relative wants introduced in chapter 5), tourist schemes where local people host visitors in their homes run the risk of radically raising material expectations. While this is not inherently a bad thing, people with higher material aspirations may exert greater pressure on natural resources in order to boost their income.

A final social factor is the issue of cultural decay. Many observers of ecotourism bemoan the exposure, particularly of the youth, to new ideas and materialism. Thus, recreational drugs can become a problem among people choosing to befriend foreign visitors. Inappropriate clothing and behavior of tourists are other commonly reported difficulties, as well as overcrowding, social tensions, resentment, and crime in the host community. In a similar vein, where the ecotourist enterprise includes "experiencing cultural diversity" there is the problem of the commodification of culture (Cohen, 1988), addressed first in chapter 2 in connection with anthropological reserves. There are, however, many ways in which ecotourist projects focusing on both biodiversity and cultural diversity can boost rather than erode cultural identity and local pride. This is most in evidence in South America, where groups like the Xavante and Kayapó began to market their cultural identity quite aggressively throughout the 1980s, often in conjunction with environmental NGO initiatives (see chapter 8). Similarly, the Tlingit Indians have entered into a productive partnership with the Sitka National Historical Park at the mouth of the Indian River in Alaska, with Native Americans playing a key role as guides, presenters, and interpreters. Tourists enjoy watching artists working and elders telling their stories, even though these activities were designed as much for the Tlingit as the tourists (Bowman, 1998).

Unsustainable tourism generally arises in situations where the influx of tourists precedes planning and management (Steele, 1995; and see Box 10.3). Accordingly, as with any other conservation strategy, the successes of ecotourism lie in planning, compromise, trade-offs, and zoning, all of which require agreements between diverse stakeholders. Planning generally entails three increasingly effective measures. At minimum, tour operators should adopt a code of practice, for instance, the slogan of the arctic, "Take nothing but photographs, Kill nothing but time, Leave nothing but footprints" (Mason, 1997, 161). Such guidelines are designed to minimize negative impacts but are not enforceable. Codes of practice become more effective as tour operators and other interest groups enter into voluntary schemes of compliance, such as the agreement among Nepalese lodge operators in the Annapurna Conservation Area. These families committed as a user group not only to heat their lodges and cook for their guests with kerosene rather than with locally depleted firewood sources, but also to maintain a trail cleaning service (see Box 2.7). At best, ecotour operators can improve compliance with such schemes by setting up a system of accreditation to win the trust (and business) of socially and ecologically responsible tourists (Wearing and Neil, 1999). Such programs improve industry standards in a highly competitive market. In some circumstances the control of ecotourism is best facilitated by a broader integrated development project, insofar as this provides a structure whereby tour operators can work together and establish rules for their use of a natural commons. In other cases ecotourism is most effectively regulated by the government: thus, at the extreme end of government intervention the government of Bhutan effectively caps the number of visitors to the nation by charging a hefty daily fee (Wells, 1993), such that in 2001 only 6,000 tourists visited the country.

National initiatives are nevertheless fragile. First, they are dependent on the current cast of political characters, and therefore by definition are volatile; for example, a National Ecotourism Strategy launched by Australia's government in 1994 was abandoned two years later following the federal election of 1996 (Wearing and Neil, 1999). In a similar vein, political instability can cause tourist numbers to plummet, as occurred at various junctures in Uganda, Zimbabwe, and most recently Madagascar. Second, given the present political climate of deregulation and the emphasis on community-based ecotouristic enterprises, there is plenty of resistance to centralized regulating agencies, since regulations protecting the environment almost inevitably reduce short-term earnings. An ideal solution is to establish a more cooperative approach to the design and implementation of ecotourism that incorporates local stakeholders from the local tourism sector and the community. It is in this context then that we can understand why it is so important for ecotouristic enterprises to obtain full participation of local communities and stakeholders in the planning, implementation, and management of tourism. Nations like Belize are particularly proactive in developing national policies to empower local communities in taking control of tourist opportunities.

In sum, given the received wisdom that in ecotourist ventures economic success and environmental impact are negatively correlated, sustainable operators must find a balance between the interests of nature and visitors, since the discerning ecotourist will no longer visit a degraded or even heavily commercialized site. Ecotourism is simply a form of development and as such it raises, in a distilled form, all the conflicts between protection and utilization that lie at the heart of this book. Without the political will to manage ecotourism in a sustainable way, ecotourism simply slides into mass tourism, as in the Galápagos, merging with the business it tried so hard to distinguish itself from (Weinberg et al., 2002). Whether self-regulation or top-down directives are most effective will inevitably vary by site, but in all

cases it seems that collaboration between government, operators, developers, tourism, and locals is critical in developing acceptable guidelines that all parties can agree upon. In the meantime, we need to remain suspicious of claims that the aggregate benefits of ecotourism outweigh the costs (see, for example, Box 9.1): benefits usually accrue to those actively involved in the tourist industry, while costs are borne by those who are least well positioned to provide services, namely the poor (Pleumarom, 1998; Stonich, 1998). In this respect ecotourism raises issues regarding social justice (chapter 7, particularly the justice associated with the ecological economics perspective discussed in chapter 9).

10.6 Integrated Conservation and Development

Far more radical than simple outreach or extension work is the integrated conservation and development (ICD) approach that springs directly from the premise that conservation cannot be achieved without development (and vice versa), a natural development of the 1970s policy shifts spurred by the Brundtland Report (see chapter 2). An ICD project emphasizes the links between development outside of protected areas with more effective protection inside (see Figure 2.4).

Ideally, the ICD approach is not conservation through development, or conservation with development, or even conservation adjoined with development; it is the achievement of conservation goals and development needs together (Sanjayan et al., 1997), and many kinds of projects have masqueraded under the label. Increasingly, however, it becomes clear that such "win-win" scenarios, where conservation really can be achieved through development, are rare, and attention shifts to more complex, multipronged interventions. Typically such projects have three objectives: improved management of natural resources within the protected area, compensation or resource substitution schemes for local communities who must stint in their use of protected resources, and appropriate social and economic development outside the protected area designed to take the pressure off the resources currently under threat. These activities can be referred to as enhancement, compensation, and alternatives (Abbot et al., 2001). A full ICD intervention incorporates into a single project each of these activities, a strategy that can lead to better coordination and less redundancy than when many smaller projects are involved.

In Abbot et al.'s (2001, 1115) words, "the oft quoted logic [of the ICD approach] looks neat: to be sustainable, development needs conservation of natural resources, and conservation, to succeed in low-income countries, requires development"—truly, in Alpert's (1996, 845) term, "a marriage of convenience." In reality, such marriages can be quite awkward. Typical tensions can be seen in Ranomafana (Box 10.4), an interesting case in part because unlike several other ICD initiatives the project puts biodiversity protection as its primary goal, in part because its long-term ecological monitoring includes assessment of human impacts on the rainforest (Wright, 1997), in part because it attempts to monitor the health and nutrition in surrounding communities (Wright and Andriamihaja, 2002), and in part because of its openness to evaluation by so many constituencies (Peters, 1998); more often NGO and foreign aid missions view researchers as meddlesome (Barrett et al., 2001), or at least an expensive luxury (Newmark and Hough, 2000).

Perhaps because the ICD strategy became such a showcase for the emerging alliance between conservation and development in the 1980s (see chapter 2), these projects drew sharp scrutiny, such that after an initial period of guarded optimism (Brandon and Wells, 1992) criticism began rolling in. Intriguingly this came on both the social and the natural science fronts. Biologists saw in the elision of conservation and development

a fatal faltering in the commitment to protect biodiversity, and concluded that nature had yet again been sold out; they supported this view with the observation that of thirty-eight ICD projects reviewed in the early 1990s, almost none had any ecological monitoring and only five showed evidence of improved wildlife numbers (Kremen et al., 1994). Thus, looking primarily at Africa (where ICD initiatives are most numerous), Alpert (1996, 852) concluded that these projects yield "tangible benefits for communities" but fewer clear successes for conservation. More disparaging reviews followed, for both Africa (Kramer et al., 1997; Oates, 1999; Newmark and Hough, 2000; Soulé, 2000) and Indonesia (Wells et al., 1999). On the other front social scientists and activists were equally unimpressed. In the ICD approach they only saw more top-down fortress conservation, with the "development" component a mere window-dressing designed to obscure the continuing colonial legacy of policing, eviction, and misanthropy, and to defuse local opposition (Neumann, 1997). Both in the prominent Zambian (ADMADE) case (Gibson and Marks, 1995) and in Ranomafana (Peters, 1998; see Box 10.4) human communities were seen as shortchanged, sold out, no doubt in part because expectations of social benefits were unduly high. Here we will look first at the reasons for each front's negative evaluations, and then turn to some brighter horizons.

Problems arise primarily from irreconcilable inconsistencies within the ICD logic (Wells et al., 1992; Barrett and Arcese, 1995; Newmark and Hough, 2000). Essentially, the links between development and conservation that lie at the heart of the ICD logic are still unproven. There are two outstanding questions: Will an economically prosperous community automatically conserve its natural resources? Does rural development depend on conservation? In other words, are conservation and development two sides of the same coin? Regarding the first question, Barrett and Arcese (1995)

Box 10.4. "A Lemur Will Have to Meet You at the Airport": The ICD Project at Ranomafana

Ranomafana National Park Project was started in 1991 to link the conservation of lemur (*Hapalemur simus* and *Hapalemur aureus*) habitat within the newly established national park to rural development. Showcasing the ICD approach, the project had the strong support of the World Bank, USAID, and the Malagasy government (Wright, 1992; 1997). The objective was to win local support for lemur conservation by demonstrating tangible gains from the conservation effort, and by bringing to the area technical assistance in agriculture and forestry projects, as well as improved education and health services. Even though funding for the ICD project itself ended in 1997 (because of the funders' more general loss of confidence in the ICD approach), Ranomafana has succeeded in attracting follow-up support, and to this day continues to address the challenge of integrating conservation and development (Wright and Andriamihaja, 2002).

As in other developing countries, the ICD approach at Ranomafana strives to meet the needs of large numbers of hungry people, poor even by Malagasy standards (Hannah et al., 1998; Kottak, 1999). The original ICD project affected approximately 27,000 Tanala and Betsileo hill rice (*tavy*) cultivators. *Tavy* is a form of slash and burn cultivation, with low yields, and *tavy* cultivators supplement their subsistence with forest products, including crayfish, bamboo, reeds, firewood, honey, medicinal plants, and forage for their livestock. After the establishment of the park in 1991, most of these forest activities were banned, as was the burning of forest for *tavy* in a 3 km buffer known as the peripheral zone. To compensate for these land use changes, and to promote sustainable development in the peripheral zone, the project provided small dams for valley bottom irrigation rice (paddy), fishponds for aquaculture, and training of Malagasy as tour guides and ecological monitors; furthermore, it guaranteed that 50 percent of park entrance fees would go to village communities for conservation projects of their choice.

In many respects the ICD project and subsequent initiatives have been successful. Most people now stay out of the park, since they have reeds, crayfish ponds, firewood, and fertilizers provided outside, and the 50 percent park fees are used wisely. Local tourist facilities have been developed to benefit from the new trade, and a

(continued)

---

conclude from African case studies that development increases consumerism, makes people less reliant on (and hence less sensitive to) variations in local resources, and weakens cultural institutions for sound environmental practices (chapter 6); as an example they note how improvements in roads and transport facilities only encourage illegal offtake for distant markets. In a similar vein, drawing primarily on quantitative data

from Latin American countries, Wunder (2001) finds that poverty reduction generally speeds up forest loss rather than slowing it down, providing a new twist to the dynamics outlined in Figure 7.2. Moving to another continent, in southern Thailand, Vandergeest (1996) points out that villagers' illegal use of protected areas has nothing at all to do with their poverty (indeed, with the rapidly growing Thai economy wages

(Box 10.4 continued)

number of Malagasy are trained in ecology and conservation. However, there are ongoing problems, recognized by the project staff at its closure, and these persist to this day (Wright, 1997):

(1) *Tavy* cultivation (and the "culture of burning"; Wright and Andriamihaja, 2002, 131) persists in the peripheral zone, such that there is extensive forest clearance within the designated buffer zone (see Figure 3.6).

(2) Those who benefit most from project activities are the rich, whose land claims render them eligible for dam and irrigation schemes. Landless (or land-poor) households are most negatively affected by the project, since they critically depend on the forest for subsistence.

(3) Local villagers have for the most part failed to profit from increased tourism to the area since they have insufficient training to enter the tourist sector; wealthier outsiders benefit.

(4) The social services promised by the project, particularly education, health, and family planning, have been sporadic because the government does not prioritize such services.

(5) There is a constant need for long-term funding since the park will probably never be self-sufficient.

Critics are appalled by the fact that funders appear to care more about lemurs than Malagasy. In the opinion of a colleague of Kottak (1999, 27): "the next time you come to Madagascar there'll be no more Malagasy. All the people will have starved to death, and a lemur will have to meet you at the airport." More concretely, Peters (1998) suggests opening a trust fund with all the money paid to expatriate and local NGO officials, and using the interest (from a high yielding offshore bank or mutual fund, see chapter 9 for environmental trust funds) to make annual payments to the most affected households; using figures from the early 1990s Peters calculates that US$131 could be paid annually to each of the 4,500 families most threatening the park's biodiversity; this is two-thirds of an average family's annual income. Of course, even if this payment was made in rice or fertilizer, there is no guarantee illegal activities would stop because of the collective action problems discussed in chapter 6.

are high even in rural areas); rather, it reflects resentment regarding loss of traditional land. As such, Vandergeest implies development, or poverty alleviation, would do nothing to change people's behavior. In China, applications of the ICD approach at the Xishuangbanna State Nature Reserve have entailed intensification of agriculture outside of protected areas, which in turn increase income and put more pressure on protected area timber for new construction and protected area water sources for irrigation (Tisdell, 1995). Finally, the conventional view based on studies in South Africa and Tanzania, that the affluent tend to be more pro-conservation than the poor (reviewed in Newmark and Hough, 2000), is not holding up. Certainly, there is evidence to believe that, at least in some circumstances, wealthier people suffer less than the poor from reduced access to natural resources and also benefit more from outreach programs, as in Ranomafana (see Box 10.4) and the Usambaras cardamom project (discussed below). But new evidence suggests other dynamics too. In Tanzania, where Holmes (2003a) focused specifically on the utilitarian value of resources in the protected area as perceived by locals living outside Katavi National Park, affluent households (effectively people with capital and time) are often the ones who most resent the protectionist measures that interfere with their small-scale business enterprises. Similarly, in an evaluation of four ICD projects in Madagascar, Marcus (2001) found no evidence that people of high socioeconomic status have more positive perceptions of protected areas than do poorer people. So the answer to the first question, does economic prosperity assure conservation, is a categorical "no."

Turning now to the second question, does rural development depend on conservation? Over the long term the answer is of course yes, but the benefits of the conservation generally kick in only far beyond the timeframes relevant to most rural inhabitants (time horizons are discussed in chapter 5). While sustainable development necessarily relies on an ecosystem that cycles nutrients, preserves water supplies, and enhances soil formation, a hungry farmer on the hills outside Ranomafana thinks less about such ecosystem services than the food in her cooking pot; unsurprisingly, politicians who represent these communities, and who can either encourage or impede policy changes, operate on equally short timeframes.

Consistent with these fundamental conceptual flaws are several sets of observations. One regards immigration. In Nigeria and the Central African Republic, immigrants have been attracted to the areas with ICD projects, precisely because of the services they provide (Oates, 1995; Noss, 1997; and see upcoming discussion of CAMPFIRE). Another regards the sustainable offtake of meat or other products (discussed further in the next section). For the most part, ICD projects based on the regular provision of meat to local communities are not sustainable, particularly where game meat can be sold outside of the community to urban markets. As Barrett and Arcese (1995; see also 1998) argue, when managers are expected to provide meat to local communities, they will find it very difficult to reduce offtake when wildlife populations are declining and most vulnerable (as in drought years when people too are hungriest), at least not without losing community support; even if the managers are local, this problem does not go away. In fact, Barrett and Arcese, discussing only the African context, go as far as to conclude that because of these problems of stochastically varying supply and constant demand, the only way to advance the conservation status of large mammals and rural human populations is to decouple human needs from wildlife—the very antithesis of the ICD approach. Indeed, as Hackel (1999) notes for Swaziland, Ethiopia, and other cases, such decoupling may become inevitable as communities are forced by demographic and social pressures to forsake wildlife conservation for more intensive agricultural development. Just as rural people adopt conservation for economic benefits, so they may reject it later when economic conditions change, indicative of the inherent slipperiness in ecological economic calculations (chapter 9).

Problems also arise from flaws in project design and implementation. A key issue regarding design is whether incentives that are offered to communities as public goods, such as schools and hospitals for the use of all (see chapter 6), can alter individual behavior. An individual's self-interested course of action (predicted from chapter 5) is to derive all the benefits from the new social services and benefit-sharing schemes, while persisting with economically advantageous activities, such as woodcutting or poaching, if the risks of getting caught and punished are not too high. An example of this is reported from Ranomafana, where former poachers hired as forest guards allegedly use their new earnings to expand their erstwhile businesses (Newmark and Hough, 2000). Implementation brings its own challenges. Project managers must, at minimum, conduct basic research, so as to be able to anticipate outcomes. For example, in Tanzania's Eastern Usambara Mountains an ICD project gave incentives to local farmers to plant the spice cardamom on their plots, with the aim of reducing illegal cultivation in the protected forest. This backfired because the managers did not know that it was primarily landless people who grow cardamom in the forest. With increased legal (farm-based) production the price of cardamom fell, forcing landless families (unable to benefit from the scheme) to clear increasingly large areas of protected forest to maintain their income (Stocking et al., 1995). Finally, even with research under wraps, implementation raises its own challenges. Insofar as ICD projects mimic earlier top-down development initiatives, typical not only of colonial days but the big integrated rural development projects of the 1970s, local people are often treated as recipients of aid rather than partners in development; as such, communities can feel disenfranchised from the project, promoting old feelings of dependency and an unsurprising urge to pursue personal benefit from the project, whenever possible. The perennial problem arises, that of the wealthy taking the cream from project activities, as reported for Tanzania's Selous Game Reserve (Songorwa, 1999). Indeed, a frequent set of complaints heard in evaluations of ICD initiatives includes scarce employment opportunities within the project, limited trickle down of benefits through the local economy, escalating consumption norms, and inflation.

What seems to have happened is that the gem at the heart of ICD approach, the idea that at least in some instances "win-win" strategies, where tangible links between conservation and development can be exploited (Getz et al., 1999), has over time mutated into a far more concrete claim: that conservation is compatible with development, that conservation depends on development, and

ultimately that conservation must be development, mirroring subtle policy shifts examined in chapter 2. As with all conservation strategies, the challenge now lies in determining the situations in which such win-win outcomes can be attained. Over the years it has become clear that ICD projects work best at sites where the threat is local and not too extreme, and that are blessed with at least some of the following features: high tourist revenues, strong national political support, high potential for sustainable extraction, low natural growth in population, low immigration rates, strong and intact communities, stable consumption norms, and a fundamental compatibility between project goals and local cultural and economic traditions; it is also clear that if the resource to be exploited is of too high value, like ivory, the project can become unstable (Freese, 1998). By contrast, ICD approaches work poorly where biodiversity is already threatened, and should not be funded unless the national government agrees to support the project's goals with strong institutional support (often difficult in weakly organized developing tropical nations; Barrett et al., 2001; and see below).

There are, however, some exciting new directions in which the ICD idea can move. First, there are success stories, albeit few and far between. Focusing on the Kilum-Ijim Forest Project, an ICD project established in 1987 in northwestern Cameroon, Abbot et al. (2001) tested a critical assumption at the heart of ICD logic, that development activities will affect the attitudes and behavior of local people in such a way as to make them more supportive of conservation measures that regulate resource use. The project's goal is to conserve a small montane forest, and to do this through increasing the productivity of lands adjacent to the forest and improving incomes from harvesting forest products (the "alternatives" and "enhancement" approaches described, respectively, above). Abbot et al.'s (2001) attitudinal study showed that the percentage of people with negative views toward the forest and boundary declined from 52 to 18 percent (averaged across villages) throughout twelve years of project activity, and those who participated most closely with the project were the most positive about forest protection. Indeed, the majority of communities adjacent to the forest have established forest management institutions, an initiative that entails significant investments in village organization, planning demarcation, conflict resolution, and meetings with neighbors and officials. Though the ecology has not been closely monitored, it appears forest clearance is much less here than in other parts of the Bamenda Highlands; as such, this case points to the potential for an ICD approach to modify behavior through attitudinal change. Another case, less closely studied, is the Batang Ai National Park in Sarawak, eastern Malaysia, formed in 1984 (Horowitz, 1998). ICD planners decided from the outset to allow local communities their customary rights within the park in return for a complete ban on some activities, such as cutting primary forest and the commercial sale of game meat. It appears that the local people are playing an important role in keeping outsiders from exploiting the park's resources, and gradually requesting further services from the project personnel in an attempt to find alternative economic activities.

A second encouraging development lies in CBNRM projects. These differ from the standard ICD approach in that instead of offering development services in exchange for a protected area, they devolve management responsibility for natural resources to local communities, the CBC initiative introduced in chapter 2. An almost iconic and very carefully crafted example of this approach is the CAMPFIRE program in Zimbabwe, with its goal of establishing communities that manage wildlife for their own interests (Box 10.5). CAMPFIRE is an ICD project that specializes in enhancing the value of the resource to be protected, and it does this by marketing trophy species to rich foreign safari hunters. By embracing the CBC approach, CAMPFIRE (and similar programs in Zambia and Namibia) aims to avert the difficulties associated with top-down directives and community disenfranchisement. While, even in the eyes of its erstwhile promoters, CAMPFIRE and other such projects have not for the most part met their objectives, many interesting lessons have been learned from varied experiences across many southern African sites, sites that can almost be thought of as experimental variants. A recent overview of these cases concludes that although devolving power to communities is a step in the right direction, its achievements have been limited, due to weak government policy, limited markets (in particular, no market for ivory), and slow pace of policy reform (Hulme and Murphree, 2001). This overview also shows that, despite all the initial enthusiasm, the possibilities of linking even large mammal conservation to human welfare are quite limited, as anticipated by Barrett and Arcese (1995); indeed, CAMPFIRE programs only work in areas where there are high densities of large-bodied wildlife, and even then there can be problems (as discussed in the next section on extractivism).

Policy amendments at the national (and international) level are a key ingredient to ICD success (Wells et al., 1992), as indeed they are to all conservation initiatives (chapter 9). We have already seen in chapter 6 how local communities are often inadequately equipped to deal with pressures that emerge outside of their community, and how an ideal common-pool resource management system lies nested within broader regional and national institutions (see Table 6.2). CAMPFIRE tries to deal with this by placing power at the district government level, but the program's long-term success still depends on broader developments at the national level, in particular, political and economic stability (see Box 10.5). In the case of ICD initiatives, national policies are required to buffer the projects from myriad external pressures, and might include limiting immigration to the ICD area, providing

Box 10.5. CAMPFIRE under Fire

Zimbabwe's CAMPFIRE aims to achieve "sustainable rural development that enables rural communities to manage, and benefit directly from indigenous wildlife" (Zimbabwe Trust 1990, cited in Derman, 1995, 202). Since 1975 owners and occupiers of land have, under the Parks and Wildlife Act of 1975, held direct custodianship rights to wildlife, fish, and plants occurring on their land. After Independence this act was amended (in 1982) to apply not only to the land of white commercial farmers but also to Africans living on so-called communal and much poorer quality lands. In effect, the Rural District Councils (RDCs) had only to register as "authorized authority" to harvest wildlife for their own benefit, and so CAMPFIRE was born (Murombedzi, 1991, 4). Its prospects as an ICD project, founded on a rich vein of wealth in the developing world, foreign safari hunting, promised a particularly favorable outcome.

The general model is this: consumptive (sport hunting) rights, and less commonly, nonconsumptive tourism rights, are leased to private-sector entrepreneurs (usually white-dominated safari hunting operators). The proceeds (in excess of US$9.3 million between 1989 and 1996) are divided such that 50 percent goes to wards (groups of villages), 35 percent to the RDC, and 15 percent to the program. While the RDC money goes toward monitoring the outside safari agents' offtake of wildlife, the ward's money is used for village projects (like flour grinding mills or schools, or even cash dividend payments to each household), and is supposed to provide local people the financial incentive to participate in the program (Metcalfe, 1994; Bond, 2001).

In reality the program operates very differently in the various districts and wards, depending on the density of wildlife, the size of the human population, and many aspects of the local community such as size, ethnic mix, and leadership (Derman, 1995). In general, two issues emerge again and again: are the financial returns from the program adequate for providing local incentives for conservation, and how should the money be allocated between district, wards, and villages? Regarding the former, Bond (2001) finds that between 1989 and 1996 payments plummeted, due to declining wildlife numbers, a growth in the number of participating wards, and, in Murombedzi's (2001) analysis, immigration of rural and urban Zimbabweans to areas of lower human population density; dividends rarely exceeded 10 percent of a typical agricultural household's income. Under these conditions, asks Bond, how can wildlife income act as an economic incentive for sustainable wildlife management and conservation? Here we have empirical data supporting Barrett and Arcese's (1998) household-based model of the fragility of linking human needs to wildlife offtake. As regards the question concerning decentralization, according to the program's "producer community principle," ward allocations should be proportional to the number of animals shot in its villages, to compensate the people who actually have to live with the wildlife and incur the most costs. It appears, however, that the RDCs, currently the big beneficiaries from CAMPFIRE, would be unwilling to let so much money go, illustrating the inherent tensions in decentralization (Murphree, 1994).

(continued)

(Box 10.5 continued)

So has CAMPFIRE attained its most celebrated objective—creation of the necessary conditions for institutional change in the management of wildlife, habitat, and other natural resources (Murphree, 1994)? There are some "all-too-rare successes" (Murphree, 2001, 190), wards with low population density and high wildlife populations where locals have become willing conservationists, or wards well enough endowed with natural beauty that revenue derived from nonconsumptive ecotourism far exceeds that derived from agriculture. Elsewhere, however, the myth of CAMPFIRE is evaporating, with reports of resentment against white safari hunt operators, who offer few jobs to black Africans and fail to acknowledge their superb tracking skills; of bitterness about the hiring of paid professionals to crop wildlife ("orgies of butchery," Murombedzi, 2001, 252) when local hunting traditions are banned; and of conservationists' frustrations, as communities use wildlife profits to finance ecologically deleterious development.

So are we back at "fences and fines" again, with a mere gloss of outreach? No. This ambitious ICD project has led the way in changing the legal status of wildlife within the national constitution. Moreover, wildlife habitat loss has been slowed, and in some areas illegal poaching brought under control (Lewis et al., 1990). Clearly, international legislation approving the sale of ivory would boost CAMPFIRE revenue, highlighting the importance of an enabling political climate discussed in chapter 9. But the outstanding problem for CAMPFIRE today lies not in the design of the program (in fact, it still stands as a fine model of an incentive structure, albeit in need of tinkering), but in population growth and the expansion of urban poor into remote, sparsely populated, and successful CAMPFIRE areas. The growing fragility of Zimbabwe's economy and political system, and the recent political appropriations of communal areas, only exacerbate these processes.

secure land tenure to the target communities, supporting prices of products sustainably produced, providing infrastructure and control of tourism (if this is part of the plan for the area), and banning access from outside (including international) logging or other extractive businesses. Indeed, the causes of ICD failure often lie right at the top. In Indonesia, for example, such projects flounder not because of local villagers' cutting of timber and hunting of wildlife, but because of the state's tacit sanctioning of the illegal logging and mining activities organized and financed by outsiders (Wells et al., 1999). Furthermore, to be maximally effective in conserving biodiversity, ICD projects need to be part of a broader landscape-wide conservation plan, protecting metapopulations across fragmented habitats (see chapter 3). Madagascar is currently exploring such an approach, eliciting the participation and collaboration of more government agencies, and focusing on forests beyond as well as within parks in order to secure watersheds and wildlife corridors (Hannah et al., 1998; Wright and Andriamihaja, 2002). Only with this kind of collaboration between community, project, and nation, formally captured in co-management schemes (see chapter 11), can the ICD approach work.

10.7 Extractive Reserves

Moving further along the continuum of increasing use, we now turn to extractivism.

The central tenet of extractivism is that one or a few species are exploited, and that this selective use is justified by the conservation of many others, particularly when the ecosystem itself is at risk (e.g., see Medley, 1993). Philosophically this is founded on the utilitarian rationale for conservation (chapter 1), employing as a means the "use it or lose it" strategy discussed in relation to commercial interventions (chapter 9). Accordingly, a key proviso of extractivism is sustainability, both social (see Box 2.5) and ecological (see Box 3.5). Often, extractive systems focus on plants, fungi, or invertebrates that are abundant, easily collected, and valuable—resources often called NTFPs, many of which are marketed in local towns or beyond. Among the most well known examples are brazil nuts (*Bertholleta excelsa*), latex (*Hevea brasiliensis*, the commercial exploitation of which predates the term *extractivism*), and açaí palm (*Euterpe oleracea*), but countless other products (including seeds, fruits, leaves, bark, vines, bamboo, gum, pollen, oil, wax, and even whole organisms) are collected under the umbrella of extractivism. Hunting for meat is often not considered in the context of extractivism, but conceptually it falls squarely within this framework. Because NTFP incomes can be earned without losing forest cover, extractive reserves have been hailed as an especially promising conservation strategy (Fearnside, 1989; Schwartzman et al., 2000a). Furthermore, the economic benefits from forest extraction can exceed those from logging or conversion to pasture in some cases (Hecht, 1992; Salafsky et al., 1993).

While generally applied to traditional, small-scale production systems using tropical forests, extractivism is not fundamentally different from some of the schemes incorporated in outreach and ICD initiatives, discussed above, but it warrants its own treatment for two reasons. First, extractive reserves are closely associated with the demand for indigenous land tenure, social justice, and economic welfare in Amazonia.

They grew out of a complex history: in Brazil, the 1920s collapse of the rubber trade, extensive rural unrest among the erstwhile *sereingueiros* (rubber tappers) throughout the 1960s and 1970s, and state- and military-sanctioned subsequent development of the Amazon, all of which led to Chico Mendez's demand for the establishment of indigenous territories with legitimate property rights over the products of the forest (Hecht, 1985; and see chapter 6). With this history extractive reserves occupy a high-profile and often politically charged place in the conservation world, epitomizing the much sought after conjunction of conservation and human interests, and testing the assumption that native peoples are good allies for conservation (chapter 8). As such, projects like the Maya Biosphere Reserve in Guatemala (Box 10.6) have drawn much international attention. Second, in terms of sheer area, indigenous (or anthropological) reserves (falling under IUCN categories V and VI, Table 2.1) have tremendous potential for conservation, particularly in the Neotropics. The numbers are astounding. Indigenous peoples have established rights to 20 percent of Brazilian Amazon (c. 1 million km^2). This is an area five times the size of all nature reserves in the Brazilian Amazon, twice the size of California, and the largest expanse of tropical forest under any form of protection anywhere. What is more, fewer than 200,000 people inhabit these areas. Nearly half of the Colombian Amazon, about 185,000 km^2, is indigenous reserve and inhabited by around 70,000 people (Schwartzman et al., 2000a). Native peoples have made substantial gains in recent years in recognition of lands rights in Peru, Bolivia, and Ecuador, and indigenous organizations are increasingly active in Venezuela and Guyana, so the conservation significance of extractive reserves will remain high for some time.

Intuitively, the first question is whether the extracted resource can be harvested sustainably (in its ecological sense, Box 3.5), in other words, at a rate lower or the same as its natural rate of growth. One might be

Box 10.6. Extraction and the Maya Biosphere Reserve

The Maya Biosphere Reserve (MBR) sits within the Peten region of Guatemala near the borders with Mexico and Belize. Gazetted in 1990 with the dual goals of conservation and improving rural livelihoods, MBR is Mesoamerica's largest protected area complex. Within this sprawling landscape are numerous protected areas, including Laguna del Tigre, Mirador-Rio Azul, Sierra del Lacandon and Tikal National Parks, two national and cultural monuments, and adjoining MBR on the Mexican side of the border is Calakmul Biosphere Reserve. Not surprisingly, this area has received tremendous international attention, recognized as a biosphere reserve, a WHS, an "endemic bird area" (Slattersfield et al., 1998), a hotspot (Myers et al., 2000), a "last wild place" (Sanderson et al., 2002), and a Ramsar wetland of international importance, and is included in the "Global 200" ecoregions (Olson and Dinerstein, 1998). In contrast to the traditional biosphere reserve model, with a pristine core area in its center, MBR is centered on an 848,440 ha "extractive reserve" or multiple-use zone covering nearly 40 percent of the overall area. At its heart, literally and figuratively, MBR is an extractive reserve.

Despite all the international attention and tens of millions of dollars in conservation funding, the MBR still faces acute threats from colonization and attendant deforestation, particularly in younger, less traditional villages (Escamilla et al., 2000; Hayes et al., 2002; Atran et al., 2002). Many of these areas also suffer from overexploitation of NTFPs, opportunistic hunting of forest wildlife, and illegal collection of macaws, which are hugely valuable for the pet trade. To complicate matters, these illegal activities sometimes occur in association with legal "extractive" trips to the forest. Nevertheless, as Gould et al. (1998) argue, these problems are more than offset by the positive relationships fostered with communities, suggesting that such goodwill has helped catalyze new conservation activities in the area.

Has MBR fared better or worse as a result of extractive use? Are there more forests? Fewer macaws? Both? Will the

Harvesting guano (*Sabal mauritiiformis* [*Areca*]) in the Bio-itza, a self-managed indigenous reserve in the buffer zone of the Maya Biosphere Reserve.
Guano is used locally for thatching traditional lowland Maya houses and as bedding in forest temporary camps. Photograph courtesy of Scott Atran.

integrity of moderately protected areas be maintained longer than if they were strictly protected? This brief example illustrates the difficulty in answering such questions. On the one hand, there seem to be clear ecological costs associated with extractive use; on the other hand, it is unlikely that the extractive areas could have been gazetted for strict protection, so they are probably afforded a greater level of protection than they would have received otherwise. Unfortunately, protected areas do not come in replicates, so we will never know, though carefully controlled comparative studies can provide insights (see text).

tempted to think of a population (e.g., of a favored game species or a useful palm) as capital whose interest can be harvested, leaving the capital untouched. But biological "interest" is not proportionate to capital (Bennett and Robinson, 2000b). Biological populations are not like bank accounts (see chapter 3). Sustainability is therefore better thought of as harvesting below or equal to production, and production can be viewed as a function of the density of the species and the rate at which it reproduces (r_{max}). Harvest can of course equal production at many different population sizes, raising the question of at what size the population can be most productively harvested (see MSY, Box 3.2). Also discussed in chapter 3 are the problems that arise once nonequilibrial systems are acknowledged, insofar as populations may be subject to density-independent shocks such as drought, fire, and unanticipated interspecific and intraspecific dynamics. Thus, even on the African grasslands, where hugely productive herds of ungulates render a standing biomass of large mammals nearly twenty times greater than in Neotropical forests, meat production can be unsustainable; Barrett and Arcese's (1995; 1998) analyses of a hypothetical savanna extractive reserve suggest that environmental variability and human population growth can quickly overwhelm the capacity for exploitation for even such a productive ecosystem, as we saw in the discussion of ICD projects.

Using considerations introduced in chapters 3 and 5 to determine the conditions under which harvesting practices (traditional or not) are sustainable, extractive reserve managers need first to know the size of a population and its reproductive rate. With such information it appears that even the harvest of brazil nuts, Brazil's principal NTFP, is unsustainable in several areas (Peres et al., 2003). It is also critical to know the unpredictable shocks to which a population might be subjected, and perhaps most important the preferences and behavioral shifts of the human harvesters (Winterhal-

der and Lu, 1997). With respect to hunting managers will also need to take account of recent comparative studies showing that hunting pressure interacts with factors such as reserve size to affect sustainability, specifically that harvested populations in small reserves are at particularly high risk of local extinction (Peres, 2001). When these and other (as yet unidentified) parameters are unknown, it is perhaps safest for managers to suggest simple conservative rules, such as no-entry areas (see chapter 11), bans on particularly at risk species, seasons in which extraction is prohibited, or (for hunting) restrictions on taking females (FitzGibbon, 1998).

Beyond biological limits there is another key factor regarding ecological sustainability—ecological interactions. What does the sustainable harvest of one species do for other species? Here we call on the discussion in chapter 3 that good conservation entails much more than the sustainable management of a single species, and the concept of cascade effects (Terborgh, 1992; Box 5.3). In this context Redford and Feinsinger (2001) introduce the idea of a "half empty forest," and note how exploitation may be sustainable from the perspective of any one species' population, but not when one considers whether or not that species continues to fulfill its ecological role. Moegenburg and Levey (2002) have recently brought experimental evidence to bear on this issue. They looked at how bird populations are affected by a common management recommendation for extractive reserves, intensified NTFP collection. Examining açaí (an Amazonian palm species, *Euterpe oleracea*, harvested for its fruits), they found in an experimental study that when only 40 percent of the fruits were removed there was no effect on bird diversity in comparison with nonharvested areas, whereas when 75 percent were taken, fruit-eating bird diversity declined. Another common recommendation is to husband NTFP products by transplanting favored species in new areas, effectively enriching the forests in the way traditional land uses may have done for millennia (see chapter 4). This intervention was also evaluated, and again species diversity was affected, this time with nonfrugivious species less abundant in enriched than unenriched areas. A related problem is the fact that when one or a few species are formally exploited others may be damaged in the very process of exploitation, for example, by trampling or tree falls, undermining the "selective" nature of extractivism (Ochoa, 1998). To the extent that these findings can be generalized to other ecosystems and resources, they demonstrate that extraction, even if sustainable, can undermine conservation objectives. There is clearly a delicate balance in extractive systems: too much harvest or intensification and the conservation value is undermined, too little and the practice is economically not viable.

Another unpredictable but essentially destabilizing factor is commercialization. Many NTFPs are harvested for their commercial value. The dangers of the market had been noted already by the end of the 1980s, when Vasquez and Gentry (1989) observed that many species had disappeared from the Iquitos market throughout the decade. Markets are particularly dangerous where the density of the species harvested is low, where resource tenure is open-access (see chapter 6), and where the communities depending on this commercial extraction have little political power to protect the forest from other uses, as in Kalimantan (Salafsky et al., 1993). Markets also provide incentives for people living in their vicinity to convert primary forest to secondary status. Demonstrating this point, Ticktin et al. (2002) found that commercial bromeliad production was highest in secondary, rather than primary, Mexican forests, and Muniz-Miret et al. (1996) note that in areas near markets, intensified açaí cultivation in home gardens is far more lucrative than collection in semi-natural secondary forests. Distance nevertheless becomes a practically irrelevant component of commercialization once we consider the roads

built by logging operators, the impact of these roads on the emergence of a trade in bushmeat across many parts of South America, Central Africa, and Asia, and the devastating effects of this trade on natural populations (see Robinson and Bennett, 2000, and references therein; also Milner-Gulland, 2002). Tropical forests can generally sustain hunting to feed about one person per square kilometer. But once roads penetrate the forest, bringing an influx of people looking for work and money and a seemingly constant stream of logging trucks between forest concessions and urban markets, the equation changes. Hunters can enter deeper into the forest, and export larger quantities of meat to more distant markets, than was ever possible on foot. Apparently, limitless markets and easy transport effectively eliminate diminishing marginal returns, in this case, the value of the next prey item taken, since the meat can always be transported and sold (see chapter 5). Accordingly, even sustainable logging practices need to be carefully scrutinized since they may conceal a wealth of illegal trade in other commodities, thereby raising a host of new management challenges (see Box 11.3).

Finally, we need to think of the social and economic sustainability of extractive reserves. Extractivism does not have to involve local communities; we saw, for example, in chapter 1 that it was promoted as conservation by Gifford Pinchot within the bureaucracy of the U.S. Forestry Department. Increasingly, however, rights to extraction of natural resources are delegated to local communities, in wildlife (IIED, 1994), forests (Ghimire and Pimbert, 1997), and fisheries (Pomeroy et al., 2001). As might be expected from the above discussion, outcomes are highly variable. While in some cases it is clear that economic benefits are appreciated, and help to compensate somewhat for the costs of stinting (see fisheries co-management, chapter 11), extractivism does not ensure satisfaction. Particularly complicated are cases where management of protected resources is perceived as a restriction of traditional use, rather than a concession from the protected area. In Thailand, for example, many communities' "traditional rights" often included commercial as well as subsistence production (Vandergeest, 1996). In this case granting communities rights to extraction for domestic but not commercial purposes is seen as an infringement on, not an enhancement of, economic opportunity.

All in all to achieve sustainable extraction several conditions must be satisfied: (1) people must practice restraint, so that commodities can be harvested indefinitely; this may require eschewing use of modern technologies; (2) extractive reserves will have to be isolated from the rising flood of people, increased demand, and the siren call of Western-style consumerism; (3) extractive reserves will also have to be isolated from markets, or at least changing markets that can produce cheaper plantation-produced or synthetic commodities that compete with natural products (Browder, 1992; Soulé, 2000). The question then is whether nations have the domestic resources or moral authority to control internal migration, to inoculate whole regions from market forces, or to freeze rural communities in "splendid bucolic isolation" (Soulé, 2000, 60).

It is also patently clear that much of the prominent contemporary debate over extractive reserves can be traced to implicit yet often widely divergent definitions of biodiversity (see chapters 1 and 3). What are extractive reserves really for? In Schwartzman et al.'s (2000a) opinion, even if local people empty the forests of large animal populations through hunting, this is not an undesired outcome; for these authors a forest is a collection of trees that sequesters carbon, does not burn, has stable hydrology and soils, and provides a home for forest dwellers—in other words, provides basic ecosystem services. For others a well-conserved forest is one that harbors ecologically

functional populations of all species within the ecosystem (e.g., Redford, 1992), thereby providing for the conservation of the full set of species, genes, and ecological relationships among these species (in the conventional sense defined in chapter 1). Debates over extractive reserves spring from these distinct criteria of success. Biologists are disturbed by analyses that list conservation and economic development as dual objectives, but focus only on economics, as with Mahpatra and Mitchell's (1997) review of extractive reserves in India. Social scientists question conservation biologists' evidence that cascade effects can arise even from light harvesting practices (Schwartzman et al., 2000b). The debate speaks to the issue of monitoring (to which we now turn), and to the need for clearly articulated goals (chapter 11).

10.8 Monitoring and Evaluation

Many of the debates in this chapter, and indeed in this book, reduce to questions about what succeeds or fails in the practice of conservation. But how do we know whether a project, entailing a particular set of strategies, is successful? This is where monitoring and evaluation come in. Monitoring can be thought of as an ongoing appraisal of how a project is progressing, and evaluation as a more infrequent (or terminal) analysis of its overall success, but in reality both activities entail the same sets of questions: determining whether all the planned project activities were carried out, whether these activities resulted in the expected outputs, whether these outputs had the anticipated results, and whether the goals of the conservation project were met. For example, as Sutherland (2000) outlines, evaluation of a fictitious wildlife club project would entail asking whether there are now clubs in all the primary schools in a particular district, whether students have put on village drama productions addressing the local depletion of certain species used for firewood, whether women are now choosing different species to harvest as a result of these dramatic productions, and ultimately whether populations of locally depleted species are rebounding. Monitoring might also entail determining whether the project has unanticipated side effects (e.g., women with large families now purchase charcoal, illegally cut in a protected area, because they have insufficient time to go to the more distant sites where less threatened species abound). Or alternatively we might need to know whether the positive results of the project are in fact due to other factors; for example, does the regeneration of locally depleted tree species merely reflect uncharacteristically heavy rainfall in that year?

Clearly, this is a massive undertaking, entailing not only the monitoring of species, suites of species, and sometimes even broader ecological processes, but also all the contingencies that link circumstance, knowledge, and practice in what could be a hugely complex causal model. Given the enormity of this challenge, it is hardly surprising that monitoring often receives short shrift. Added to this, few projects (conservation or other) want their failures to be known, so evaluations tend to be in-house and minimal. Increasingly, however, both government and nongovernmental organizations are realizing the need to be accountable, and are struggling to develop methods that measure effectiveness. For example, Redford and Taber (2000) argue for a cultural shift in the conservation world, toward a "safe-fail" environment, where frank discussion of trial and error will lead to more innovative and adaptive conservation interventions, building on ideas of adaptive management (Box 3.4). In this section we look at developments in monitoring, first with respect to ecological monitoring, and then with the monitoring of social and economic outcomes.

Looking first on the ecological side, there has been a shift from not doing any monitoring at all, to everyone paying lip service to the critical role of monitoring, to a recognition that effective monitoring is both extremely difficult and exorbitantly expensive. The justification for not monitoring is expense, but in fact it may be even more expensive to pour resources into ineffective interventions. As Kremen et al. (1994) stress, ecological monitoring is key, both for determining the success of conservation projects and for adapting and improving management strategies. However, almost half of the thirty-six ICD projects they examined worldwide in 1994 lacked ecological monitoring altogether, and only two contained a comprehensive ecological monitoring component.

Biodiversity monitoring should ideally document the baseline conditions of both the natural area and adjacent human-dominated landscapes, and then determine changes over time in both the distribution and abundance of species. In reality, baseline data are rarely available, and data collection begins after the project has started. At this stage we can identify two kinds of monitoring, biodiversity and impact monitoring (Kremen et al., 1994). With respect to biodiversity monitoring, conservationists would like to inventory the distribution and abundance of all species in the area and, more generally, ecosystem health. Because in most cases this would be extraordinarily costly and take a very long time, they use indicators of biodiversity or surrogates of ecosystem health (see Box 3.6). Ideal candidates are those species whose populations are sensitive to disturbance (or to fluctuations in the populations of the other species), and which are easy to observe and count (Caro and O'Doherty 1999). In reality, though, even where such useful indicators exist, the research leading to their identification might be inordinately expensive, and the scope of their applicability limited to a particular site, threat, or taxonomic group (e.g., Brown, 1997; Lawton et al., 1998).

An alternative strategy suggested by Kremen, impact monitoring, is relevant where certain species are known (or hypothesized) to be at risk from human impacts, either directly or indirectly, and is therefore most useful in buffer, peripheral, or multiple-use zones, areas that lie in the front lines for both conservation and development. Using this method in Madagascar Kremen et al. (1998) measured not only the relative importance of various useful plant and animal species to the local human population (to determine which species were most at risk) but assigned species to groups of substitution products; an example might be a group of different kinds of grasses that can all be used interchangeably for thatching. Substitution groups can provide early warnings of nonsustainable use. For example, if a survey shows that nonpreferred species within the thatching substitution list are being used, this would suggest that the preferred species are declining locally and action must be taken. Similarly, it is also possible to monitor the size or age classes of exploited species (e.g., the size of a buffalo's horns, a bull elk's antlers, or the length of sawn timber). Thus, by focusing on substitution or declining quality, impact monitoring signals overharvesting, in some cases long before a change in density can be detected.

Increasingly, ecologists are recognizing the difficulty and expense entailed in a good monitoring system, even one pared down to indicator and impact species (as described above), and are looking for new tools. Salafsky and Margoluis (1999) propose a "threat reduction index." Effectively, project personnel identify threats to the biodiversity of a region, rank them, and then monitor over the years the project's progress in reducing these threats. No specific quantitative methods are used, and proponents of these methods acknowledge that evaluations of threat reduction may be biased or inaccurate. Analyses based on these "soft indicators" are nevertheless much less expensive, and can be quite useful in comparative studies

(Salafsky et al., 2001). On the other hand, many biologists remain frustrated when no quantitative measures of population numbers are available. Another useful monitoring tool, not mentioned by Kremen, recognizes that resource users spend much more time "sampling" than do managers or scientists, and that this behavior can yield important insights about the state of resources. As described in chapter 5, foragers are often cognizant of "catch per unit effort." Just as the trend in size or quality of a resource can inform monitoring, so can changes in return rates or the distances resource collectors must travel to fill their larder. Indeed, foragers and resource extractors can usefully collect the kind of data scientists need (see Box 5.5), and the significance of this is discussed in relation to citizen science in chapter 11.

This brings us to the evaluation of a project's success in securing social and economic objectives, where the principal method has been attitudinal studies. There are in fact now quite a large number of studies where researchers have visited households in areas affected by a conservation program, measured people's attitudes toward conservation with questions like "What do you think of the national park? How would you feel if it were extended? Do you receive sufficient compensation?" and so on, and then examined the causes of variability within villages or communities. What these studies show is first, great variability among sites. Holmes's (2003b) review shows that in eight of eighteen cases (in Bolivia, Nepal, Uganda, Cameroon, Tanzania, Mozambique, and South Africa) people living adjacent to protected areas hold generally positive attitudes and are opposed to the abolition, or agricultural conversion, of parks; reasons commonly nominated are watershed protection, the generation of foreign exchange, park outreach programs, and local climatic stabilization (see also Newmark and Hough, 2000). Positive attitudes are commonest among individuals who live close to the protected areas, who have rights of access to park resources, who are educated or literate, or who are wealthy (Abbot et al., 2001). In seven of the eighteen cases reviewed by Holmes, negative attitudes predominated, sometimes even despite the receipt of tangible benefits from the project, as in Nepal's Koshi Tappu Wildlife Reserve (Heinen, 1993); restrictions on grazing and fuelwood collection, wildlife damage to crops and livestock, unpleasant contacts with park personnel, a feeling of inadequate compensation, limited services, and dashed expectations were common candidate reasons in this category. In the three remaining cases attitudes were very mixed. Ambiguity arises sometimes from a discrepancy in views toward different parts of the project (e.g., popular community forestry and unpopular wildlife protection schemes; e.g., Mehta and Kellert, 1998), and sometimes from a greater fondness for the abstract idea of conservation than the reality of nasty encounters with game scouts. Describing such dynamics, and there is of course great variability between sites, is extremely important because only with an understanding of the differing views among various stakeholders and intended beneficiaries can effective conservation compromises be reached. Effectively, these attitudinal surveys can serve as a baseline from which to both guide management decisions and determine the effectiveness of management actions (Jacobson, 1995).

Attitudinal studies are interesting evaluative tools because they get at the interstices between knowledge, values, and (ultimately) behavior. As such, they permit measurement of the effects of conservation education campaigns and incentive programs (such as those described for outreach and ICD projects) on behavior, and hence conservation outcomes. With attitudinal studies the core assumption to most modern conservation projects—that development activities will affect local peoples' views toward conservation in ways that make them more supportive of conservation measures regulating resource use and thus more likely to stint—can be evaluated. Insofar as restraint is gen-

erally not in an individual's self-interest (see chapter 5), such attitudinal shifts are critical to a project's ultimate success in conserving biodiversity, as we also pointed out in the earlier section on outreach and education.

There are four critical steps whereby attitudinal studies can demonstrate such links: (1) that the people or communities primarily targeted by conservation and development projects are more conservation-minded than those given less attention; (2) that projects demonstrably shift people's attitudes toward conservation over time; (3) that what is known as "linkage" occurs, in other words, that people change their attitudes toward conservation precisely because of their recognition that the economic benefits of a project are contingent on their restraint; (4) that changes in behavior and practice ensue. Let us look at each of these steps in turn.

First, are people who are receiving benefits from a project more conservation-minded than those who receive fewer or no benefits? There is considerable evidence from different parts of the world that this is the case, whether benefits are measured in terms of legitimate access rights to the park's resources or receipt of outreach services, as reviewed above (Newmark and Hough, 2000; Abbot et al., 2001; Holmes, 2003b); causality is difficult to determine, but it could be that projects are more likely to be set up in conservation-minded communities. Turning to the second step, and effectively from a cross-sectional to a longitudinal perspective, Abbot et al. (2001) explored attitudinal changes over time in response to the establishment of an ICD project at the Kilum-Ijim Forest Project in Cameroon (albeit with retrospective data) and found increased support of forest conservation since the project's inception (as reported earlier in this chapter), contingent primarily on an increased water supply.

Evidence for the third question, does linkage occur, is harder to pin down. The key idea here is that if there is no explicit recognition that the benefits received from a project are contingent on restraint, none of the approaches reviewed above will be sustainable; resentment will persist and infringement will continue unabated, irrespective of the services received. Such linkage is difficult to prove. In evaluating four different ICD sites in Madagascar, Marcus showed that villagers who feel these projects are helping them are more likely to have a positive view of protectionism than those who do not see this help (a positive score for step 1, above). Closer investigation, however, through focus group discussion in local communities, suggests that while people appreciate the odd windfall to their village—a well, a doctor, or a school—they do not (somewhat surprisingly) link the projects' more substantive agricultural and training programs to the existence of a national park in their area. In Marcus's opinion, if people only associate gifts with the park, and not broader regional development, the logic and efficiency of an ICD approach is severely diluted. A similarly puzzling situation exists at an ecotourism project linked to the conservation of Komodo dragons (*Varanus komodoensis*) in Indonesia's Komodo National Park; even though villagers with positive attitudes to tourism support conservation (again in line with the first logical step posited above), those whose actually earn their income from tourism are negatively disposed to conservation (Walpole and Goodwin, 2001). This rather surprising result suggests not only that linkage is absent, but that the project has generated some unanticipated dynamics; possibly there are tensions between conservation authorities and some of the less responsible elements in the tourist industry, reminiscent of the Galápagos story told earlier in the chapter.

Finally, there is the tricky issue of whether attitudinal changes affect behavioral outcomes. Most attitudinal studies to date fail to consider associations between reported attitudes and what people actually do. There are, however, some positive indications that attitudinal changes precipitate behavioral change. In a study that looked at how perceptions of an outreach program at Tanzania's Katavi National Park are linked to sustainable patterns of wood use, Holmes (2003b) found that households more interested in seeing the park degazetted exhibited less sustainable wood extraction methods than households supportive of the park. Such correlative studies at the individual level are interesting as regards linkage but, as Holmes notes, they still raise issues of causality. Similarly concerned with the effects of attitudes on behavior, Adams and Infield (2001) examined a community conservation project in Uganda's Mgahinga Gorilla National Park that entails education, benefit-sharing, the funding of the community project, and some extractive use. They found a marked reduction in the rate of illegal entry and resource use by local Ugandans since the project's inception, but as the authors note, such data are also open to various interpretations. Numbers of arrests or snares removed could reflect three things: the deterrent effect of stronger patrolling, lackadaisical policing (depressing figures for arrests not illegal behavior), or local people's acquiescence to the new conservation measures. The last interpretation, which would support the claim that attitudes affect behavior, receives some, albeit weak, support from participatory research (see chapter 11), where villagers were asked to draw Venn diagrams of how the various institutions associated with the park and conservation program were linked to the village, as helpers or hindrances. On the whole, villagers saw the conservation program as a helping influence (though there was variation between villages), and this viewpoint was linked to an appreciation of the economic benefits of being a park neighbor.

For the most part we can conclude that the effects of attitudinal changes on behavior are only weakly attested to by empirical data. With so little data to adjudicate on this issue, some investigators simply continue to assume that attitudes are a useful surrogate for behavior while others emphasize the need to study (and change) behavior rather than attitudes. In the latter camp, and on the basis of his experiences in participatory rural development, Chambers (1997, 233) argues "it is easier, quicker and less dominating to provide opportunities for new behavior and experiences, than frontally changing belief systems." While it is possible that changes in behavior subsequently influence beliefs and attitudes, this could be a dangerous strategy given that conservation cannot always be made to yield economic payoffs (chapter 9).

In short, attitudinal studies are a useful tool in evaluating the success of the development aspects of a conservation project, and hence its social sustainability over the long term. Such surveys usefully identify who does (and who does not) benefit from a particular strategy, and thereby serve to pinpoint effective and ineffective approaches. What they do not tell us much about yet is whether development and conservation are linked in people's minds, or how well attitudes map onto behavior, and hence measurable conservation outcomes.

Monitoring and evaluation are much more straightforward if the goals of conservation are clearly defined (Salafsky et al., 2002). Success cannot be claimed if there are no concrete objectives, or if these goals are not shared. For instance, earlier in this chapter we saw that national parks do a good job conserving various assemblages of mammals and vegetation in Tanzania (Caro et al., 1998; Caro, 1999; Pelkey et al., 2000), while Schroeder (2002) claims that the Tanzania National Park's outreach programs are wholly inadequate, at least in comparison to the participatory programs set up by private safari operators in the less heavily protected areas. Schroeder is not reversing the Tanzanian story, but rather he has broadened the currency of success from technical measures of biodiversity to the equitable sharing of wildlife benefits (an indicator of social and environmental justice; see chapter 7). The debate over extractive reserves discussed earlier betrays the simi-

larly unshared criteria of success. We discuss the specification of goals and targets in the next chapter.

10.9 Conclusion

This chapter looked at various conservation strategies ordered loosely in terms of their emphasis on protectionism and utilization. Strict protectionism may be the only solution for certain critically threatened resources or ecosystems, but can usually be made more socially sustainable through outreach work of various kinds. Conservation education can be phenomenally effective, but needs to be combined with other programs that improve local livelihoods and guarantee environmental justice. Ecotourism has saved some areas from development, and has brought economic benefits to communities that act as stewards of these resources, but incurs its own risks. If few tourists come there is too little cash generated to improve significantly the livelihoods of many people; alternatively, if the site becomes an international magnet, ecological damage and political maneuvering can be devastating. Integrating conservation and development, the great hope of the 1980s, is now seen as having far narrower application than once dreamed of; even on the African savannas, with their great potential for linking large mammal populations to human livelihoods through sustainable offtake, the prognosis is poor. Extractive reserves designed on a similar logic face a delicate balance between maximizing resource values and undermining biodiversity value. Whether the consensus for any particular site lies in pure protectionism, outreach, ecotourism, integrated conservation and development, or extractive reserves, none of these strategies will work without national enabling legislation, or without international treaties and support (chapter 9), whether these address migration, commercial operators, taxation, property rights, or international trade.

There are two outstanding issues. First,

given that each approach has its own pitfalls and problems, conservation projects must combine elements of different approaches. Second, as we proceeded through the different strategies, capacity-building becomes ever more critical. It is one thing to advocate ecotourism (or any other strategy), but if nobody in the local community knows accounting, how to census animals, or even how to drive, these strategies are untenable in the short term and unsustainable in the long. In the following, and final, chapter, we look at some of these novel combinations of ideas and also initiatives that involve local people in the science and management of a project.

CHAPTER 11

Red Flags: Still Seeing Things in Black and White?

11.1 Introduction

Conservation is all about choices, and in the opening chapter we showed how these choices reflect ethical judgment. Such judgment structures distinct rationales for conserving biodiversity, particularly emphasis on the intrinsic versus the instrumental value of nature, and divide self-proclaimed conservationists into often mutually suspicious factions, with the former dismissing the latter as shallow materialists bowing to political expediency and the latter viewing the former as unrealistic, unscientific, and sometimes too radical in their methods. Though ethical conviction spurs energy, ardor, and commitment in conservationists of every hue, we see also inherent dangers: the tendency to conflate the objectives and goals of conservation with its means, and the tendency for practitioners on both sides of debates to see everything in black and white. In this final chapter we break away from the polarized viewpoints that characterize conflicts in the world of conservation, its literature and practice, and we look to novel perspectives and experiments that fracture conventional alignments of tools and ideas. We argue that there are no monolithic "right" answers, and even if they did exist, conservation in practice is shaped by the realities and constraints of particular places, people, and ecosystems, rather than theoretical assertions, no matter how right these might be. Solutions are needed that combine centralization with decentralization, protectionism with utilization, NGO activities with state policy changes, and professionals with inspired gifted amateurs, solutions that combine elements from different ideological and disciplinary perspectives.

We start out by reviewing briefly the major polarities that divide contemporary conservationists, effectively providing a summary of principal themes within each of the preceding chapters and their links (11.2). From there, we explore some novel and integrative solutions emerging in the conservation world (11.3) and use these examples as a foundation for a more general discussion of scale (11.4), of the need to specify clearly the goals and targets of a conservation project, and of the importance of keeping these separate from the means of achieving these goals (11.5). We conclude with a call for more interdisciplinary training, research, and action (11.6).

11.2 No Development without Tears, and Other Debates

The challenge to protect nature raises numerous questions that have structured preceding chapters. Most centrally there is the debate between conservation and development. In chapter 2 we saw how sharp attacks on preservationism over the last thirty years have swung the pendulum back again toward utilitarianism. Given the intensity of threats to biodiversity, escalating global inequalities between rich and poor nations, and the growing influence of a huge bureaucracy of government departments and NGOs on environmental policy and practice, the old intrinsic/instrumental divide has become dressed up in development politics and ideology. At the center of the debate is the relationship between conservation and development, founded on the unsteady fulcrum of sustainable development with its conjoint objectives of protecting the environment at the same time as alleviating poverty, building capacity, and instilling social justice and empowerment. There may be intellectual justification to the assumption underlying sustainable development, that just as development depends on the conservation of the environment so conservation depends on development, but many conservation practitioners emphasize the difficulties in striving simultaneously for two such antithetical ends. Conservationists prefer to highlight the inevitable trade-offs that must be made depending on whether conservation or development is the primary objective, echoing the sentiment of a colonial administrator in Tanzania as he contemplated delivery of an eviction order in the name of economic advancement—"development will not take place without tears" (Neumann, 1998, 70).

While it has been fashionable among conservationists to stress this inherent contradiction, not all conservation-development conflicts are necessarily zero-sum games (situations where parties gain only from the losses of their opponents). In other words, sustainable development should not be abandoned as a goal simply because it cannot substitute as a goal in many of the cases where biodiversity is critically under threat. A very large-scale and quite successful project in this respect, combining economic development with the eradication of invasive exotic vegetation, comes from South Africa (Box 11.1). Even in cases where development does not contribute directly to conser-

vation outcomes, it still has a role to play in building a constituency for conservation. Recognizing this, many people are unwilling to give up on the conservation-development alliance, despite its difficulties. Indeed, although the much maligned World Summit on Sustainable Development, held in Johannesburg in 2002, produced only a toothless international agreement and a very small commitment of funds, it does seem to be inspiring a number of international scientific organizations to fund research centers in Asia, the Pacific, Africa, Latin America, and the Caribbean to promote the best science for sustainable development. The overall lesson here then is that despite the inherent contradictions in combining conservation and development, the goal must still be to seek development alternatives that minimize environmental degradation and conservation solutions that minimize economic losses. In short, we argue not for the elusive "win-win" situation (even though such cases can exist, as in the South African water project), but a strategically calculated set of trade-offs that can be designed across a landscape scale (see below).

While the conservation-development debate took hold in policy circles, changes were occurring in the discipline of ecology, with the recognition of the complexity of ecosystems (chapter 3). The (relatively) simple solutions from equilibrial models can no longer be called on in the design of healthy ecosystems: accordingly, policies such as destocking are being examined more critically, the value judgments for prioritizing conservation actions are becoming more explicit, more flexibility is now built into management plans, and broader regional factors (both ecological and economic) are brought into conservation planning, directing greater attention to the neighbors of protected areas.

But how local communities should be incorporated into conservation strategies is still at issue. Debates within anthropology over the last decade have served to dispatch the excessively romantic concept of an eco-

Box 11.1. Working for Water in South Africa

Post-apartheid South Africa faces a serious water supply crisis as demand increases from rural areas and rapidly growing cities. New dams are controversial and unlikely to satisfy demand. Intriguingly, it is the alien species (such as pines from Europe and wattles from Australia, which have invaded much of the country) that lie behind this water crisis. Invasive exotics use 3.3 billion m^3 of water in excess of that used by indigenous vegetation every year, almost 7 percent of the runoff of the country. In the Western Cape fynbos biome, an area designated as a biological hotspot, estimates suggest that within the next 100 years invasive aliens will result in the loss of more than 30 percent of the water supply to the city of Cape Town (Le Maitre et al., 1996). Exotics also threaten biodiversity, constitute a serious fire hazard, and aggravate erosion.

To meet this water crisis, an ambitious Working for Water program was initiated in 1995 to involve local communities in the destruction of invasive alien species across the vast tracts of the country, and in so doing to improve the water supply (Binns et al., 2001). But the program is much more than an invasive eradication scheme. Launched, funded, and administered by twelve different government ministries, it has been designed to play a major role in alleviating poverty, creating jobs, and empowering locals, at the same time as improving ecological integrity, protecting biodiversity, and counteracting fire, erosion, and flooding. Capitalizing on the heavy work entailed in clearing alien vegetation, the program directly addresses poverty, since anyone living in a catchment area targeted for alien clearance can apply for contract work for teams of up to twenty workers. Equity is encouraged by ensuring 60 percent of the workers are women, 20 percent youth, and 2 percent disabled, and by making childcare available. Social sustainability is promoted through providing vocational training, environmental education, primary health care and HIV-AIDS awareness, and empowerment. Gradually the program is devolving to local hands; in the words of one coordinator, "Now people are starting to own the project . . . [they] are no longer passive, they are decision-makers" (quoted in Binns et al., 2001, 351). Finally, on the ecological front, there is considerable evidence in some areas that water supplies are recovering.

The project does not work well everywhere. There are instances of corruption and of crippling bureaucratic delays, for example, where recently cleared areas are denied timely follow-up and have to be cleared from dense regrowth all over again. There are also serious issues concerning sustainability, both ecological and social. Regarding the former, clearance of large areas of forest can easily lead to increased erosion and silting of water courses and reservoirs. Although the program aims to reseed with indigenous species soon after clearance, any delay opens further the window of risk. In addition, the broader microclimatic effects of cutting pine and wattle forests are also as yet undetermined. On the social side, too, there are sustainability issues. Insofar as the very goal of the program is to eradicate invasive aliens, at some time in the future, if the project succeeds, there will be no aliens left to clear. What will people trained primarily in the secondary industries associated with wood extraction do once the glut of raw material is exhausted? Follow-up and more extensive training will be necessary, although this does not lie within the remit of the program.

logically noble savage (chapter 4), since archaeological and ethnographic data show that humans have always had significant impacts on their environments. The extent to which these effects are countered by the role local communities can play as sustainable harvesters of resources, or ardent defenders of their traditional homelands, is still debated. Emphasis on conservation benefits from lands under the management of traditional peoples reaches its extreme formulation in a recent IWGIA statement where Gray et al. (1998) ask why, if conservation organizations now recognize indigenous territories, it is still necessary to have protected areas. On the other side, some biologists find conservation based on indigenous land management disingenuous, arguing either that it is irresponsible to expect resource users to be resource conservers (the idea of putting the fox in charge of the henhouse) or that it is inappropriate to expect relatively powerless people to stop the economically and politically powerful forces behind deforestation. As biocentrics, Noss and Cooperider (1994) make the extraordinary claim that they are only interested in the survival of those cultures that are compatible with conservation of biodiversity, whereas advocates of anthropocentricism (Alcorn, 1993) argue that conserving biodiversity can only occur in collaboration with local peoples. Clearly, there is a wide spectrum of opinion.

Humans have always had significant impacts on their environments, and this raises the possibility that human presence may in some cases be critical to the structure of ecosystems designated for conservation, at least if they are to remain in their present state. If this is the case, it will be necessary to incorporate historical and indigenous land use patterns into management plans, for example, by allowing burning or grazing at certain times of year. To some extent this is a problem, since there are numerous studies, primarily from the Neotropics, which show that hunters' behavior is driven more by efficiency than conservation considerations. Though human impacts on plant species diversity are less well understood, low population densities and limited extractive technology, not conservation ethics, keep rural communities from depleting many of their natural resources (chapter 5). Understanding the importance of time discounting and property tenure in relation to a particular set of resources is key to predicting the effects of shifting management to a community-based structure; certainly, such a shift by no means guarantees successful conservation. This presents an awkward trade-off to managers of anthropogenic landscapes—exclude people from an area slated for pure protectionism only then to mimic their anthropogenic effects through management regimes. But what is the alternative? While it is more politically satisfying to restore an ecological system with its human component intact, in some cases the price may be too high. Almost every human community lives in the ambit of growing markets and affluence norms, and there is no guarantee that people will exercise restraint in their harvest and use of nature.

Which brings us to the issue of community regulatory institutions. Among scholars the challenge of conservation and resource management is often understood as a dilemma of collective action. Remarkable advances over the last twenty years in our understanding of local institutions and their role in contributing to sustainable natural resource management (chapter 6) have both stimulated and been buttressed by the policy shift toward decentralization. Decentralization is based on the conviction that local bodies will be more responsive to conservation initiatives than national governments. Local communities are thought to have greater knowledge of the intricate dynamics of their natural resource base than outsiders, and also to have more incentive to sustain it over time; they are also presumed to have lower discount rates than, for example, a commercial intruder. There are other assumptions behind this conviction (Wyckoff-Baird et al., 2000): first, that the devolution of authority, responsibility, and funding to local institutions and organizations gives greater power over natural resource management to those most in contact with these resources; and second, that these people have the wish to both promote conservation and reduce threats to biodiversity. As we already saw in chapters 4 and 5, many of these views have only limited validity. Decentralization in itself probably does little to ensure positive conservation outcomes. Rather, its success depends largely on the nature of the institutions to which power is devolved and the conditions antecedent to the decentralization process. All this points, unsurprisingly, to the fact that policy initiatives, such as decentralization, do not obliterate history, bringing us to the broader frame of political ecology.

In theory, political ecology provides a useful framework for looking at the less proximal causes of environmental degradation (chapter 7). It asks who has the power to pollute a lake, to fell a forest, or to restrict resource harvesting at a given locale, seeking answers in history, politics, and economics, at least insofar as these are played out at the local level. It also probes explanations that attribute biodiversity loss to such gross concepts as population pressure, exploring instead the more precise mechanisms whereby population does (or in some cases does not) occasion undesirable environmental change. These insights are intensely valuable insofar as they can suggest more specific avenues for policy intervention. They also shed light on how various actors competing for a natural resource construe the situation, again with practical implications. However, as the scale of political ecological analysis expands, its focus becomes diminishingly ecological; perhaps for that reason it has been largely ignored by conservation biologists. While political ecologists, for the most part, fail to examine ecological and conservation outcomes in the quantitative comparative manner conservation biologists seek, conservation biologists themselves rarely take full consideration of the historical, political, and social contexts of the problem

they address. The scope for productive collaboration here is enormous. A model would seem to lie in ecologists' and economists' studies that examine conservation options within broader national, even international, economic frameworks, such as for Kenya (Box 9.1). Political ecology could add the missing political and social dimensions to such analyses.

A political ecological perspective on the alliances between nongovernmental conservation organizations and local communities (chapter 8) reveals both the potential and pitfalls for such initiatives. One issue that becomes clear from reviewing experiences in this area is that often the safest approach lies in working with local communities to protect traditionally owned land and water from unrestrained infusions of the market and immigrant settlers; in some cases there may be options for helping such communities defend locally established protected areas, or for compensating them directly for their setting aside of conservation areas (see below). Here the interests of the conservationists and the local communities largely overlap. More risky are the strategies that seek to offer economic incentives for saving nature, such as pharmaceutical prospecting and green marketing. Apart from the difficulties in regulating the often insatiable appetite of international markets as they creep into communities that are marginal even to their own nation's economy, introducing conservation projects as income-generating projects opens the possibility for bitter conflict. This is not to say such projects are doomed, only that they must be implemented with enormous care.

Once conservation problems are seen in their broader political context, the critical role of international markets and international regulations comes to the fore. In chapter 9 we saw how business corporations, many of them with budgets larger than those of individual nations, are necessary partners in conservation efforts, and yet how difficult these are to regulate. Clearly, we are at a critical juncture in history, with respect to bringing nations and multinational corporations within a global regulatory framework, whether this is over the control of water, fish, carbon stores, or biodiversity. Similar battles are simultaneously being fought in the international realm over trade, the protection of human rights, and the right to use military force.

As we have seen again and again in this book, promoters of different conservation agendas tend to define their approach in opposition to the other's perceived failure. This style of argument may be arresting, but encourages an unnecessary polarization of views among conservationists. It obscures the considerable areas of agreement between protagonists, accentuates differences, serves to conceal the fact that rarely can any single approach work in isolation, and promotes the idea that there is a single "best" approach to any problem. In short, it restricts vision and curbs imagination. In the next lengthy section we pick up the theme of chapter 10, which looked at more conventional conservation solutions, and highlight novel approaches that we believe are necessary for conservation to be successful in the multifaceted, idiosyncratic, and complex landscapes of the twenty-first century. These cases form the backdrop for our return to the bigger picture at the end of the chapter.

11.3 Integrative and Novel Solutions

Just as parks cannot carry the entire burden for biodiversity conservation, local people cannot take the sole responsibility for political viability of protected areas (Brandon, 1998). Given these realities, many have recognized the need for conservationists to engage a multistranded approach, combining different elements of different strategies to best suit the local conditions and the specific challenge. Here we explore what we see as exciting developments, particularly insofar as they break down the old tendencies of seeing everything in black and white, categorical thinking that waves red flags in the

face of conservationists and affected populations alike. Some combine traditionally separate approaches to form a more comprehensive strategy for conservation, others provide novel ways of implementing conservation actions, and still others mark a revival of older methods for conservation long presumed dead. What links these cases is their innovative nature and a willingness to craft new, locally appropriate solutions to conservation problems.

Community-Based Protected Areas

Local groups are beginning to take the initiative in establishing protected areas. In an analysis of the sustainability of different species hunted for both commercial and subsistence purposes in lowland Peru, Bodmer (2000) shows how indispensable unhunted source populations are to the maintenance of productivity from areas close to roads and villages, areas that are clearly (in Bodmer's analyses, see Table 4.2) becoming depleted of wildlife. These set-asides of nonhunted areas serve as source populations that can replenish adjacent areas that are heavily hunted (Novaro, 1995; and see chapter 5). As such, communities can ensure within their territories the sustainability of hunting (and other forms of extraction) for subsistence and monetary needs. Bodmer (2000, 288) sees much potential for community-based protectionism in the Amazon, suggesting that "local communities . . . in the vast expanses of western Amazonia naturally recognize the value of setting aside non-hunted areas as source populations," and presents evidence of a successful case in Peru where the community established the Reserva Comunal Tamshiyacu-Tahuayo, which contains its own fully protected zone (Bodmer and Puertas, 2000). This strategy offers the advantage of being grassroots, of providing full protection, and of being much less expensive than state-run protectionist programs. Furthermore, by some estimates nonhunted areas set aside by local communities may allow the 10 percent goal of global protection to be reached.

There are some budding examples of community-based protected areas that at least approach the ideal in some respects. The case from Madre de Dios, Peru, is instructive (Cincchón, 2000). In 1990 the government created the Tambopata Candamo Reserved Zone from 1.5 million ha of pristine rainforest and rural landholdings. Initially, local people reacted negatively, primarily because they had not been consulted and feared loss of their lands. However, over the following two years local communities were drawn into a planning process, and they ended up zoning a people-free core protected area (effectively a park) in the center of the reserved zone. According to Cincchón (2000, 1368), the "concept that a large portion of land should be set aside to maintain key ecological processes was debated and accepted among the local population, whose main sources of food originate in the natural forests." Later these initiatives were wrecked by the opening up of the Peruvian Amazon for oil and gas exploration, and the contracting of an oil block in the proposed national park. The leader of the families to have negotiated the Madre de Dios plan was among the first to object: "It has been hard work to conserve this region. We have asked miners to leave, we have denounced illegal logging and we also have refrained from logging in order to conserve this region" (Victor Zambrano, cited in Cincchón, 2000, 1369). In 1996 the government established the Bahuaja Sonene National Park to compensate the local communities for their loss of the original park, and under Peruvian law national park lands enjoy the highest degree of protection from oil development. Clearly, in this case local people were instrumental in effecting national park designation because it was in their own interests, both to secure their supply of forest foods and to protect their indigenous lands, although the outcome was not at all as they had anticipated.

Another example is the Mamirauá Reserve in Brazil's western Amazon. Here, although the impetus was again initially external, the idea of setting up a reserve was inspired by the existence of local protectionist institutions (Box 11.2). Founded on a history of communities flagging their lakes against the entry of outsiders' fishing vessels, this project has succeeded in drawing local communities into accepting total no-entry zones that function as reserves for their own fish stocks.

With the support of environmental NGOs, community-based protected areas such as those being experimented with in Peru and Brazil may yield valuable conservation benefits. As emphasized by Bodmer and Puertas (2000), success will depend on local communities availing themselves of the technical assistance and extension services of outside experts, with respect to ensuring both the sustainability of harvest and the health of no-take zones; locals will need training in these skills (see below) insofar as the scale of a zoned regional conservation initiative extends far beyond the ambit of traditional ecological knowledge. The strategy is most suited to areas not impacted by high population growth or strong migration, and dovetails closely with the move toward indigenous sovereignty and territorial integrity, as discussed for the Kuna and Miskito in chapter 8. Ultimately, however, the application may be wider and reach beyond remote, sparsely populated tropical forests. For instance, locally established no-entry areas can be extended to marine reserves, where there is now considerable evidence that well-monitored no-take zones can enhance overall fishing success, thereby compensating for the harvest forgone within the no-take zone (Roberts and Hawkins, 2000; and see Box 11.5). Thus, in the Soufrière Marine Management Area on the Caribbean Island of St. Lucia, within three years of the designation of a protected area, biomass tripled within the no-take zone and doubled in the fishing areas adjacent to the no-take zone;

Box 11.2. Flagging the *Várzea* in Brazil's Mamirauá Reserve

In a remote area of the western Brazilian Amazon, the Mamirauá Ecological Station (MES) was set up in 1990. The area is inhabited by local communities who are dependent on a highly productive fishery, abundant flora and fauna, timber extraction, and subsistence agriculture in the seasonally flooded forest (or *várzea*; Röper, 2000), and is of considerable commercial interest to outside logging and fishing concerns based in nearby towns like Tefé. *Várzea* water levels rise 12 m annually with the white water Andean runoff, and being full of sediment harbor enormous biodiversity.

Recognizing the importance of the area to local communities for both subsistence and commerce, the Brazilian ecologist who founded the research station began working toward a reclassification of a 1.2 million ha area as a sustainable development reserve (SDR; Ayres et al., 1999). This new category, created by the State of Amazonas for this particular project, allows for the coexistence of the human population with the protection of local biodiversity, and has three objectives: preservation of natural heritage, research on biodiversity, and promotion of sustainable development to combat poverty (Lima, 1999). Between 1992 and 1996 a draft management plan was developed through negotiations between residents' associations and environmental authorities, specifying various levels of lake zoning for different purposes, for example, preservation lakes, maintenance lakes, and commercialization lakes, with two entirely protected zones at the center of the reserve. A critical development was the speed with which fish stocks recovered, giving communities quick, economically valuable positive feedback. User groups (both residents with subsistence and commercial needs and nonresidents with commercial interests in the reserve) participated in the developments at all stages, from planning to voting. A negotiated compromise among families and communities with different economic interests was reached (Lima, 1999; Gillingham, 2001), consistent with the goal of avoiding environmental sectarianism (Ayres et al., 1999).

So in what sense are the zones set aside for the total protection of lakes and forests an example of a community-based protected area? Certainly the idea of the SDR did not originate locally. It was Ayres, the biologist who founded MES, who saw the potential for protectionism in what he believed to be a centuries-old practice of *várzea* inhabitants flagging their lakes against the entry of outsiders. While little is known of the customs of early *várzea* settlers, it is clear that after the rubber boom, when patrons (local bosses) controlled the mouths of the lakes, the current system of informal territories came into being, with local settlements controlling their adjacent lakes and forests. With the growth of Amazonian towns in the 1960s and the increasing demand for fish, this system came under strain. The Catholic Diocese of Tefé intervened with programs that not only installed a system of democratic political representation for *vargeiros* (floodplain inhabitants) but formally institutionalized "procreation" and "maintenance" lakes (Lima, 1999). However, it was only with the Mamirauá SDR that these practices became legally enforceable; policing is done locally, backed up by the national environmental agency based at Tefé.

(continued)

(Box 11.2 continued)

Protected area management therefore springs from a history and an ecology dictating territorial access. Mamirauá SDR has become a national showcase project, and lake protection now occurs in most of Brazil's *várzea*. The project remains under intense ecological study; there has been a two-thirds drop in the amount of fish sold in Tefé from Mamiruá and a reduction in the hunting of caiman (*Caiman crocodilus* and *Melanosuchus niger*), manatee (*Trichechus inunguis*), and endangered Amazonian turtles (*Podocnemis* spp.), although timber is proving harder to control (Ayres et al., 1999). Additionally, local community members are involved in participatory action research (see later in this chapter).

additionally, biodiversity increased 20 percent in the entire management area. Other marine examples come from South Africa, Kenya, Tasmania, and Australia. Though such schemes are not always initiated by the local community (though see the case for Thailand, discussed below; Johnson, 2001), the favorable effects of total protection on the economic potential of local fisheries makes the strategy very attractive.

Ultimately, community-based protected areas can become institutionalized. Institutionalization can favor conservation interests, as we saw with the sustainable development reserve in Mamirauá, or broader political goals. As evidence of the latter, in Australia an Indigenous Protected Area (IPA) program, initiated in the mid 1990s, focuses more on social justice than conservation, and endorses the fact that Aboriginal people will not necessarily value conservation over commercial exploitation (Muller, 2003).

Dealing with Commerce

As the previous cases illustrate, significant conservation benefits can be gained by working with local communities that control large and significant areas of land or water. But what happens when these areas are controlled, at least in a de facto sense, by corporations with large logging concerns

on which local communities are heavily dependent for wages? This may seem an intractable situation, but some novel ideas for thinking "outside the box" (or, more literally, outside the park) can have big payoffs for conservation.

Thinking along the lines of an extractive reserve (chapter 10), logging companies can, in principle, run sustainable enterprises, at least where regeneration rates are relatively rapid. This is because logging for tropical hardwoods is distinctly different from temperate logging of softwood species. Only one to three trees are taken from each hectare. Though this selective logging has significant effects on the botanical structure and composition of forests, a case can be made for using some forests in this way, following the "use it or lose it logic" of chapter 9, particularly when such an enterprise is well regulated and receives green certification (see Box 9.2). Unfortunately, however, there is an unanticipated consequence to tropical forest logging, emerging with a vengeance in remote forests across the world (Rao and McGowan, 2002). With an influx of people, an associated demand for food under circumstances where agricultural production is limited or impossible, and logging trucks moving between forest concessions and large urban markets, all the pieces are in place for a lucrative bushmeat trade (see chapter 10) and an eventual wild-

TABLE 11.1
Multiple Effects of Production on Biodiversity[a]

| | Effect(s) on Biodiversity | |
	Obligate	Nonobligate
Direct effect	Very difficult to reconcile conflicts (A)	May be possible to mitigate the effect if production practices can be changed (C)
Indirect effect	May be possible to mitigate the effect somewhere in the indirect pathway (B)	More tractable problems because production methods can be changed, or the indirect pathway can be disrupted (D)

[a] Some examples from a hypothetical harvesting of parrots for the pet trade illustrate the differences between the cells. If the taking of parrots directly threatens the population's chances for survival, and the parrots themselves are targets of the harvest (i.e., they are an obligate part of production), then there are few options other than to reduce or eliminate the harvest (A). Contrast this situation with one where the parrot harvest itself is sustainable (as far as the parrots are concerned), but the harvest has an indirect effect on nest-cavity breeding species because the reduced parrot population is excavating too few nest holes to provide for other cavity-nesting species (B). This conflict can be dealt with by establishing next boxes for the cavity-nesting species. Another tractable situation might arise if hunters engaging in an otherwise sustainable parrot harvest were felling trees at an unsustainable rate to gain access to parrot nests. In this case, felling trees is not an obligate part of production, since the harvesters could climb the trees (C). Changing production practices by providing ropes and tree climbing gear could solve the problem. The final and most tractable scenario is when a nonobligate practice has an indirect negative effect (D). If tree felling for parrots eliminated the number of nest holes for other species, we could either change production (stopping the felling of trees) or mitigate the indirect effect (nest boxes).
    Source: With thanks to David Wilkie who proposed this table.

life disaster. Unsustainable hunting has the potential not only of emptying logging concessions of their wildlife, but of siphoning off many species from adjacent unhunted areas, no matter what their level of protection. As such, the impact of unregulated hunting associated with logging far outweighs the direct consequences of selectively cutting down valued hardwood species.

The situation can be portrayed in a 2 × 2 matrix (Table 11.1) illustrating the four possible combinations of obligate/nonobligate and direct/indirect effects of harvesting on biodiversity, and presents a starting point for reconciling conservation and human use. Looking at the four quadrants, the upper left, where an essential part of the production process has a direct impact on a conservation target (a species, or an ecological community), there are few options for reconciling the conflict other than giving up on production or the conservation target.

But in the other cells new options emerge, some linked directly to changing production practices and others to averting the ancillary effects associated with production. It is here that the Wildlife Conservation Society (WCS) took advantage of a problem falling in the nonobligate, indirect cell (Box 11.3). By restricting the export of bushmeat from logging concessions, managing the spatial distribution of legal hunting, and developing other sources of protein, the WCS and the local logging company, Congolaise Industrielle du Bois, have improved the situation for northern Congo's wildlife, which is partially protected in Nouabalé Ndoki National Park but ranges into adjacent logging concessions. In fact, some species like gorillas, bongo, and elephant may now actually benefit from logging because it opens the forest canopy and allows more light to penetrate to the forest floor to produce herbaceous vegetation, a key forage resource.

Box 11.3. Even in the Last Place on Earth: Exploitation and Conservation in Nouabalé Ndoki

The forests of the northern Congo Republic cover approximately 30,000 km² and harbor intact populations of many large mammals (Blake et al., 1995; Fay and Agnagna, 1991), including important populations of forest elephants (*Loxodonta africana*), western lowland gorillas (*Gorilla gorilla*), chimpanzees (*Pan troglodytes*), and bongo (*Tragelaphus euryceros*). The region has extremely low human population density (less than 1/km²), and until recently has been largely isolated from modernizing influences. Recognizing the value of this area to loggers, the Congolese government established Nouabalé-Ndoki National Park. This area has been called "The Last Place on Earth" (Chadwick, 1995), but even Nouabalé Ndoki is not insulated from human influence. Even as the park was being established, it was clear that wildlife inside the park was significantly affected by what was going on in the logging concessions surrounding it.

Unfortunately, the logging enterprise was supplying a vibrant trade in bushmeat (see chapter 10) to the urban centers of Yaoundé and Douala. In the aftermath of years of civil war in Congo Republic, leaving the country bereft of infrastructure, the German-owned Congolaise Industrielle du Bois (CIB) is effectively the de facto state in this part of the country, providing health care, education, infrastructure, and law enforcement. Clearly, any conservation strategy for the area would have to engage CIB or immediately become irrelevant. Against this background the WCS struck up a relationship with CIB, which has led to a reduction in the export of bushmeat. By restricting the sales of meat from logging concessions, managing the spatial distribution of legal hunting, reducing illegal hunting through enforcement of existing hunting regulations, and developing other sources of protein, WCS and CIB have improved the situation for northern Congo's wildlife. Initially, the WCS–CIB relationship was labeled *greenwashing*, a sellout or getting in bed with the

(continued)

Nouabalé Ndoki is by no means completely safe, but this deal struck between conservationists and loggers has provided a new model for conservation in tropical production forests, one that has already been replicated elsewhere in Congo and in Cameroon, the Central African Republic, and Gabon.

Agreements among commercial producers may also be needed at national levels. Evidence of an inverse relationship between oceanic fisheries harvest and decline of wild mammals in Ghanaian national parks suggests that market forces and demand for wild meat are closely linked (Brashares, pers. comm.). This indicates that fish and

bushmeat are substitutes and that a decline in one will result in increased consumption of the other, and that at least in West Africa fish and wildlife should be managed together; interestingly, this pattern does not hold for lowland Amerindian societies in Central and South America (Wilkie and Godoy, 2001).

Direct Payments

There is, however, a more aggressive, and more preemptive, method for dealing with commercial interests, that of direct pay-

(Box 11.3 continued)

devil (see chapter 9), but both parties soon realized that the most serious effect of logging, the hunting, was not an obligate part of the logging process. Specifically, they recognized that the pathway by which logging affected wildlife was indirect, greatly enhancing the prospects for finding a solution (see Table 11.1).

Public education campaigns may also play a role in the future, if there is corroboration of the suggestion that severe diseases like HIV-AIDS and Ebola (hemorrhagic fever) cross the human-wildlife interface via hunting, butchering, and consumption of primates by human hunters. Alternatively, such evidence may be construed as scare tactics designed by conservationists to keep locals from hunting.

The WCS–CIB "model" has now been incorporated in the management planning process for all logging concessions in Congo, and other Central African countries are following suit. What is more, a deal was brokered whereby "naïve chimps" (chimpanzees living in an area

Ecoguards Checking a Logging Truck in Central Africa.
Picture courtesy of WCS-PROGEPP Congo.

so remote that they have no fear of humans because they have never been harassed or hunted) could be protected; the forests in which they live were removed from the logging estate and incorporated in Nouabalé Ndoki National Park. The lessons learned from Nouabalé-Ndoki may even reach beyond Congo and tropical timber. Similar business-NGO relationships could be forged in other countries and surrounding other frontier industries, like petroleum development.

ments, which is best suited to problems falling in the upper left quadrant of Table 11.1. Direct payment can be thought of a form of outreach (see chapter 10) insofar as in its simplest form it entails giving something, often money, directly back to disenfranchised communities to compensate them for lost access to resources (Wilkie et al., 2001). Essentially, conservationists identify priority areas for conservation and pay those who control these areas to protect the ecosystem from degradation, with payments explicitly tied to conservation performance. The basic principle, and why this method is referred to as "direct" payments, is that

the best way of getting something you want is to pay for what you want (e.g., protected rainforest), rather than pay for something indirectly related to it (e.g., development; Ferraro and Kiss, 2002).

The idea of compensating people for their role in maintaining resources of global value is not new (Swanson, 1995). The best known conservation contracts of this type are known as easements; governments compensate farmers through cash awards, tax deductions, and mitigation credits, in return for adhering to habitat conservation plans requiring restraint in the use of their land (Wilhere, 2002). For instance, in Europe fourteen nations paid an estimated $11 billion (1993–97) to farmers, diverting over 20 million ha of agricultural land into set-asides for environmental services or forestry (Ferraro, 2001). Not only the state but also NGOs use direct payment approaches; for example, Defenders of Wildlife reward U.S. landowners for occupied wolf dens on their property. Such programs are still rare outside high-income countries, with the exception of Costa Rica, where landowners are rewarded for setting up private reserves to protect ecosystem services (water capture, biodiversity protection, scenic beauty, and carbon sequestration), with money raised through fuel taxes and levies on hydroelectric plant operators (Kiss et al., 2002; similar programs are under development in El Salvador, Guatemala, and Ecuador; Langholz, 2002). Clearly, different challenges require different payment schedules: in some cases environmental damage can be pre-empted by a simple payment scheme; in other cases, for example, where wildlife are agricultural pests or dangerous to humans, supplements for damage awards may be required. In yet other cases payments need to be contingent on good conservation outcomes, for instance, evidence that numbers or densities of threatened species are increasing ("performance payments"; see Ferraro, 2001).

Innovative experiments with the direct payments method are just beginning in the developing world (Hardner and Rice, 2002), taking the forms of land leases, or concessions to certain rights, such as those to logging. In this respect they differ from the outright purchase of land as reserves discussed in chapter 9. For example, in Guyana, CI has leased the timber rights to 81,000 ha of pristine forest at $1.25 per hectare per year with a twenty-five-year renewable contract, and in Kenya the Wildlife Foundation is securing migration corridors on private land through conservation leases at $4.00 per acre per year (Ferraro and Kiss, 2002). In Bolivia, too, logging rights to a 45,000 ha concession were purchased for a one-off $100,000 payment, securing forest that was subsequently integrated into the adjacent Madidi National Park. Such concessions or lease arrangements avoid the resentment often associated with outright land purchases overseas (see chapter 9). Ideally, these arrangements should compensate for the benefits that would have come from logging, with the fee jointly compensating the government for lost taxation, the region for forfeited infrastructural development, and local communities for forgone jobs, with a fair portion of the fee directed to the local level to create employment and social programs. As such, conservation concessions provide governments with economic reasons for supporting protected areas beyond parks, and corporations with ways to offset the environmental costs and economic risks associated with their operations (Hardner and Rice, 2002).

As Ferraro (2001) outlines, direct payments are preferable to ICD in several respects. First, where payments depend on performance the explicit linkage between conservation and personal (or community) benefits is unambiguous, thereby circumventing the linkage problems associated with the ICD approach (discussed in chapter 10), and encouraging local residents to take an active role in conserving ecosystems as collaborators not adversaries. Second, performance payments can eliminate the open-access problem associated with many ICD projects, for example, the flood of immigrants into project areas to reap the benefits, since conservation contracts can be reserved only for long-term residents of the area. Third, in terms of the dangers of privatization discussed in chapter 7, performance payments finesse personal greed by compensating individuals only for pursuing strategies that are of broader social and ecological value. In comparison with ecotourist initiatives (chapter 10), the money from direct payments principally stays in the local area rather than trickling away to multinationals. Direct payments are also not dependent on fickle or insatiable markets for wildlife products and the consumerism that so easily subverts otherwise sustainable utilization projects (chapters 5 and 8). And, finally, direct payments are estimated as being more cost-effective than indirect development interventions in securing conservation outcomes.

The direct payments approach has exciting potential, but is not immune from some of the more general dangers associated with other conservation strategies. First, time discounting: in chapter 5 we saw how from an actor's point of view future benefits may seem very insignificant compared to present costs; under these conditions payments would have to be large, continuous, and renegotiable. This latter point is important since the opportunity cost of, for example, an unlogged forest will depend on the employment level, roads, markets, sawmills, technology, and other infrastructural factors that easily change over time. Second, in the case of performance payments, the resources to be protected must be carefully monitored, and if protection appears to be failing, payments must be withdrawn. Even under sound management, the vagaries of "natural" fluctuations (see chapter 3) could give an erroneous impression of failure and an unjustified withdrawal, generating serious resentment and causing the protected area to revert to open access. Third, the direct payments initiative would work better if linked to interest-paying environmental trust funds, so that one-off payments or deals can be transformed into a permanent funding mechanism, as defined in chapter 9. Fourth, there is the persistent question of who receives the payment, and how it is distributed among stakeholders. As we saw in chapter 7, communities are generally heterogeneous with respect to who suffers the cost of conservation. For just this reason it will be difficult to identify who should reap financial compensation, a problem already thrown into relief in the discussion of bioprospecting, where we saw how awkward it can be to decide whether the herbal specialist, the village, the broader community, or the na-

tion should be rewarded for potentially valuable pharmaceutical compounds. There is clearly much scope for powerful citizens to influence the allocation of conservation contracts in self-serving ways.

Last, but not least, direct payments open many possibilities for coercion: communities can blackmail conservationists to extract the maximum surplus from those willing to pay the subsidy; individuals could feign interest in converting lands that would not have been converted in the absence of payments; and conservationists can effectively force an impoverished group within a split community into accepting hard-to-resist but unethically low cash payments (Ferraro, 2001). As we saw with indigenous NGOs among the Maasai of Tanzania (chapter 8), throwing money at communities exacerbates, even incites, conflict.

Many of these problems are inherent in other strategies, and are not insuperable. It is the "direct" aspect of the payments, concessions and leases that is so attractive, because what is being purchased or leased is exactly what is wanted—an intact ecosystem.

Co-management

A conclusion of chapter 6 was that decentralization depends ultimately on the nature of the institutions to which power is devolved. Ostrom (1990) recognized this in her vision of a hierarchical nesting of institutions, with the administrative scope of each suited to the problem at hand. This conclusion points to the critical role of broader institutional arrangements, effectively the state, in any conservation strategy. While we have already looked at how the state (chapter 7) and multiple states (chapter 9) struggle with ecological crises, here we turn to a particularly productive formula for thinking about power sharing between state and local levels—co-management.

Already in the early 1990s, co-management was being promoted as a strategy emphasizing "the sharing of power and responsibility between government and local resource users" (Berkes et al., 1991, 12; see also Murphree, 1994). In this way it offers a powerful management compromise that blends the strengths of state and communal property regimes (see Table 6.1). Intellectually co-management arises from the critical observation that local autonomy can persist only if legitimated by the state, and that accordingly the integrity and function of local-level organizations depend on an umbrella of higher government structures. Co-management thereby substantiates Ostrom's celebrated framework of nested hierarchy (chapter 6), and has much in common with McNeely's (1995) emphasis on partnership among institutions, be these local, national, nongovernmental, or even commercial. Its focus is on reaching decisions over regulations and resource use by working with representatives of user groups, appropriate government agencies, and ideally research institutions. The idea is that different parties bring different tools to the crafting of workable solutions—government (its administrative assistance and enabling national legislation), local resource users (their TEK and locally adapted management practices), and scientists (their expertise and predictions regarding the conditions of sustainability). In an ideal world co-management encourages consensus-based decision making, and permits negotiation rather than litigation by providing a range of institutions for mitigating and resolving conflicts, as can be seen for Australia's showcase at Kakadu National Park (Box 11.4). In practice, as we also see in the Kakadu case, conflicts persist, but at least there is a structure within which to negotiate.

In the developing world the successes of co-management are as yet limited (Nepal, 2002), with Zimbabwe's CAMPFIRE program being an innovative but perhaps unsustainable forerunner (Box 10.5). There are promising and unexpected developments in China, where until very recently the centralized system of government had virtually ruled out local participation in park management. Now, for example, Xishuangbanna State Nature Reserve's management plan rests on a system of negotiations to balance the needs of local Dai and Han communities with the goals of conservation. Perhaps unsurprisingly, given the increase in population in the area, economic needs often eclipse protection, with farmers offered land within the reserve to offset their losses from desisting shifting cultivation in ecologically sensitive areas (Tisdell, 1995; Albers and Grinspoon, 1997; and see chapter 10), but at least the institutions for reaching future compromises are in place. In Nepal, too, political reforms over land and forest management have boosted indigenous empowerment and control, although in reality the changes in the way a park like Sagarmatha is run have been quite limited (see Box 6.6). By contrast, in places like northern Thailand, where evictions and relocations of hill tribes persist until this day (see chapter 2) and mistrust characterizes the relationships between communities, protected area authorities, and the highly centralized government, co-management is still a nonstarter. In a recent review of these cases Nepal (2002) concludes that only countries with effective political and economic reform can start down the road to co-management, and that the transition from partial to full co-management must be gradual.

Another thing that is clear is that some kinds of resources are more suited to devolved systems of co-management than others. Making proposals for the management of Mongolia's semiarid rangelands, Fernández-Giménez (2002) argues how well the residentially mobile and fluid social groupings of the Mongolian herders respond to a multitiered co-management system. This is because people who subsist primarily on domestic herds keenly appreciate how key resources, be these wells, seasonal pastures, or territory, because of differences in scale, demand different institutions for possession and management (see also Naughton-Treves, 1999). A similar case has been made for

Box 11.4. Co-management in Australia's Kakadu National Park

Kakadu National Park, in Australia's Northern Territory, is one of the most celebrated examples of the co-management (or joint management) of areas designated for biodiversity protection, serving as a model across Australia and worldwide. Forged out of a brew of intense conflicts among indigenous peoples, conservationists, government agencies, ranchers, and mining interests (areas rich in uranium), the Kakadu plan was established in 1978, following the 1976 Aboriginal Land Rights (Northern Territory) Act recognizing freehold title. It aimed to provide for biodiversity protection, while simultaneously maintaining the land's value for its traditional owners. Its key elements are: (1) the land is owned by traditional Aboriginal custodians as inalienable freehold title; (2) the land is leased by the Aboriginal owners to the government to be managed as a national park; (3) the lessors receive an annual rent from the government; and (4) the Aboriginal owners constitute a majority of the board of management, the locus of decision making (de Lacy, 1994).

Under the terms of the agreement the primary goal is the protection of biodiversity in Kakadu, although the lease can be terminated if the Aboriginal owners feel issues cannot be resolved in their favor. Additionally, Aboriginals are trained in park management, are employed under conditions that recognize their special needs and culture, own the principal tourist facilities, and are involved in the development of management plans. Co-management provides an institutional structure for incorporating TEK (see Box 4.2) into park management; in an attempt to keep this knowledge alive and vibrant, elders participate in educational initiatives.

De Lacy (1994) attributes the success of Australia's co-management schemes to the legal recognition of Aboriginal land rights. Hill and Press (1994) also note how the project has benefited (ironically) from its constant attacks from the Northern Territories government: this opposition has encouraged the traditional owners and the federal Australian National Parks and Wildlife Service to join together. Nonetheless, co-management has not been simple, and Kakadu has stayed in the political limelight since its establishment, with intense struggles among various sets of serried interests: uranium mining consortia, conservationists, Aborigines, tourists, and ranchers. There have been debates, for example, over the handling of introduced feral buffalo (which Aboriginals had come to depend on quite heavily for protein, but which conservationists see as exotics). There have also been problems over the management of sacred sites and caves, from which archaeologists and anthropologists allegedly take illegal collections. Tourism is another point of contention, despite its substantial revenue: visitors damage cultural sites, recreational fishing infringes on Aboriginal subsistence, and the massive numbers of tourists intrude into the lives of Aboriginal residents. For example, Aboriginal hunters do not like to hunt or gather within view of tourists for both public relations and safety reasons (see also Furze et al., 1996, for Uluru-Kata Tjuta).

Particularly hot have been the debates over fire. Aboriginals set closely controlled fires across different habitats, burning at different times of year and at varying frequencies and intensities (Lewis, 1989). These practices have explicit cultural, religious, and ecological significance. For example, they are used to "clean" the habitat of high grass (thereby facilitating walking around the landscape and access

(continued)

(Box 11.4 continued)

to game), to trap game, and to demonstrate continuity with ancestors by renewing spiritual sites, in other words, for both ecological and more diffuse cultural purposes. Many argue that mosaic burning helps to prevent the late dry season fires that currently sweep across vast tracts of land and that cause considerable damage to some ecological communities, and there is some experimental and comparative evidence to support this position (Woinarski et al., 2001, Yibarbuk et al., 2001). Accordingly, managers at Kakadu and elsewhere use traditional fire management techniques as preventative burns to control late dry season wildfires. Others concerned with broader indicators of biodiversity are more hesitant over such policies (e.g., Orgeas and Andersen, 2001); in their view traditional fire management does not enhance biodiversity as defined in chapter 1. Much of the debate turns on whether conservationists are trying to conserve the biodiversity that prevailed immediately prior to European settlement, or to re-create tentatively defined conditions that existed prior to human occupation (Bowman, 1998; and see chapter 7). Also at issue is the utility of Aboriginal burning patterns in a habitat so dramatically altered by the grazing of introduced buffalo and the arrival of fast-growing invasive African grasses.

The contention surrounding fire is not simply that resource managers in state conservation agencies do not easily accept Aboriginal ecological knowledge unless it conforms to their own scientific constructs (de Lacy, 1994), but rather that scientists themselves feel alienated from a management system that seems so unresponsive to scientific input (Andersen et al., 1998). The standoff at Kakadu has not been aboriginals versus colonialists but managers (black and white) versus scientists. Though these issues will not be easily settled, the benefit of co-management is that it at least offers a framework for conflict resolution, and does this through reducing the negative social and cultural consequences associated with protectionism; furthermore, in recent years it is fostering new collaboration with researchers (A. Petty, pers. comm.).

Aboriginal Burning.
Photo courtesy of Aaron Petty.

Alaska and Canada, where the success of co-management has been attributed in part to the fact that renewable natural resources (in particular, waterfowl and fish that migrate between Alaska, Canada, the lower forty-eight U.S. states, and Central America) are claimed by a variety of users (indigenous and not) with poorly defined use rights (Chase et al., 2000; see Box 7.6).

It is, therefore, not surprising that fisheries are so well suited to co-management. As with herders' pastures and water sources, pelagic resources are also often available in different places at different times of the year, and harvested by poorly defined sets of users; indeed, as outlined in chapter 6, rivers and oceans are classic common-pool resources subject to subtractability and difficulties of exclusion. It is perhaps for this reason that in many parts of the world co-management is flourishing in the regulation of fisheries. Pomeroy and Berkes (1997) describe numerous examples of fisheries triggered into successful co-management either by resource depletion or by apparently insoluble conflicts among and between resource users and various government agencies. The case of the San Salvador Island fishery stands out (Box 11.5), where local initiatives have thrived under a 1991 Philippines law aimed specifically at decentralizing government resource management functions to the municipal level (Katon et al., 1999) and subsequent laws governing broader bodies of water (Fernandez et al., 2000).

Naturally each case of fisheries co-management proceeds differently, and involves different degrees of power devolution, depending largely on the extent to which the nation-state endorses, or alternatively can tolerate, local autonomy. In some rather remarkable cases fishing communities simply take the initiative to implement and enforce rules, for example, outlawing the use of explosives and trawl nets; only subsequently do they gain the support of government and NGOs through co-management. This happened in southern Thailand, where, after villagers' nets and stationary gear were being torn up by outside operators, they instigated a total no-entry zone (Johnson, 2001). In other cases co-management cannot get off the ground without the passing of new legislation, as with the law permitting villager participation in Tanzania's Mafia Island Marine Park (Ngoile et al., 1995). In other cases they are modeled on preexisting communal institutions, as with the fisheries of Lake Kariba that draw on Zimbabwe's CAMPFIRE program. And in yet other cases (parallel to what we see in North America; see Notzke, 1995, and Box 7.6), devolved management is linked to constitutionally

granted land claims. More generally, Pomeroy and Berkes's review of international co-management fisheries worldwide underscores the importance of patience: community organizing can take anywhere from three to ten years, and even more in the absence of direct governmental interventions for institution-building. Agencies should not be too quick in evaluating outcomes or withdrawing funding.

Co-management has emerged as a promising complement to the classical model of national parks. It has the potential to mitigate the social and cultural impacts of conservation, to access and implement TEK, and to preempt or at least reconcile conflicts that arise over access to natural resources. Co-management also confers resilience and legitimacy to local institutions; the breakdown of community political institutions, systems of land tenure, and rights allocation often lie at the root of ecological degradation (as was concluded in chapter 4; Colchester, 1997; see also Zimbabwe's CAMPFIRE experiment, Box 10.5). While co-management faces inevitable challenges in institution-building, mobilization, prevention of free-riding, and reconciling inconsistencies between the objectives of the full suite of managers (external and internal), its key contribution lies in the agreement among various levels of governance, from village to international, to tackle different aspects of the management challenge, depending essentially on the scale of the specific problem. Finally, as the North American experience shows, co-management schemes built to counter territorial threats or consolidate land claims, in line with self-determination movements (see chapter 8), are more successful than those designed around externally perceived stock crises; as Pike et al. note, "hunters hate being told what to do" (cited in Notzke, 1995, 198).

Participation and Capacity-Building

Capacity-building, defined as enhancing the capability of an organization or community

Box 11.5. Fisheries Co-management in the Philippines: The Case of San Salvador

San Salvador Island lies off the western coast of Luzon, with a population of about 1,600 people, mainly fishermen making their living on its fringing reef. Since the late 1960s the island's coral reef has been dying and fish yields have plummeted. There are two reasons for this: first, an influx of migrants from the central Philippines who introduced destructive fishing methods, including blast fishing, sodium cyanide poisoning (to stun and catch fish for aquarium and up-market restaurant trade), and fine mesh nets (that indiscriminately catch large and small fish); and second, integration of the village economy into the international aquarium trade. Fish catch per unit effort on San Salvador dwindled from 20 kg in the 1960s to 1 to 3 kg in the late 1980s (Katon et al., 1999), mirroring the situation throughout much of Indonesia, where only 5 percent of all reefs are considered to be in good health. Homemade explosives leave vast tracts of pulverized reef, countless nontarget species and juveniles dead in the water, and some badly injured fishermen.

Throughout the 1990s the central government, local authorities, and NGOs took practical action, capitalizing on a strong national drive toward decentralizing natural resource management. Typically these initiatives established "no take" zones or sanctuaries, marked by buoys and guarded by locally manned watchtowers, and were accompanied by regulations on fishing gear and methods, catch size, and closed seasons. One of many successful projects can be seen on San Salvador Island.

Here an external agent initiated resource management measures—a Peace Corps volunteer who in 1987 had surveyed the reef, and held informal dialogues with the villagers. In 1989, with the support of the Haribon Foundation (a local NGO), the Marine Conservation Project for San Salvador (MCPSS) was formed. Its goals were to establish a marine reserve and a coral reef fish sanctuary, to develop and implement a management strategy, and to enhance institutional capacities (Katon et al., 1999). Despite the absence of a tradition of collective action among the island's heterogeneous inhabitants, the project succeeded in mobilizing the residents. Within less than a year the islanders had set up a sanctuary that banned all fishing and a reserve that allowed only nondestructive methods. From the start illegal fishing was monitored, and violators of the restrictions (many of whom did not live on the island) were sanctioned, fined, relieved of their equipment, or even imprisoned. By July 1989 the municipal government had passed legislation that allowed for sanctuary management and for the legal apprehending of rule breakers. In 1993 the Haribon Foundation turned the project over to a village-based fishers' organization it had helped to create, and remarkably the conservation activities continued, even escalated. The municipality, for example, demonstrated its continuing support by constructing a new guardhouse and providing food and radios for the marine guards, the village council started to pay honoraria for the guards, and fishermen continued to patrol the waters on a twenty-four-hour basis. Rigorous monitoring shows remarkable improvements in coral cover and fish abundance (see table), and socioeconomic and attitudinal surveys attest to general satisfaction.

(continued)

(Box 11.5 continued)

TABLE 11.5 BOX
Comparative Biological Data on San Salvador Island, Philippines

Indicator	1988	1991	1998
Live coral cover	23%	No data	57%
Fish yield	7 tons/km^2/yr	14 tons/km^2/yr	162 tons/km^2/yr
Species richness	126 species belonging to 19 families	No data	138 species belonging to 28 families

Co-management in the Philippines remains a challenge because coastal communities are so poor, and residents are forced out of desperation to use methods that they know are destructive to eke out a meager living. It is hindered also by local dialects that inhibit communication even within small areas. Another problem is that many islands are hard to reach, and this discourages consultation among local and national parties to the various agreements. Nevertheless, the remarkable success of many projects like that at San Salvador speaks to the appropriateness of co-management for protecting coastal resources that are seriously at risk of depletion.

to identify and solve its own problems, is increasingly recognized as a major goal of any conservation initiative (World Bank, 1998). It is essential if decisions are to be devolved to the local level, if local skills and capital are to be used in the management of natural resources, and if the project is to persist once external support is cut. Generally, capacity-building implies top-down training (in such skills as planning, management, law, finance, agriculture, and forestry), often usefully preparing local leaders to interact with external institutions and outside initiatives from conservationists and others.

A first step lies in securing local participation, since capacity can hardly be built among an alienated populace. While there is a huge literature on the importance of local participation in development interventions, it is only recently that we have tangible evidence of the importance of participation in conservation projects. Salafsky et al. (2001) evaluated twenty-nine CBC project sites established across Asia and the Pacific, each of which focused on an income-generating enterprise, such as collecting marine samples to test for pharmaceutical compounds, distilling essential oils from wild plant roots, ecotourism, and handicrafts. After four years the success of these projects, measured in terms of the degree to which the threat to particular focal resources in each area had been reduced, was determined, and three hypotheses were tested: that conservation success would be greatest where (1) the enterprise was most profitable, (2) the revenue benefits were greatest, and (3) local people had most input. Interestingly, neither enterprise success, nor the percentage contribution of the enterprises to average household income, had any effect on conservation success, whereas measures such as the degree of local management, the extent to which locals owned the enterprises, and the degree of community policing were all clearly linked with conservation success. The pattern compellingly suggests that community involvement is a critical component in any conservation strategy.

An important route to securing participation is though integrating local leaders or functionaries into the broader conservation project. We have already looked at one apparently successful example from Nepal, where deforestation within Sagarmatha National Park was slowed during the period that traditional forest guards were reinvested with authority by the national park (Box 6.6). Increasingly, such initiatives are being experimented with at broader levels. Take, for example, the new national parks in the Republic of Guinea (West Africa), where the policy pursued by European donors (and the national forestry agencies they support) is to have no park guards, and to work instead through traditional hunters' societies (Leach and Fairhead, 2002). Such "brotherhoods" are longstanding features of rural Mande and other West African villages, and are founded on initiation, apprenticeship, and shared ceremony (Croll and Parkin, 1992). Hunters know the natural history of their forests well, and in this cultural area are often viewed as providers, healers, and protectors of the community against evil; they also have wide supernatural powers, and recognize taboos linking people to the bush and nature more generally. National park authorities, with donor support, are striving to promote these hunters' societies, harnessing them for conservationist ends. In fact, hunting brotherhoods are now formed at district, regional, and national levels in a federated structure mirroring the national administration, and are invested with responsibilities to respect the hunting calendar, protect endangered species, fight bush fires, and reinforce natural resource management programs. Innovative as such initiatives are, they raise concern among some who fear that outsiders have inflated the social importance of hunters in Mande communities (Leach and Fairhead, 2002). The argument here is not simply that a Mande hunter is no ecologically noble savage, but that he may be less a pillar of local political life than a powerful (perhaps even unaccountable) actor on its margins. This case points to the value not simply of working with local institutions (such as

hunters' brotherhoods), but of closely understanding the dynamics of these institutions. More generally, success in working with local leaders is likely to vary as a function of how well the outsiders avoid co-opting local office holders into the service of conservation as opposed to forging genuine collaboration.

A broader, novel, and still more democratic development in this area of participation and capacity-building is citizen science, which arose from the critique of the role of science in colonial, fascist, and democratic regimes. Citizen science questions who has the legitimacy to identify research questions, and how these should be answered; it promotes the idea that this right is not confined to specialists, laboratories, or development programs but should be open to public participation. Citizen science emerged as the public began to engage with science, its agenda, and even its methods, and unsurprisingly took hold primarily in Northern industrial societies (Irwin, 1995). Lay citizens began to contest expert science by conducting their own research and experiments, most preeminently with the "popular epidemiology" surrounding issues of toxic waste pollution (Fischer, 2000), where activists were simply not willing to accept at face value scientists' claims that communities exposed to toxins were not at risk.

While citizen science thrives primarily in developed nations with widespread educational institutions, it has huge potential in the developing world, and has already taken hold under other labels, such as Participatory Action Research (PAR, e.g., Parkes and Panelli, 2001). Through PAR local communities define research questions, collect data, use data for management decisions, and present findings to a broader public. As such, local people can take more control of research agendas, methods, analysis, and outcomes. To date most PAR has been used in environmental health and natural resource management projects, with, for example, Malian farmers diagnosing the quality of their soils and developing appropriate farm management strategies through resource flow models (Defoer et al., 1998) and Zambians using GIS to map their villages with respect to resolving disputes over land use and planning within the ADMADE program (Lewis, 1995). Application to nature conservation issues is less prevalent, although we saw some aspects of PAR with the Aché (Box 5.5), where Aché systematically census animal numbers along transects. One of the great strengths of participatory research is that the researchers, as community members, are the eyes and ears that identify ecological problems in their area. This enlarges the scale of data collection, both temporarily and spatially, as well as enormously reducing costs. Another strength is that community education about research objectives occurs throughout the process of data collection, and even publication (Bawa Village Community, 1997). In similar developments local communities have begun mapping their lands and threatened natural resources, thereby focusing attention on as yet unanticipated problems and empowering them as decision-making bodies with regards to land and wildlife (Poole, 1995; and chapter 8).

Citizen science has the potential to strengthen conservation practice in the developing world. Most likely its research tradition will join forces with that surrounding TEK (see chapter 4), a research field that explores local expertise on health, welfare, and nature and that has as yet largely failed to engage critically with "expert" science. As developing world communities become more distrustful of science, or at least of development rhetoric (e.g., Guha, 1997), local and indigenous knowledge will be brought to challenge outsiders' scientific conclusions, as we saw with the social ecology movement in India (Box 7.5). Citizen science is also being incorporated into adaptive management (see Box 3.4), such as in Australia's Kakadu National Park, where in the late 1980s both the park managers and local Aboriginals agreed that fire is key to maintaining diversity of habitats within the park. On the other side of the coin, in challenging

expert science citizen science may derail current conservation practice, for better or worse. Returning again to the Australian outback, park managers may have agreed over the importance of using fire as a management tool, but disagree deeply over how (see Box 11.4 again); while conventional managers plan fires in accordance with a system of "controlled burns" based on scientific measurements, local Aboriginals burn according to traditional rules of thumb (when the bush has become "dirty"; Lewis, 1989, 949).

Important as participation, citizen science, PAR, and other strategies are in engaging local communities, we need to recognize that these initiatives are driven by two arguments: the practical need to develop nonreductionist approaches to environmental issues by contextualizing expert knowledge, and the moral need to democratize science among all citizens. Steps toward the former goal will ensure progress on the latter goal. But a question remains. Under what circumstances must lay knowledge be superseded by that of specialists? Can lay knowledge be privileged simply to amend the previous imbalances and injustices? This is a matter we have met before in the discussion of ecological economics (chapter 9), where we asked why people are willing to promote contingency valuation (or lay viewpoints) in the execution of environmental policy but not bridge building or back surgery. If citizen science and other forms of knowledge participation evolve within the worldwide practice of conservation (as indeed they should), Berkes is right to caution that "traditional knowledge is complementary to Western scientific knowledge, not a replacement for it" (Berkes, 1999, 109).

11.4 Spatial Scale and Integrating Conservation and Development

Many of the cases reviewed above illustrate another principle: that the potential for reconciling land use conflicts increases with spatial scale. A brief thought experiment illustrates this idea. Imagine a protected area manager's universe that only includes the protected area. If there are land use conflicts within the protected area, the options for reconciling the conflict are limited to those activities or levels of extraction that do not harm the biodiversity of the reserve. However, if we expand the spatial scale slightly, areas outside the reserve come into play, expanding the options available to the reserve manager and the other stakeholders. For example, Abbot and Homewood (1999) examine a situation where fuelwood is collected within a small protected area in Malawi. If fuelwood collection is detrimental to the biodiversity of the park, a rational response might be to provide an alternative source of fuel, perhaps through establishment of a woodlot. However, working only within the protected area, planting a woodlot is not a viable option for wildlife (because to be productive, woodlots are often monocrops of exotic species). However, looking across a broader landscape, the potential for an intensively managed woodlot would be much greater. Economists are well aware of these "comparative advantages," because they drive specialization and trade, both of which tend to operate at larger spatial scales.

A related point is that political will for conservation becomes much greater when thinking and working at large scales. For example, at the national level, nearly everyone agrees that protected areas are important and should be established and protected. The widely cited target for countries is that 10 to 15 percent of their terrestrial area should be under protection specifically for biodiversity. That leaves 85 percent for other activities, which may or may not incorporate biodiversity conservation. This is a very modest (some say under-ambitious and inadequate) target, and in general, politicians, managers, and the general public willingly accept this level of commitment (see Table 2.2). Moving down to smaller scales, however, protected areas, especially large, ecologically viable ones, often occupy more than 10 to 15 percent of a local region or district, making this larger commitment less palatable. And of course, a very small area completely embedded in a protected area has few to no options outside of those compatible with conservation, so this may be the most difficult sale of all.

The issue of spatial scale also sheds some light on the links between conservation and development, and when or how they are mutually reinforcing. Clearly, at national scales the links between economic development and conservation are most straightforward; for example, economic opportunities in the rural or urban sector allow people to forgo resource collection in protected areas, pursue economic alternatives, or even move to another region with a strong job market. In other words, a strong industrial economy physically separated from areas of remaining biodiversity can actually help in their conservation. But when these activities are physically closer, and the dependencies of people and wildlife are tighter, more difficult trade-offs become apparent, as we saw in the discussion of ICD projects.

Although there is clearly no "right" or "wrong" scale, only more and less appropriate ones for a given question, the scale at which a system, process, or problem is viewed will determine the pattern observed. Indeed, central to ecologists' enterprise is to understand how local interactions at one level of biological organization (small scale) influence phenomena or patterns at higher levels of organization (large scale), and to determine how different processes may be entrained at different scales. As we have seen again and again in these chapters, a relationship or process that seems clear at one scale may become less clear, disappear, or even reverse at larger spatial or longer temporal scales.

Take tropical forest clearance for agriculture and timber as an example of this point. Foreign debt can be a key factor precipitating deforestation at an international level (chapter 9) but is less relevant to regional processes driven by the availability of labor and capital and their interactions, as we saw in the analysis of Latin American deforestation (Figure 7.2), and almost irrelevant (or at least extremely distal) when looking at village by village patterns, where considerations stemming from the marginal returns to competing activities (chapter 5), as well as time discounting and property rights (chapter 6), become most salient. Similar issues arise in considering the significance of state property law and land use policy, and their effects on the rates of vegetation change that vary between neighboring countries, as in the comparison of land degradation between China, Mongolia, and Russia (chapter 6). Dramatic as these differences can be, they do not necessarily preclude more conventional explanations at different scales. Thus, Homewood et al.'s (2001) demonstration of the significance of property rights in driving different rates of land conversion and game densities between Kenya and Tanzania does not exclude the possibility that stocking densities may still affect ecological outcomes more locally (see Box 3.3).

A final example comes from the role of local populations in protecting wildlife. In chapter 4 we recognized that local populations often appear to hunt certain prey species beyond sustainable levels, depleting or rendering locally extinct their prey basis, whereas later in that chapter (and in chapter 8) we argued that indigenous and local populations may offer the best landscape-level solution to forest protection in some parts of the world. This conclusion is not drawn from ethical criteria (fundamental though these are; chapter 2), but rather from the potential for large tribally managed territories to offer protection to source-sink structured populations, metapopulations, and buffer zones, as we saw with the no-take or no-hunting zones discussed earlier in this chapter. The key point, then, is that we need to be constantly vigilant about arguments or conservation approaches that, intentionally or not, slide between different geographic scales, and that inappropriately infer mech-

anisms and dynamics from one level to the next. In a similar vein we need to guard against arguments that slide between different temporal scales, as we discuss in Box 3.5 in relation to strategies that look sustainable over long timeframes despite being unfeasible in the short term.

11.5 Means and Ends: The Many Orthogonal Axes of Conservation in Practice

The thrust of this final section is that there are no recipes for "conservation." A recipe assumes that we are all trying to produce the same dish. The truth is that conservationists serve up many different dishes, starting with diverse ingredients and cooking tools, and must make their offerings palatable to vastly different clientele. Here, to wind up our arguments, we point to two pitfalls that bedevil so many conservation projects, and to some resolution: the failure to specify the targets and goals of a conservation endeavor, and the tendency to conflate these ends with the means of achieving them. Our arguments follow in the spirit of a number of recent overviews of conservation policy (Redford and Richter, 1999; Mace et al., 2000; Salafsky et al., 2002; Redford et al., 2003), but focus particularly on general recommendations with respect to the persistent tensions over conservation versus development.

We start with a consideration of conservation targets and goals. An important characteristic of the more successful projects outlined in this book is that they clearly articulate conservation targets and maintain a focus on achieving as a goal some level of protection or restoration of these targets (e.g., a critical population size or density of individuals). Unfortunately, remarkably few conservation projects define their targets, and even fewer their goals. This is not surprising given that, as we saw in chapter 3, these targets can range from populations, species, and species assemblages (or communities), to ecosystems, ecosystem health, and evolutionary processes (Box 3.6). Biodiversity, the hopeful neologism scrutinized in chapter 7, usefully captures the common biological essence among these targets, but its simple label should not trick us into thinking there is any single strategy that will succeed in any of these very distinct challenges, or that strategies will be compatible. Targets also vary from being truly wild or scenic places (natural features) to human-dominated (and sometimes culturally valued) landscapes, in other words, according to their anthropogenic component. To make things worse, the very rationale for conservation, whether we value biodiversity for instrumental or intrinsic reasons, has swung pendulum-like across the years, with various periods (and distinct lobbies within these periods) emphasizing a different cocktail of these ideologies. Accordingly, targets have expanded to include sustainable use by human populations, equitable sharing, and human welfare. Against this history it is not surprising targets vary dramatically; indeed, from a set of twenty-one different conservation approaches promulgated by thirteen conservation organizations Redford et al. (2003) identified twelve distinct targets.

Many of the same issues hold for goals. A conservation goal specifies what precisely is aimed at with respect to the chosen target, for example: how many populations? of what size? how many species? how many ecosystems? Goal specification is extraordinarily difficult, in part because of the diversity of targets, as discussed directly above, and in part because a judgment must be made as to how much conservation is enough. As we suggested in chapter 1, the choice of goals is driven by values that reflect historical, sociological, philosophical, and personal preferences; these values betray, ultimately, a commitment to either an instrumental or inherent philosophical rationale for conservation. Layered on top of this are the policy shifts in the late twentieth century (see chapter 2) that make some choices more fashionable than others, with sustainable

development introducing a whole new set of criteria—poverty alleviation, capacity-building, social equity, justice, and empowerment. As a result, when biodiversity conservation is stated as a goal in the early twenty-first century, it has many different meanings (Table 2.3) and can be used in the service of very different agendas.

With such a smorgasbord of targets and goals, stakeholders are drawn to the modern conservation arena, all looking for very different outcomes. Such circumstances inevitably spark conflicts. Also, arguably, they conspire to encourage NGOs, governments, stakeholders, and other parties to leave their goals somewhat unspecified and loose (Redford and Richter, 1999). For instance, some may choose not to specify targets for fear of alienating stakeholders, whereas others may eschew identifying goals too closely to avoid an overt failure to meet these goals, thus jeopardizing funding sources that are likely to dry up after hearing of failure (see chapter 10). In such an arena of inexplicit assumptions, confused objectives, and naïve expectations, it is hardly surprising that conflict, mistrust, and disappointment are so common.

There are several advantages to having clearly specified targets and goals. First, they force conservationists to develop a precise conceptual framework for their activities, within which "conservation problems" become tractable and empirical issues; without being forced to clarify what each intervention is intended to accomplish it is easy to adopt seemingly sensible activities that have tenuous or nonexistent links to conservation targets. A second advantage is that with precise goals, monitoring and evaluation (see chapter 10) become more straightforward (Salafsky et al., 2002); indeed, many of the current "hot debates" in conservation (and this book!) would not exist if the intervening parties had clearly stated their targets and goals at the outset and monitored their progress toward them. Without this, academic arguments about the efficacy of benefit-sharing, conservation education, or strict protection can drag on endlessly, without reaching resolution.

There are also costs associated with not specifying goals. If in the interests of avoiding alienation, stakeholders are misled as to the real goals of a project, this can backfire and create greater problems in the future, even a complete withdrawal of cooperation once the stakeholders recognize the deception (see chapter 10). In this vein Robinson and Redford (2004) characterize ICD projects as "Jacks of all trades, and masters of none." Recognizing the span of mutually incompatible interventions, they advocate acknowledging that there really exist two kinds of projects, Conservation Projects with Development (CPwD) and Development Projects with Conservation (DPwC).

We turn now to the dangers of conflating ends (by which we mean targets and goals) with the means of achieving these ends. Consider the arguments among proponents of CBC and those of protectionism. Protectionism is criticized on many counts. Parks are expensive, are often poorly protected, commonly have weak institutional and local support, and sometimes override the human rights of erstwhile inhabitants; they may also fail to protect complete ecosystems, or the ecosystems most in need of protection (chapter 2). CBC has its own warts, mostly associated with its assumption that resources can be conserved through consumption (its "use or lose it" philosophy), as well as its failure to recognize that some resources must be put off limits, to acknowledge that population and demand are not constant, and to acknowledge competition and conflict within communities (chapters 7 and 10).

This debate easily accrues political (even personal) innuendo. Thus, in the eyes of its critics CBC is a palliative for conservation agencies eager to leave behind their environmentally checkered pasts, to put a politically correct spin on protecting biodiversity, and to attract new funding streams (Young, 1993). From the other side, conservation biologists are seen as authoritarian, meddling patricians, blinded to injustice by a love of biology (Guha, 1997). As such, we have two mutually suspicious factions. Driving these suspicions are the stereotypical characterizations of positions at either end of a perceived spectrum. A typical depiction of this continuum would place at one end the biologist with a deeply held siege mentality desperately goading governments or international donors into protecting wild places from the legions of disenfranchised poor, and at the other end, the naïve social scientist trusting that if everyone benefits from wilderness and wildlife, sustainable use and effective conservation will be the outcome. These views imply a single philosophical axis on which each conservation project (and conservationist) must jostle for a spot. This axis is shown in Figure 11.1. Collapsed along this single axis are rationales for conservation (utilitarian versus intrinsic), goals (utilitarianism versus preservationism), targets (anthropogenic landscapes versus wilderness areas), methods (emphasizing utilization and sustainable extraction versus non use), and style of management.

This simple picture confounds numerous dimensions. Most critically, for those concerned with the conflicts between conservation and development, there is a conflation of goals with means. The policy shift from top-down centralized strategies toward the bottom-up, grassroots, decentralized approaches that occurred in the 1980s (chapter 2) coincided with the shift away from the ideology embracing the inherent value of nature to that of its instrumental value, or from preservationism to utilitarianism. The means and goals of conservation are nevertheless independent, and collapsing them into the single spectrum shown in Figure 11.1 minimizes the space for novel solutions. Indeed, centralization and level of consumption need not co-vary; consumption can be high or low at any level of centralization. This is the crux of an alternative conceptual framework presented in Figure 11.2. To visualize this idea, consider cen-

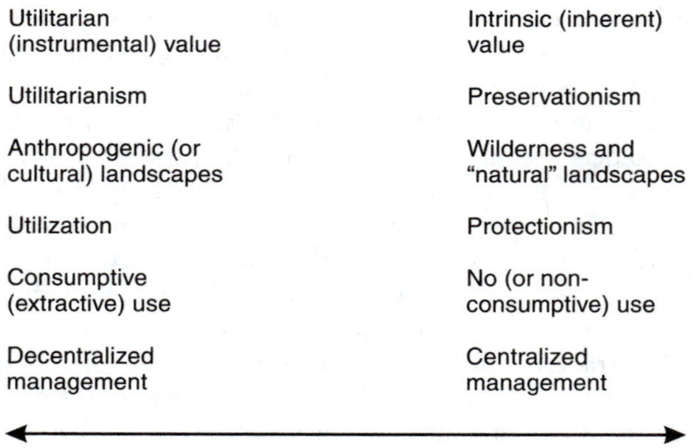

Utilitarian (instrumental) value
Intrinsic (inherent) value
Utilitarianism
Preservationism
Anthropogenic (or cultural) landscapes
Wilderness and "natural" landscapes
Utilization
Protectionism
Consumptive (extractive) use
No (or non-consumptive) use
Decentralized management
Centralized management

Figure 11.1. Single Axis Combining Means and Ends.

Centralized
Decentralized
Utilitarianism
Preservation
1
2
3
4
5
6
7
8
9
10
vi
v
iv
iii
ii
ib
ia
TC
PF
Ramsar
RR
NP
NWH
IUCN I
BRT
BRB
BRC
C
PR

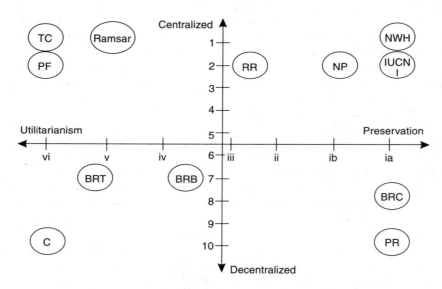

Figure 11.2. Orthogonal Axes of Ends and Means.

The horizontal axis represents the management philosophy, and its values range from 1 to 6, corresponding to the IUCN categories of protected areas (see Table 2.1). The vertical axis represents the degree of centralization with values ranging from 1 to 10 corresponding to Murphree's categories; essentially, these range from 1, multinational agreements or actors, to districts (4), to communities (7), to individuals (10). Well-known conservation strategies described in the text are plotted along each axis, albeit approximately. BRB, biosphere reserve buffer; BRC, biosphere reserve core; BRT, biosphere reserve transition; CPR, commons with common-property regime; ER, extractive reserve; IUCN I, strict nature reserve; NF, national forest (managed for both production and wilderness protection); NP, national park; OA, commons with complete open access; PF, production forests (merging in social forestry [SF] as becomes more decentralized); PR, private reserves; TC, timber certification schemes (see Box 9.2); WH, world heritage site (Natural).

tralization and utilitarianism as orthogonal axes that can be depicted graphically (note that we retain the old term *preservationism* here, recognizing that this strategy does not exclude as a target the conservation of dynamic evolutionary forces nor anthropogenic landscapes, following Meffe and Carroll, 1997). The vertical axis represents the degree of centralization, and the horizontal axis is level of consumption (or protection). The collapsed single axis from Figure 11.1 would run diagonally from the lower left to the upper right. As shown in the figure, different conservation strategies fall in all four quadrants.

There are several points to note in this picture. First, different components of a single conservation project can fall in different parts of the spectrum portrayed in Figure 11.2. Thus, a biosphere reserve's core, where little or no extractive use is allowed, should fall in the lower right quadrant, while its buffer and transitional zones (allowing increased levels of consumptive use; Batisse, 1986) fall to the left side of the central axis. Second, different aspects of a project may fall in different places along either axis; thus, the extent to which a project is centralized will depend on whether we are concerned with how it was initiated, its land tenure arrangements, the involvement of the local population in policy or management decisions, or simply the flow of benefits from conservation (Stevens, 1997b). Third, note that co-management schemes consist, by definition, of hierarchically nested institutions that cover the full gamut from centralized to decentralized management, dependent in part on how much participation and capacity-building is entailed; they are not shown in the figure. Fourth, with respect to other novel schemes discussed in this chapter, we can see that community-based protected areas set up in Amazonia would fall in the bottom right quadrant, as would direct payments to set nature "off limits" (at least if payments are made to local communities and administrative bodies), and the collaboration with logging concessions

over controlling bushmeat hunting. South Africa's Working for Water program stretches even this disaggregation of axes by simultaneously promoting employment and eradicating exotics.

Thinking along multiple axes in this way frees us from some of the doctrinaire, uncritical applications of conservation tools, and encourages us to cobble together mixed strategies to meet the specific demands and threats confronting the local situation. Furthermore, considering more openly and experimentally how different conservation tools suit different problems throws into relief the value of adaptive management, the idea of learning from doing, and of integrating design management and monitoring to create a framework whereby we can learn from mistakes (see Box 3.4).

We plot strategies on Figure 11.2 not to pigeon-hole projects, but to demonstrate the broader claim that whatever philosophical motivations lie behind personal valuations of nature, this should not constrain the full range of potential strategies for conserving biodiversity. While history shows that specific conservation means have traditionally been linked with certain ends (chapter 1), associations between utilitarianism and decentralization, and between preservationism and centralization, are certainly not obligate. Recognizing this may reveal a wider variety of means to achieve various conservation targets and goals, and may help guide us toward new solutions—solutions that integrate different tools in different combinations, such as some of the novel approaches reviewed in this chapter. Once this wider range of solutions appears on the conservationists' radar, the challenge of matching problems to solutions arises. Consistently in this book we have noted that, so to speak, different horses are suited to different courses—parks to remote areas with little threat, ICD projects to nations with strong central policies regarding economic development and immigration, biosphere reserves to more threatened ecosystems, game ranching to contexts with strong commercial infrastructure, ecotourism to situations with political stability, and so on. Despite innovative qualitative work in this area (Brandon, 1998; Terborgh et al., 2002) there is now a need for quantitative meta-analyses. Great steps have already been taken in both designing projects in such a way as to yield comparative prospective data for testing hypotheses (Salafsky et al., 2001) and in reviewing case studies strategically selected on the basis of certain key variables (such as the extent of decentralization; Wyckoff-Baird et al., 2000). Far more work is needed in this area, building on the model of comparative studies of common-pool resource management introduced in chapter 5.

11.6 Conclusion

This chapter began with a warning: conservation is about choices and usually difficult ones at that. When these choices are made consciously and strategically, significant conservation successes can be gained. Confronting these choices out in the open allows conservationists and other stakeholders to identify where and when different actors' objectives are consistent, and where they are in conflict. Often these opportunities and challenges do not map onto our expectations, and presumed links (like those between conservation and development) may break down. But in other cases, opportunities may present themselves. In this chapter we have looked at the chance to protect gorillas, elephants, and bongo through logging concessions in Congo, the no-entry zones set up by local communities in Amazonia, the eradication of exotics in South Africa in the service of development, the use of payments made to local communities and administrative bodies to ensure protection, and some successful programs built on co-management and participatory action. None of these are "silver bullets," but each is working rather well in achieving its specific goals.

Conservation solutions clearly demand a nimble mind. While in all likelihood biodiversity will continue to mean different things to different people, and conservation biologists, economists, anthropologists, and social activists will never entirely agree on an overall rationale for conserving nature, we are by no means at an impasse. An important advance lies in recognizing all the different dimensions to a conservation problem, whether these pertain to scale, targets, goals, or means. The breaking apart of conventional axes, and the expansion of scale, opens up different niches for the collaboration of different agencies, scientists, stakeholders, and amateurs, with each bringing complementary skills to different parts of the problem—for example, what target to focus on, at what scale to intervene, what degree of consumption to strive for, and with what levels of decentralization. Approaches that seem incompatible at one scale, or from the perspective of a single axis, emerge as complementary, or at least nested, when viewed in this broader framework. Thinking about conservation and people in this way highlights the need for a multidisciplinary training, or at least a tolerance of interdisciplinary dialogue. This book has tried to entice the reader down such a path.

Bibliography

Abbot JIO, and Homewood K (1999) A history of change: Causes of miombo woodland decline in a protected area in Malawi. Journal of Applied Ecology 36:422–433.

Abbot JIO, and Mace R (1999) Managing protected woodlands: Fuelwood collection and law enforcement in Lake Malawi National Park. Conservation Biology 13:418–421.

Abbot JIO, Thomas DHL, Gardner AA, Neba SE, and Khen MW (2001) Understanding the links between conservation and development in the Bamenda highlands, Cameroon. World Development 29:1115–1136

Abler DG, Rodriguez AG, and Shortle JS (1998) Labor force growth and the environment in Costa Rica. Economic Modelling 15:477–499.

Acheson JM (1988) The lobster gangs of Maine. Hanover, NH: University Press of New England.

Acheson JM (1998) Lobster trap limits: A solution to a communal action problem. Human Organization 57:43–52.

Acheson JM, and Wilson J (1996) Order out of chaos: The case for parametric fisheries management. American Anthropologist 98:579–594.

Adams W, and Hulme D (2001) Conservation and community: Changing narratives, policies and practices in African conservation. In D Hulme and M Murphree (eds.), African wildlife and livelihoods: The promise and performance of community conservation. Oxford: James Currey, pp. 9–23.

Adams W, and Infield M (2001) Park outreach and gorilla conservation: Mgahinga Gorilla National Park, Uganda. In D Hulme and M Murphree (eds.), African wildlife and livelihoods: The promise and performance of community conservation. Oxford: James Currey, pp. 131–147.

Agarwal A, and Narain S (1991) Global warming in an unequal world: A case of environmental colonialism. New Delhi: Centre for Science and Environment.

Agarwal B (1992) The gender and environment debate: Lessons from India. Feminist Studies 18:119–158.

Agrawal A (1994) Rules, rule-making and rule-breaking: Examining the fit between rule systems and resource use. In E Ostrom, R Gardner, and J Walker (eds.), Rules, games and common-pool resources. Ann Arbor: University of Michigan Press, pp. 413–439.

Agrawal A (2002) Common resources and institutional sustainability. In E Ostrom, T Dietz, N Dolšak, P Stern, S Stronich, and E Weber (eds.), The drama of the commons. Washington, DC: National Academy Press, pp. 41–85.

Agrawal A, and Gibson CC (1999) Enchantment and disenchantment: The role of community in natural resource conservation. World Development 27:629–649.

Agrawal A, and Yadama GN (1997) How do local institutions mediate market and population pressures on resources? Forest Panchayats in Kumaon, India. Development and Change 28:435–465.

Aiken SR (1994) Peninsular Malaysia's protected areas coverage, 1903–92: Creation, rescission, excision, and intrusion. Environmental Conservation 21:49–56.

Albers HJ, and Grinspoon E (1997) A comparison of access restriction between Xishuangbanna Nature Reserve (China) and Khao Yai National Park (Thailand). Environmental Conservation 24:351–362.

Alcorn JB (1993) Indigenous peoples and conservation. Conservation Biology 7:424–426.

Alcorn JB (1994) Noble savage or noble state?

Northern myths and southern realities in bio-diversity conservation. Etnoecologica 2:7–19.

Alcorn JB (1995) Comment on Alvard. Current Anthropology 36:789–818.

Alcorn JB, and Molnar A (1996) Deforestation and human-forest relationships: What can we learn from India. In LE Sponsel, TN Headland, and RC Bailey (eds.): Tropical deforestation: The human dimension. New York: Columbia University Press, pp. 99–121.

Alcorn JB, and Toledo V (1998) Resilient resource management in Mexico's forest ecosystems: The contribution of property rights. In F Berkes and C Folke (eds.), Linking social and ecological systems. Cambridge: Cambridge University Press, pp. 216–249.

Alford RA, and Richards SJ (1999) Global amphibian declines: A problem in applied ecology. Annual Review of Ecology and Systematics 30:133–165.

Allen W (2001) Green Phoenix: Restoring the tropical forests of Guanacaste, Costa Rica. Oxford; New York: Oxford University Press.

Alpert P (1996) Integrated conservation and development projects: Examples from Africa. Bioscience 46:845–855.

Alroy J (2001) A multispecies overkill simulation of the end-Pleistocene megafaunal mass extinction. Science 292:1893–1896.

Alston LJ, Libecap GD, and Mueller B (2000) Land reform policies, the sources of violent conflict, and implications for deforestation in the Brazilian Amazon. Journal of Law, Economics and Organization 39:162–188.

Alvard MS (1993) A test of the ecologically noble savage hypothesis: Interspecific prey choice by neotropical hunters. Human Ecology 21:355–387.

Alvard MS (1994) Conservation by native peoples: Prey choice in a depleted habitat. Human Nature 5:127–154.

Alvard MS (1995a) Shotguns and sustainable hunting in the neotropics. Oryx 29:58–66.

Alvard MS (1995b) Intraspecific prey choice by Amazonian hunters. Current Anthropology 36:789–818.

Alvard MS (1998a) Indigenous hunting in the neotropics: Conservation or optimal foraging. In T Caro (ed.), Behavioral ecology and conservation biology. New York: Oxford University Press.

Alvard MS (1998b) Evolutionary anthropology and resource conservation. Evolutionary Anthropology 7:62–74.

Alvard MS (2000) The potential for sustainable harvests by traditional Wana hunters in Morowali nature reserve, Central Sulawesi, Indonesia. Human Organization 59:428–440.

Alvard MS, and Nolin DA (2002) Rousseau's whale hunt? Coordination among big-game hunters. Current Anthropology 43:533–559.

Andersen AN, Braithwaite RW, Cook GD, Corbett LK, Williams RJ, Douglas Michael M, Gill AM, Setterfield SA, and Muller WJ (1998) Fire research for conservation management in tropical savannas: Introducing the Kapalga fire experiment. Australian Journal of Ecology 23:95–110.

Anderson AB (1997) Prehistoric Polynesian impact on the New Zealand environment: Te Wjenua Hou. In PV Kirch and TL Hunt (eds.): Historical ecology in the Pacific islands. New Haven: Yale University Press, pp. 271–283.

Anderson AB, and Posey DA (1989) Management of a tropical scrub savanna by the Goritire Kayapo of Brazil. Advances in Economic Botany 7:159–173.

Andrewartha HG, and Birch LC (1954) The distribution and abundance of animals. Chicago: University of Chicago Press.

Arias M, and Chapin M (1998) Panama: The research project for the management of wilderness areas in Kuna Yala. In IWGIA (ed.), From principles to practice: Indigenous peoples and biodiversity conservation in Latin America. Copenhagen: International Work Group for Indigenous Affairs, pp. 237–280.

Arizpe L, and Velazquez M (1994) The social dimensions of population. In L Arizpe, MP Stone, and DC Major (eds.), Population and environment. Boulder: Westview Press, pp. 15–40.

Armsworth PR, and Roughgarden JE (2001) An invitation to ecological economics. Trends in Ecology and Evolution 16:229–234.

Arrow KJ (1994) Methodological individualism and social knowledge. American Economic Review 84:1–9.

Atran S, Medin D, Ross N, Lynch E, Coley J, Ucan Ek' E, and Vapnarsky V (1999) Folkecology and commons management in the Maya Lowlands. Proceedings of the National Academy of Sciences 96:7598–7603.

Atran S, Medin D, Ross N, Lynch E, Vapnarsky V, Ucan Ek' E, Coley J, Timura C, and Baran M (2002) Folkecology, cultural epidemiology, and the spirit of the commons: A garden experiment in the Maya lowlands, 1991–2001. Current Anthropology *43*:421–450.

Aungur R (1992) The nutritional consequences of rejecting food in the Ituri forest of Zaire. Human Ecology *20*:263–291.

Aungur R (1994) Are food avoidances maladaptive in the Ituri forest of Zaire? Journal of Anthropological Research *50*:277–310.

Auzel P, and Wilkie DS (2000) Wildlife use in northern Congo: Hunting in a commercial logging concession. In JG Robinson and EL Bennett (eds.), Hunting for sustainability in tropical forests. New York: Columbia University Press, pp. 413–426.

Axelrod R, and Hamilton WD (1981) The evolution of cooperation. Science *211*:1390–1396.

Ayensu E, Claasen DV, Collins M, Dearing A, Fresco L, Gadgil M, Gitay H, Glaser G, Lohn CL, Krebs J, Lenton R, Lubchenco L, McNeely JA, Mooney HA, Pinstrup-Andersen P, Ramos M, Raven P, Reid WV, Samper C, Sarukhan J, Schei P, Tundisi JG, Watson RT, Xu GH, and Zakri AH (1999) International ecosystem assessment. Science *286*:685–686.

Ayres JM, Alves AR, Queiroz HL, Marmontel M, Moura E, Lima DM, Azevedo A, Reis M, Santos P, da Silveira R, and Masterson D (1999) Mamirauá: The conservation of biodiversity in an Amazonian flooded forest. In C Padoch, JM Ayres, M Pinedo-Vasquez, and A Henderson (eds.), Várzea: Diversity, conservation and development in Amazonia's whitewater floodplains. New York: Botanical Garden Press, pp. 203–216.

Ayres JM, Lima DM, de Souza ME, and Barreiros JLK (1991) On the track of the road: Changes in subsistence hunting in a Brazilian Amazonian village. In J Robinson and K Redford (eds.), Neotropical wildlife use and conservation. Chicago: University of Chicago Press, pp. 82–92.

Bahn PG, and Flenley J (1992) Easter Island, Earth Island. New York: Thames and Hudson.

Balée WL (1989) The culture of Amazonian forests. In DA Posey and WL Balée (eds.), Resource management in Amazonia: Indigenous and folk strategies. New York: The New York Botanical Garden, pp. 1–21.

Balée WL (1994) Footprints of the forest: Ka'apor ethnobotany: The historical ecology of plant utilization by an Amazonian people. New York: Columbia University Press.

Balée WL (1998) Advances in historical ecology. New York: Columbia University Press.

Balick MJ, Elisabetsky E, and Laird SA, eds. (1996) Medicinal resources of the tropical forest. New York: Columbia University Press.

Balmford A, Bruner A, Cooper P, Costanza R, Farber S, Green Rhys E, Jenkins M, Jefferiss P, Jessamy V, Madden J, Munro K, Myers N, Naeem S, Paavola J, Rayment M, Rosendo S, Roughgarden J, Trumper K, and Turner RK (2002) Economic reasons for conserving wild nature. Science *297*:950–953.

Balmford A, Mace GM, and Ginsberg JR (1998) The challenges to conservation in a changing world: Putting processes on the map. In GM Mace, A Balmford, and JR Ginsberg (eds.), Conservation in a changing world. Cambridge: Cambridge University Press, pp. 1–28.

Balmford A, Gaston KJ, Rodrigues ASL, and James A (2000) Integrating costs of conservation into international priority setting. Conservation Biology *14*:597–605.

Balmford A, Moore JL, Brooks T, Burgess N, Hansen LA, Williams P, and Rahbek C (2001) Conservation conflicts across Africa. Science *291*:2616–2619.

Balmford A, and Whitten T (2003) Who should pay for tropical conservation, and how could the costs be met? Oryx *37*:238–250.

Bandyopadhyay J (1992) Sustainability and survival in the mountain context. Ambio *21*:297–302.

Bannister K, and Barrett K (2001) Challenging the status quo in ethnobotany. Cultural Survival Quarterly *24*:10–13.

Barber BR (1995) Jihad vs. McWorld. New York: Times Books.

Barbier EB (1992) Economics for the wilds. In TM Swanson and E Barbier (eds.), Economics for the wilds: Wildlife, diversity, and development. London: Earthscan, pp. 15–33.

Barbier EB (1998) The economics of environment and development. Cheltenham, UK: Edward Elgar.

Barbier EB, Burgess JC, Bishop JT, and Aylward BA (1994) The economics of the tropical timber trade. London: Earthscan.

Barraclough SL, Ghimire K, and United Nations (2000) Agricultural expansion and tropical deforestation: Poverty, international trade and land use. London: Earthscan.

Barrett CB, and Arcese P (1995) Are integrated conservation-development projects (ICDPs) sustainable? On the conservation of large mammals in Sub-Saharan Africa. World Development *23*:1073–1084.

Barrett CB, and Arcese P (1998) Wildlife harvest in integrated conservation and development projects: Linking harvest to household demand, agricultural production, and environmental shocks in the Serengeti. Land Economics 74:449–465.

Barrett CB, Brandon K, Gibson C, and Gjertsen H (2001) Conserving tropical biodiversity amid weak institutions. Bioscience 51:497–502.

Barrett CB, and Grizzle R (1999) A holistic approach to sustainability based on pluralism stewardship. Environmental Ethics 21:23–42.

Barrett CB, and Lybbert TJ (2000) Is bioprospecting a viable strategy for conserving tropical ecosystems? Ecological Economics 34:293–300.

Barrow E, and Murphree M (2001) Community conservation: From concept to practice. In D Hulme and M Murphree (eds.), African wildlife and livelihoods: The promise and performance of community conservation. Oxford: James Currey, pp. 24–37.

Basu K (1996) Methodological individualism: Resurrecting controversy. Economic and Political Weekly 31:269–270.

Bates DM (1985) Plant utilization: Patterns and prospects. Economic Botany 39:241–265.

Batisse M (1982) The biosphere reserve: A tool for environmental conservation and management. Environmental Conservation 9:101–111.

Batisse M (1986) Developing and focusing the biosphere reserve concept. Nature and Resources 22:3–11.

Batisse M (1993) The Silver Jubilee of Mab and its revival. Environmental Conservation 20:107–112.

Bawa Village Community (1997) Mozambique's Tchuma Tchato initiative of resource management on the Zambezi: A community perspective. Society & Natural Resources 10:409–413.

Beckerman S, and Valentine P (1996) On native American conservation and the tragedy of the commons. Current Anthropology 37:659–661.

Bedoya E (1996) The social and ecological causes of deforestation in the Peruvian Amazon basin: Natives and colonists. In M Painter and WH Durham (eds.), The social causes of environmental destruction in Latin America. Ann Arbor: University of Michigan Press, pp. 217–246.

Begossi A (1998) Resilience and neo-traditional populations: The caicaras (Atlantic Forest) and caboclos (Amazon, Brazil). In F Berkes and C Folke (eds.), Linking social and ecological systems. Cambridge: Cambridge University Press, pp. 129–157.

Behnke RH (2000) Equilibrium and non-equilibrium models of livestock population dynamics in pastoral Africa: Their relevance to Arctic grazing systems. Rangifer 20:141–152.

Beinart W (1990) Empire, hunting, and ecological change in southern and Central Africa. Past and Present 128:162–186.

Bell R (1987) Conservation with a human face: Conflict and reconciliation in African land use planning. In D Anderson and R Grove (eds.), Conservation in Africa: People, policies and practice. Cambridge: Cambridge University Press, pp. 63–101.

Belovsky GE (1988) An optimal foraging-based model of hunter-gatherer population dynamics. Journal of Anthropological Archaeology 7:329–372.

Bennett EL (1996) A trip to the Che-Wong. In M Borgerhoff Mulder and W Logsdon (eds.), I've been gone far too long. Oakland: RDR Books, pp. 141–150.

Bennett EL, Nyaoi AJ, and Sompud J (2000) Saving Borneo's bacon: The sustainability of hunting in Sarawak and Sabah. In JG Robinson and EL Bennett (eds.), Hunting for sustainability in tropical forests. New York: Columbia University Press, pp. 305–324.

Bennett EL, and Robinson JG (2000a) Hunting for sustainability: The start of a synthesis. In JG Robinson and EL Bennett (eds.), Hunting for sustainability in tropical forests. New York: Columbia University Press, pp. 499–519.

Bennett EL, and Robinson JG (2000b) Hunting for the Snark. In JG Robinson and EL Bennett (eds.), Hunting for sustainability in tropical forests. New York: Columbia University Press, pp. 1–9.

Bennett JW (1976) The ecological transition: Cultural anthropology and human adaptation. Oxford, England: Pergamon Press.

Bennett JW (1993) Human ecology and human behavior: Essays in environmental and development anthropology. New Brunswick: Transaction Publishers.

Benthall J (1995) The greening of the purple. Anthropology Today 11:18–20.

Berkes F (1996) Social systems, ecological systems, and property rights. In S Hanna, C Folke, and K-G Maler (eds.), Rights to nature: Ecological, economic, cultural, and political principles of institutions for the environment. Washington, DC: Island Press, pp. 87–107.

Berkes F (1998) Indigenous knowledge and resource management systems in the Canadian subarctic. In F Berkes and C Folke (eds.), Linking social and ecological systems. Cambridge: Cambridge University Press, pp. 98–128.

Berkes F (1999) Sacred ecology: Traditional ecological knowledge and resource management. Philadelphia, PA: Taylor & Francis.

Berkes F, Colding J, and Folke C (eds.) (2000) Rediscovery of traditional ecological knowledge as adaptive management. Ecological Applications 10:1251–1262.

Berkes F, Feeny D, McCay BJ, and Acheson JM (1989) The benefits of the commons. Nature 340:91–93.

Berkes F, Folke C, and Colding J (1998) Linking social and ecological systems: Management practices and social mechanisms for building resilience. Cambridge: Cambridge University Press.

Berkes F, Folke C, and Gadgil M (1995) Traditional ecological knowledge, biodiversity, resilience and sustainability. In CA Perrings, K-G Maler, C Folke, and CS Holling (eds.), Biodiversity conservation. Dordrecht: Kluwer Academic Publishers, pp. 281–299.

Berkes F, George PJ, and Preston RJ (1991) Co-management: The evolution in theory and practice of the joint administration of living resources. Alternatives 18:12–18.

Berrill M (1997) The plundered seas: Can the world's fish be saved? San Francisco: Sierra Club Books.

Bezanson K, and Mendez R (1995) Alternative funding: Looking beyond the nation-state. Futures 27:223–229.

Bilsborrow RE (1987) Population pressures and agricultural development in developing countries: A conceptual framework and recent evidence. World Development 15:182–203.

Binns JA, Illgner PM, and Nel EL (2001) Water shortage, deforestation and development: South Africa's working for water programme. Land Degradation & Development 12:341–355.

Biodiversity Support Program (2001) An ounce of prevention. Washington, DC: Biodiversity Support Program, WWF.

Birdsell J (1958) On population structure in generalized hunting and collecting populations. Evolution 12:189–205.

Blaikie PM, and Brookfield HC (1987) Land degradation and society. London; New York: Methuen.

Blake S, Rogers E, Fay JM, Ngangoue M, and Ebeke G (1995) Swamp gorillas in northern Congo. African Journal of Ecology 33:285–290.

Blanchard KA (1995) Reversing population declines in seabirds on the north shore of the Gulf of St. Lawrence, Canada. In SK Jacobson (ed.), Conserving wildlife: International education and communication approaches. New York: Columbia University Press, pp. 51–63.

Blomquist W (1994) Changing rules, changing games: Evidence from groundwater systems in southern California. In E Ostrom, R Gardner, and J Walker (eds.), Rules, games, and common-pool resources. Ann Arbor: University of Michigan Press, pp. 283–300.

Blum E (1993) Making biodiversity conservation profitable—a case study of the Merck Inbio agreement. Environment 35:16+.

Bodmer RE (2000) Integrating hunting and protected areas in the Amazon. In A Entwistle and N Dunstone (eds.), Priorities for the conservation of mammalian diversity: Has the panda had its day? Cambridge: Cambridge University Press, pp. 277–290.

Bodmer RE (1994) Managing wildlife with local communities in the Peruvian Amazon: The case of the Reserva Comunal Tamshiyacu-Tahuayo. In D Western and RM Wright (eds.), Natural connections. Washington, DC: Island Press, pp. 113–134.

Bodmer RE, and Puertas P (2000) Community-based co-management of wildlife in the Peruvian Amazon. In JG Robinson and EL Bennett (eds.), Hunting for sustainability in tropical forests. New York: Columbia University Press, pp. 395–409.

Bohn H, and Deacon R (2000) Ownership risk, investment, and the use of natural resources. American Economic Review 90:526–549.

Bohnet I, and Frey BS (1999) The sound of silence in prisoner's dilemma and dictator games. Journal of Economic Behavior & Organization 38:43–57.

Bond I (2001) CAMPFIRE and the incentives for institutional change. In D Hulme and M Murphree (eds.), African wildlife and livelihoods: The promise and performance of community conservation. Oxford: James Currey, pp. 227–243.

Bongaarts J, and Bulatao RA (2000) Beyond six billion: Forecasting the world's population. Washington, DC: National Academy Press.

Bonner R (1993) At the hand of man: Peril and hope for Africa's wildlife. New York: Alfred A. Knopf.

Bookbinder MP, Dinerstein E, Rijal A, Cauley H, and Rajouria A (1998) Ecotourism's support of biodiversity conservation. Conservation Biology 12:1399–1404.

Borgerhoff Mulder M, and Ruttan LM (2000) Grassland conservation and the pastoralist commons. In LM Gosling, WJ Sutherland, and M Avery (eds.), Behavior and conservation. Cambridge: Cambridge University Press.

Borgerhoff Mulder M, and Sellen DW (1994) Pastoralist decision making: A behavioral ecological perspective. In E Fratkin, KA Galvin, and EA Roth (eds.), African pastoralist systems: An integrated approach. Boulder: Lynne Reinner Publications, pp. 205–229.

Boserup E (1965) The conditions of agricultural growth: The economics of agrarian change under population pressure. London: G. Allen & Unwin.

Boss D, Grootenhuis GG, and Prins HHT (2000) The economics of wildlife tourism: Theory and reality for landholders in Africa. In HHT Prins, JG Grootenhuis, and TT Dolan (eds.), Wildlife conservation by sustainable utilization. Boston: Kluwer Academic Publishers, pp. 277–294.

Bowler PJ (1993) The Norton history of the environmental sciences. New York: Norton.

Bowman DMJS (1998) The impact of Aboriginal landscape burning on the Australian biota. New Phytology 140:385–410.

Bowman S (1998) Parks in partnership. National Parks January/February: 30–33.

Boyd R, and Richerson PJ (1985) Culture and the evolutionary process. Chicago: University of Chicago Press.

Boyd R, and Richerson PJ (1988) The evolution of reciprocity in sizeable groups. Journal of Theoretical Biology 132:337–356.

Boyd R, and Richerson PJ (1992) Punishment allows the evolution of cooperation (or anything else) in sizable groups. Ethology and Sociobiology 13:171–195.

Brandon K (1997) Policy and practical considerations in land-use strategies for biodiversity conservation. In RA Kramer, C van Schaik and J Johnson (eds.), Last stand: Protected areas and the defense of tropical biodiversity. New York: Oxford University Press, pp. 90–114.

Brandon K (1998) Perils to parks: The social context of threats. In K Brandon, KH Redford, and SE Sanderson (eds.), Parks in peril: People, politics, and protected areas. Washington, DC: Island Press, pp. 415–439.

Brandon K, Redford KH, and Sanderson SE (1998) Introduction. In K Brandon, KH Redford, and SE Sanderson (eds.), Parks in peril: People, politics, and protected areas. Washington, DC: Island Press, pp. 1–23.

Brandon K, Redford KH, and Sanderson SE, eds. (1998b) Parks in peril: People, politics, and protected areas. Washington, DC: Island Press.

Brandon K, and Wells M (1992) Planning for people and parks: Design dilemmas. World Development 20:557–570.

Brashares JS, Arcese P, and Sam MK (2001) Human demography and reserve size predict wildlife extinction in West Africa. Proceedings of the Royal Society Biological Sciences Series B 268:2473–2478.

Bray W (2000) Ancient food for thought. Nature 408:145–146.

Brechin SR, West PC, Harmon D, and Kutay K (1991) Resident peoples and protected areas: A framework for inquiry. In PC West and SR Brechin (eds.), Resident peoples and national parks. Tucson: University of Arizona Press, pp. 5–28.

Breslin P, and Chapin M (1984) Conservation Kuna-style. Grassroots Development 8:26–36.

Brewer SW, Rejmanek M, Johnstone EE, and Caro TM (1997) Top-down control in tropical forests. Biotropica 29:364–367.

Brightman RA (1993) Grateful prey: Rock Cree human-animal relationships. Berkeley: University of California Press.

Brockington D, and Homewood K (2001) Degradation debates and data deficiencies: The Mkomazi Game Reserve, Tanzania. Africa 71:449–480.

Brokensha D, Warren DM, and Werner O (1980) Indigenous knowledge systems and development. Washington, DC.: University Press of America.

Bromley D, ed. (1992) Making the commons work: Theory, practice and policy. San Francisco: ICS Press.

Brosius JP (1997) Endangered people, endangered forest: Environmentalist representations of indigenous knowledge. Human Ecology 25:47–69.

Brosius JP (1999) Analyses and interventions: Anthropological engagements with environmentalism. Current Anthropology 40:277–309.

Brosius JP, Tsing AL, and Zerner C (1998) Representing communities: Histories and politics of community-based natural resource management. Society & Natural Resources 11:157–168.

Browder JO (1992) The limits of extractivism. BioScience 42:174–182.

Brown K (1998) The political ecology of biodiversity, conservation and development in Nepal's Terai: Confused meanings, means and ends. Ecological Economics 24:73–87.

Brown KS (1997) Diversity, disturbance and sustainable use of Neotropical forests: Insects as indicators for conservation monitoring. Journal of Insect Conservation 1:25–42.

Brown LH (1971) The biology of pastoral man as a factor in conservation. Biological Conservation 3:93–100.

Bruce JW, Fortmann L, and Nhira C (1993) Tenures in transition, tenures in conflict: Examples from Zimbabwe social forest. Rural Sociology 58(4):626–642.

Bruner AG, Gullison RE, Rice RE, and da Fonseca GAB (2001) Effectiveness of parks in protecting tropical biodiversity. Science 291:125–128.

Brush SB (1996a) Whose knowledge, whose genes, whose rights? In SB Brush and D Stabinsky (eds.), Valuing local knowledge. Washington, DC: Island Press, pp. 1–21.

Brush SB (1996b) Is our common heritage outmoded? In SB Brush and D Stabinsky (eds.), Valuing local knowledge. Washington, DC: Island Press, pp. 143–164.

Bryant RL (1992) Political ecology: An emerging research agenda in Third-World studies. Political Geography 11:12–36.

Bryant RL (1997) Beyond the impasse: The power of political ecology in Third World environmental research. Area 29:1–17.

Bryant RL, and Bailey S, eds. (1997) Third world political ecology. London: Routledge.

Budianski S (1995) Nature's keepers. New York: Free Press.

Bullard RD (1994) Unequal protection: Environmental justice and communities of color. San Francisco: Sierra Club Books.

Bunting BW, Sherpa MN, and Wright M (1991) Annapurna Conservation Area: Nepal's new approach to protected area management. In PC West and SR Brechin (eds.), Resident peoples and national parks. Tucson: University of Arizona Press, pp. 160–172.

Burney DA (1993) Recent animal extinctions: Recipes for disaster. American Scientist 81:530–541.

Butler PJ (1995) Marketing the conservation message: Using parrots to promote protection and pride in the Caribbean. In SK Jacobson (ed.), Conserving wildlife: International education and communication approaches. New York: Columbia University Press, pp. 87–102.

Callenbach E (1977) Ecotopia: The notebooks and reports of William Weston. New York: Bantam Books.

Callicott JB (1994) Earth's insights: A survey of ecological ethics from the Mediterranean basin to the Australian outback. Berkeley: University of California Press.

Callicott JB, and Mumford K (1997) Ecological sustainability as a conservation concept. Conservation Biology 11:32–40.

Caro TM (1999) Densities of mammals in partially protected areas: The Katavi ecosystem of western Tanzania. Journal of Applied Ecology 36:205–217.

Caro TM, Borgerhoff Mulder M, and Moore M (2003) Effects of conservation education on reasons to conserve biological diversity. Biological Conservation 114:143–152.

Caro TM, and Laurenson MK (1994) Ecological and genetic factors in conservation: A cautionary tale. Science 263:485–486.

Caro TM, and O'Doherty G (1999) On the use of surrogate species in conservation biology. Conservation Biology 13:805–814.

Caro TM, Pelkey N, Borner M, Campbell KLI, Woodworth BL, Farm BP, Ole Kuwai J, Huish SA, and Severre ELM (1998) Consequences of different forms of conservation for large mammals in Tanzania: Preliminary analyses. African Journal of Ecology 36:303–320.

Caro TM, Rejmánek M, and Pelkey N (2000) Which mammals benefit from protection in East Africa? In A Entwistle and N Dunstone (eds.), Priorities for the conservation of mammalian diversity: Has the panda had its day? Cambridge: Cambridge University Press, pp. 221–238.

Carpenter SR, and Turner M (2000) Opening the black boxes: Ecosystem science and economic valuation. Ecosystems 3:1–3.

Carrillo E, Wong G, and Cuaron AD (2000) Monitoring mammal populations in Costa Rican protected areas under different hunting restrictions. Conservation Biology 14:1580–1591.

Carson R (1962) Silent spring. New York: Fawcett Crest.

Cashdan E (2001) Ethnic diversity and its environmental determinants: Effects of climate, pathogens, and habitat diversity. American Anthropologist 103:968–991.

Catlin G (1990) An artist proposes a national park. In R Nash (ed.), American environmentalism: Readings in conservation history. New York: McGraw-Hill, pp. 31–35.

Caughley G (1994) Directions in conservation biology. Journal of Animal Ecology 63:215–244.

Ceballos G, Rodriguez P, and Medellin RA (1998) Assessing conservation priorities in megadiverse Mexico: Mammalian diversity, endemicity, and endangerment. Ecological Applications 8:8–17.

Chadwick DH (1995) Ndoki: Last place on earth. National Geographic 188:2–45.

Chambers R (1997) Whose reality counts? Putting the first last. London: Intermediate Technology.

Chapin M (1991) Losing the way of the great father. New Scientist 131:40–44.

Chapin M (1994) Recapturing the old ways: Traditional knowledge and western science among the Kuna Indians of Panama. In CD Kleymeyer (ed.), Cultural expression and grassroots development: Cases from Latin America and the Caribbean. Boulder: Lynne Reinner Publishers, pp. 83–101.

Chapin M (2001) The struggle over land on Central America's last frontier. In S Lobo and S Talbot (eds.), Native American voices: A reader. Upper Saddle River, NJ: Prentice Hall, pp. 401–411.

Chapman M (1985) Environmental influences on the development of traditional conservation in the south Pacific region. Environmental Conservation 12:217–230.

Chase A (1987) Playing god in Yellowstone: The destruction of America's first national park. San Diego: Harcourt Brace.

Chase LC, Schusler TM, and Decter DJ (2000) Innovations in stakeholder involvement: What's the next step? Wildlife Planning and Management 28:208–217.

Chernela J (1987) Endangered ideologies: Tukano fishing taboos. Cultural Survival 11:50–52.

Child B (1996) The practice and principles of community-based wildlife management in Zimbabwe: The CAMPFIRE Programme. Biodiversity and Conservation 5:369–398.

Chjatterjee N (1995) Social forestry in environmentally degraded regions of India: Case-study of the Mayurakshi Basin. Environmental Conservation 22:20–30.

Choudhary K (2000) Development dilemma: Resettlement of Gir Maldharis. Economic and Political Weekly: 2662–2668.

Cincchón A (2000) Conservation theory meets practice. Conservation Biology 14:1368–1369.

Cincotta RP, and Engelman R (2000) Nature's place: Human population and the future of biological diversity. Washington, DC: Population Action International.

Cincotta RP, Wisnewski J, and Engelman R (2000) Human population in the biodiversity hotspots. Nature 404:990–992.

Ciriacy-Wantrup SV, and Bishop RC (1975) "Common property" as a concept in natural resources policy. Natural Resources Journal 15:713–727.

Clark CW (1973) The economics of overexploitation. Science 181:630–634.

Clay J (1984) Organizing to survive. Cultural Survival Quarterly 8:2–5.

Clay J (1991) Cultural survival and conservation: Lessons for the past twenty years. In ML Oldfield and JB Alcorn (eds.), Biodiversity: Culture, conservation and ecodevelopment. Boulder: Westview Press, pp. 248–273.

Clay J (1992) Buying in the forests: A new paradigm to market sustainably collected tropical forest products protects forests and forest residents. In KH Redford and C Padoch (eds.), Conservation of neotropical forests: Working from traditional resource use. New York: Columbia University Press, pp. 400–415.

Clutton-Brock TH, and Parker GA (1995) Punishment in animal societies. Nature 373:209–216.

Cohen E (1988) Authenticity and commoditization in tourism. Annals of Tourism Research 15:371–386.

Colchester M (1993) Pirates, squatters and poachers: The political ecology of dispossession of the native peoples of Sarawak. Global Ecology and Biogeography Letters 3:158–179.

Colchester M (1997) Salvaging nature: Indigenous peoples, protected areas and biodiversity conservation. In KB Ghimire and MP Pimbert (eds.), Social change and conservation: Environmental politics and impacts of national parks and protected areas. London: Earthscan, pp. 97–130.

Colchester M (2000) Self-determination or environmental determinism for indigenous peoples in tropical forest conservation. Conservation Biology 14:1365–1367.

Colchester M, and Gray A (1998) Foreword. In IWGIA (ed.), From principles to practice: Indigenous peoples and biodiversity conservation in Latin America. Copenhagen: International Work Group for Indigenous Affairs, pp. 10–17.

Colding J, and Folke C (2001) Social taboos: "Invisible" systems of local resource management and biological conservation. Ecological Applications 11:584–600.

Collar NJ, and International Council for Bird Preservation (1992) Threatened birds of the Americas: The ICBP IUCN red data book, part 2, third edition: Smithsonian Institution Press in cooperation with International Council for Bird Preservation.

Commoner B (1971) The closing circle: Nature, man, and technology. New York: Knopf.

Conklin BA (2002) Shamans versus pirates in the Amazonian treasure chest. American Anthropologist 104:1050–1061.

Conklin BA, and Graham LR (1995) The shifting middle ground: Amazonian Indians and eco-politics. American Anthropologist 97:695–710.

Connell JH (1978) Diversity in tropical rain forests and coral reefs. Science 199:1302–1309.

Corry S (1993) The rainforest harvest: Who reaps the benefit? The Ecologist 23:148–153.

Costanza R (1989) What is ecological economics? Ecological Economics 1:1–8.

Costanza R, d'Arge R, de Groot R, Farber S, Grasso M, Hannon B, Limburg K, Naeem S, O'N eill RV, Paruelo J, Raskin RG, Sutton P, and van den Belt M (1997) The value of the world's ecosystem services and natural capital. Nature 387:253–260.

Costanza R, and King J (1999) The first decade of ecological economics. Ecological Economics 28:1–9.

Croll EJ, and Parkin DJ, eds. (1992) Bush base: Forest farm: Culture, environment, and development. New York: Routledge.

Crooks KR, and Soule ME (1999) Mesopredator release and avifaunal extinctions in a fragmented system. Nature 400:563–566.

Cuello C, Brandon K, and Margoluis R (1998) Costa Rica: Corvocado National Park. In K Brandon, KH Redford, and SE Sanderson (eds.), Parks in peril: People, politics, and protected areas. Washington, DC: Island Press, pp. 143–191.

Cummings BJ (1990) Dam the rivers, damn the people: Development and resistance in the Amazonian jungle. London: Earthscan.

Daily GC, and Ehrlich PR (1996) Socioeconomic equity, sustainability, and earth's carrying capacity. Ecological Applications 6:991–1001.

Daily GC, Soderqvist T, Aniyar S, Arrow K, Dasgupta P, Ehrlich PR, Folke C, Jansson A, Jansson BO, Kautsky N, Levin S, Lubchenco J, Maler KG, Simpson D, Starrett D, Tilman D, and Walker B (2000) The value of nature and the nature of value. Science 289:395–396.

Daily GC, and Walker BH (2000) Seeking the great transition. Nature 403:243–245.

Dasmann RF (1988) Towards a biosphere consciousness. In D Worster (ed.), The ends of the earth: Perspectives on modern environment history. Cambridge: Cambridge University Press, pp. 277–288.

Davidson J, and Andrewartha HG (1948) The influence of rainfall, evaporation and atmospheric temperature on fluctuations in the size of a natural population of Thrips imaginis (Thysanoptera). Journal of Animal Ecology 17:200–222.

Davis K (1963) The theory of change and response in modern demographic history. Population Index 29:345–366.

Dawes R, van de Kragt AJC, and Orbell JM (1988) Not me or thee, but we: The importance of group identity in eliciting cooperation in dilemma situations: Experimental manipulations. Acta Psychologica 68:83–97.

de Beer JH, and McDermott MJ (1989) Economic value of non-timber forest products in south-east Asia. The Netherlands: Council for the International Union of the Conservation of Nature.

de Lacy T (1994) 50,000 years of Aboriginal law and land management changing the concept of national parks in Australia. Society and Natural Resources.

De Leo GA, Rizzi L, Caizzi A, and Gatto M (2001) Carbon emissions: The economic benefits of the Kyoto Protocol. Nature 413:478–479.

Deacon RT, and Murphy P (1997) The structure of an environmental transaction: The debt-for-nature swap. Land Economics 73:1–24.

Defoer T, De Groote H, Hilhorst T, Kante S, and Budelman A (1998) Participatory action research and quantitative analysis for nutrient management in southern Mali: A fruitful marriage? Agriculture Ecosystems & Environment 71:215–228.

Demsetz H (1967) Toward a theory of property rights. American Economic Review 57:347–359.

Denevan WM (1992) The pristine myth: The landscape of the Americas in 1492. Annals of the Association of American Geographers 82:369–385.

Denmer J, Godoy R, Wilkie D, Overman H, Taimur M, Fernando K, Gupta R, McSweeney K, Brokaw N, Sriram S, and Price T (2002) Do levels of income explain differences in game abundance? Biodiversity and Conservation 11:1845–1868.

Derman B (1995) Environmental NGOs, dispossession, and the state: The ideology and praxis of African nature and development. Human Ecology 23:199–215.

DeWalt BR (1994) Using indigenous knowledge to improve agriculture and natural resource management. Human Organization 53:123–131.

Diamond J (1986) Overview: Laboratory experiments, field experiments, and natural experiments. In J Diamond and TJ Case (eds.), Community ecology. New York: Harper and Row, pp. 3–22.

Diamond JM (1997) Guns, germs, and steel: The fates of human societies. New York: Norton.

Diamond PA, and Hausman JA (1994) Contingent valuation: Is some number better than no number? Journal of Economic Perspectives 8:45–64.

DiSilvestro RL (1993) Reclaiming the last wild places: A new agenda for biodiversity. New York: Wiley.

Dixon JA, and Sherman PB (1990) Economics of protected areas: A new look at benefits and costs. Washington, DC: Island Press.

Dobson A (2000) Green political thought. London: Routledge.

Dobson AP, Rodriguez JP, Roberts WM, and Wilcove DS (1997) Geographic distribution of endangered species in the United States. Science 275:550–553.

Dobson AP, Bradshaw AD, and Baker AJM (1997) Hopes for the future: Restoration ecology and conservation biology. Science 277:515–522.

Dombrowski K (2002) The praxis of indigenism and Alaska native timber politics. American Anthropologist 104:1062–1073.

Dove MR (1993) A revisionist view of tropical forest and development. Environmental Conservation 20:17–24, 55–56.

Dove MR (1995) The theory of social forestry intervention: The state of art in Asia. Agroforestry Systems 30:315–340.

Dove MR (1996) Center, periphery, and biodiversity: A paradox of government and a developmental change. In SB Brush and D Stabinsky (eds.), Valuing local knowledge. Washington, DC: Island Press, pp. 41–67.

Dugatkin LA (1998) Game theory and cooperation. In LA Dugatkin and HK Reeve (eds.), Game theory and animal behavior. New York: Oxford University Press, pp. 38–63.

Duke JA (1996) The role of medicinal plants in health care in India. In MJ Balick, E Elisabetsky, and SA Laird (eds.), Medicinal resources of the tropical forest. New York: Columbia University Press, pp. 266–277.

Durham WH (1996) Political ecology and environmental destruction in Latin America. In M Painter and WH Durham (eds.), The social causes of environmental destruction in Latin America. Ann Arbor: University of Michigan Press, pp. 249–264.

Durning AT (1989) Poverty and the environment: Reversing the downward spiral. Washington, DC: Worldwatch Institute.

Durning AT (1993) Guardians of the land: Indigenous peoples and the health of the earth. Washington, DC: Worldwatch Institute.

Dye T, and Steadman DW (1990) Polynesian ancestors and their animal world. American Scientist 78:207–215.

Earnshaw A, and Emerton L (2000) The economics of wildlife tourism: Theory and reality for landholders in Africa. In HHT Prins, JG Grootenhuis, and TT Dolan (eds.), Wildlife conservation by sustainable utilization. Boston: Kluwer Academic Publishers, pp. 315–334.

Easterlin RA (1995) Will raising the incomes of all increase the happiness of all? Journal of Economic Behavior & Organization 27:35–47.

Edelman M (2000) Review commentary: A conservation history of Costa Rica. Human Ecology 28:651–660.

Edwards PJ, and Abivardi C (1998) The value of biodiversity: Where ecology and economy blend. Biological Conservation 83:239–246.

Edwards R, and Kumar S (1998) Dust to dust. New Scientist: 18–19.

Ehrenfeld DW (1976) The conservation of non-resources. American Scientist 64:648–656.

Ehrenfeld DW (1978) The arrogance of humanism. New York: Oxford University Press.

Ehrlich PR (1968) The population bomb. New York: Ballantine Books.

Ehrlich PR, and Ehrlich AH (1990) The population explosion. New York: Simon and Schuster.

Ehrlich PR, Ehrlich AH, and Daily GC (1995) The stork and the plow. New Haven: Yale University Press.

Ellen RF (1986) What Black Elk left unsaid: On the illusory images of Green primitivism. Anthropology Today 2:8–12.

Ellis JE, and Swift DM (1988) Stability of African pastoral ecosystems: Alternate paradigms and implications for development. Journal of Range Management 41:450–459.

Elster J (1986) Introduction. In J Elster (ed.): Rational choice. New York: New York University Press, pp. 1–33.

Ensminger J (1996) Culture and property rights. In SS Hanna, C Foke, and KG Maler (eds.), Rights to nature. Washington, DC: Island Press, pp. 179–203.

Ensminger J, and Knight J (1997) Changing social norms: Common property, bridewealth, and clan exogamy. Current Anthropology 38:1–24.

Entwistle A (2000) Convention on biological diversity, CoP V: Moving from concepts to action? Oryx 34:338–340.

Escamilla A, Sanvicente M, Sosa M, and Galindo-Leal C (2000) Habitat mosaic, wildlife availability, and hunting in the tropical forest of Calakmul, Mexico. Conservation Biology 14:1592–1601.

Escobar A (1995) Encountering development: The making and unmaking of the Third World. Princeton, NJ: Princeton University Press.

Escobar A (1999) After nature: Steps to an anti-essentialist political ecology. Current Anthropology 40:1–30.

Evans D (1997) A history of nature conservation in Britain. London: Routledge.

Fairhead J, and Leach M (1996) Misreading the African landscape: Society and ecology in a forest-savanna mosaic. Cambridge: Cambridge University Press.

Fairhead J, and Leach M (1998) Reframing deforestation: Global analyses and local realities with studies in West Africa. London; New York: Routledge.

Faith DP, Walker PA, Ive JR, and Belbin L (1996) Integrating conservation and forestry production: Exploring trade-offs between biodiversity and production in regional land-use assessment. Forest Ecology and Management 85:251–260.

FAO (1997) The state of world fisheries and aquaculture. Rome.

Farnsworth NR, Akerele O, Bingel AS, Soejarto DD, and Guo Z (1985) Medicinal plants in therapy. Bulletin of the World Health Organization 63:965–1170.

Fay JM, and Agnagna M (1991) A population survey of forest elephants (Loxodonta-Africana-Cyclotis) in northern Congo. African Journal of Ecology 29:177–187.

Fearnside PM (1989) Extractive reserves in Brazilian Amazonia. Bioscience 39:387–393.

Feeny D, Berkes F, McCay BJ, and Acheson JM (1990) The tragedy of the commons: Twenty-two years later. Human Ecology 18:1–19.

Fehr E, and Gachter S (2000a) Cooperation and punishment in public goods experiments. American Economic Review 90:980–994.

Fehr E, and Gachter S (2000b) Fairness and retaliation: The economics of reciprocity. Journal of Economic Perspectives 14:159–181.

Fehr E, Fischbacher U, and Gachter S (2002) Strong reciprocity, human cooperation, and the enforcement of social norms. Human Nature 13:1–26.

Feit H (1973) The ethno-ecology of the Waswanipi Cree: Or how hunters can handle their resources. In BA Cox (ed.), Cultural ecology. Toronto: McClelland and Stewart Ltd, pp. 115–125.

Fernandez PR, Matsuda Y, and Subade RF (2000) Coastal area governance system in the Philippines. Journal of Environment and Development 9:341–369.

Fernández-Giménez ME (2002) Spatial and social boundaries and the paradox of pastoral land tenure: A case study from postsocialist Mongolia. Human Ecology 30:49–74.

Ferraro JP (2001) Global habitat protection: Limitations of development interventions and a role for conservation performance payments. Conservation Biology 15:990–1000.

Ferraro PJ, and Kiss A (2002) Direct payments to conserve biodiversity. Science 298:1718–1719.

Firth R (1959) Economics of the New Zealand Maori. Wellington, New Zealand: Government Printer.

Fischer F (2000) Citizens, experts, and the environment: The politics of local knowledge. Durham: Duke University Press.

Fisher WH (1994) Megadevelopment, environmentalism, and resistance: The institutional context of Kayapó indigenous politics in central Brazil. Human Organization 53:220–232.

Fisher WH (2000) Rain forest exchanges: Industry and community on an Amazonian frontier. Washington, DC: Smithsonian Institution Press.

FitzGibbon C (1998) The management of subsistence harvesting: Behavioral ecology of hunters and their mammalian prey. In T Caro (ed.), Behavioral ecology and conservation biology. New York: Oxford University Press, pp. 449–473.

FitzGibbon CD, Mogaka H, and Fanshawe JH (1995) Subsistence hunting in Arabuko-Sokoke Forest, Kenya, and its effects on mammal populations. Conservation Biology 9:1116–1126.

Folke C, and Costanza R (1996) The structure and function of ecological systems in relation to property rights regimes. In S Hanna, C Folke, and K-G Maler (eds.), Rights to nature: Ecological, economic, cultural, and political principles of institutions for the environment. Washington, DC: Island Press, pp. 13–34.

Ford J (1971) The role of trypanosomiasis in African ecology. Oxford: Oxford University Press.

Fortmann L, Antinori C, and Nabane N (1997) Fruits of their labors: Gender, property rights, and tree planting in two Zimbabwe villages. Rural Sociology 62:295–314.

Fox SR (1985) The American conservation movement: John Muir and his legacy. Madison: University of Wisconsin Press.

Frank RH (1988) Passions within reason: The strategic role of the emotions. New York: Norton.

Frank RH (1999) Luxury fever: Why money fails to satisfy in an era of excess. New York: Free Press.

Franke RW, and Chasin BH, eds. (1992) Seeds of famine: Ecological destruction and the development dilemma in the West African Sahel. Lanham, MD: Rowman and Littlefield.

Frazer JG (1955) The golden bough: A study in magic and religion. New York: St. Martin's.

Freese CH, ed. (1997) Harvesting wild species: Implications for biodiversity conservation. Baltimore: Johns Hopkins University Press.

Freese CH (1998) Wild species as commodities: Managing markets and ecosystems for sustainability. Washington, DC: Island Press.

Fromm O (2000) Ecological structure and functions of biodiversity as elements of its total economic value. Environmental & Resource Economics 16:303–328.

Furze B, De Lacy T, and Birckhead J (1996) Culture, conservation and biodiversity. Chichester: Wiley.

Gadgil M (1987) Diversity: Cultural and biological. Trends in Ecology and Evolution 2:369–373.

Gadgil M, Berkes F, and Folke C (1993) Indigenous knowledge for biodiversity conservation. Ambio 22:151–156.

Gadgil M, and Guha R (1992) This fissured

land: An ecological history of India. Berkeley: University of California Press.

Gadgil M, and Guha R (1995) Ecology and equity: The use and abuse of nature in contemporary India. New Delhi: Penguin Books India.

Gadgil M, Hemam NS, and Reddy BM (1998) People, refugia and resilience. In F Berkes and C Folke (eds.): Linking social and ecological systems. Cambridge: Cambridge University Press, pp. 30–47.

Gadgil M, and Thapar R (1990) Human ecology in India: Some historical perspectives. Interdisciplinary Science Reviews 15:209–223.

Gadgil M, and Vartak VD (1976) The sacred groves of Western Ghats in India. Economic Botany 30:152–160.

Galaty JG (1994) Rangeland tenure and pastoralism in Africa. In E Fratkin, KA Galvin, and EA Roth (eds.), African pastoralist systems: An integrated approach. Boulder: Lynne Reinner Publications, pp. 185–204.

Gall S (2001) The bushmen of southern Africa: Slaughter of the innocent. London: Chatto & Windus.

Gardner GT, and Stern PC (1996) Environmental problems and human behavior. Boston: Allyn and Bacon.

Gascon C, and Lovejoy TE (1998) Ecological impacts of forest fragmentation in central Amazonia. Zoology (Jena) 101:273–280.

Gatto M, and De Leo GA (2000) Pricing biodiversity and ecosystem services: The never-ending story. Bioscience 50:347–355.

Gedicks A (1995) International native resistance to the new resource wars. In T Bron (ed.), Ecological resistance movements: The global emergence of radical and popular environmentalism. Albany: SUNY Press, pp. 89–107.

Geist HJ, and Lambin EF (2002) Proximate causes and underlying driving forces of tropical deforestation. BioScience 52:143–150.

Getz WM, Fortmann L, Cumming D, du Toit J, Hilty J, Martin R, Murphree M, Owen-Smith N, Starfield AM, and Westphal MI (1999) Conservation: Sustaining natural and human capital: Villagers and scientists. Science 283: 1855–1856.

Ghimire KB, and Pimbert MP (1997) Social change and conservation: An overview of issues and concepts. In KB Ghimire and MP Pimbert (eds.), Social change and conservation, environmental politics and impacts of national parks and protected areas. London: Earthscan, pp. 1–45.

Giannini IV (1996) The Kikrin do Cateté Indigenous Area. In KH Redford and JA Mansour (eds.), Traditional peoples and biodiversity conservation in large tropical landscapes. Arlington, VA: The Nature Conservancy, pp. 115–136.

Gibson CC, and Koontz T (1998) When "community" is not enough: Institutions and values in community-based forest management in southern Indiana. Human Ecology 26: 621–647.

Gibson CC, and Marks SA (1995) Transforming rural hunters into conservationists: An assessment of community-based wildlife management programs in Africa. World Development 23:941–956.

Gilles JL, and Jamtgaard K (1981) Overgrazing in pastoral areas: The commons reconsidered. Nomadic Peoples 10:1–10.

Gillingham S (2001) Social organization and participatory resource management in Brazilian Ribeirinho communities. Society and Natural Resources 14:803–814.

Ginsberg JR, and Milner-Gulland EJ (1994) Sex-biased harvesting and population dynamics in ungulates: Implications for conservation and sustainable use. Conservation Biology 8:157–166.

Gintis H (2000a) Strong reciprocity and human sociality. Journal of Theoretical Biology 206: 169–179.

Gintis H (2000b) Beyond Homo economicus: Evidence from experimental economics. Ecological Economics 35:311–322.

Gobbi JA (2000) Is biodiversity-friendly coffee financially viable? An analysis of five different coffee production systems in western El Salvador. Ecological Economics 33:267–281.

Godoy R, and Bawa KS (1993) The economic value and sustainable harvest of plants and animals from the tropical forest: Assumptions, hypotheses, and methods. Economic Botany 47:215–219.

Godoy R, and Contreras M (2001) A comparative study of education and tropical deforestation among Lowland Bolivian Amerindians. Economic Development and Cultural Change 49:555–574.

Godoy R, Brokaw N, and Wilkie D (1995) The effect of income on the extraction of non-timber tropical forest products: Model, hypotheses, and preliminary findings from the Sumu Indians of Nicaragua. Human Ecology 23: 29–52.

Godoy R, Kirby K, and Wilkie D (2001) Tenure security, private time preference, and use of natural resources among lowland Bolivian Amerindians. Ecological Economics 38:105–118.

Godoy R, Lubowski R, and Markandya A (1993) A method for the economic valuation of non-timber tropical forest products. Economic Botany 47:220–233.

Godoy R, Wilkie D, Overman H, Cubas A, Cubas G, Demmer J, McSweeney K, and Brokaw N (2000) Valuation of consumption and sale of forest goods from a Central American rain forest. Nature 406:62–63.

Gonzalez N (1992) We are not conservationists: Interview conducted by Celina Chelala. Cultural Survival Quarterly 16:43–45.

Goodland RJ, Herman ED, and Serafy SE (1993) The urgent need for rapid transition to global environmental sustainability. Environmental Conservation 20:297–309.

Goodwin HJ, and Leader-Williams N (2000) Tourism and protected areas: Distorting conservation priorities towards charismatic megafauna? In A Entwistle and N Dunstone (eds.), Priorities for the conservation of mammalian diversity: Has the panda had its day? Cambridge: Cambridge University Press, pp. 257–275.

Gordon HS (1954) The economic theory of a common property resource: The fishery. Journal of Political Economy 62:124–142.

Gould K, Howard Andrew F, and Rodriguez G (1998) Sustainable production of non-timber forest products: Natural dye extraction from El Cruce Dos Aguadas, Peten, Guatemala. Forest Ecology & Management 111:69–82.

Grabowski R (1988) Theory of induced institutional innovation: A critique. World Development 16:385–394.

Gray A (1990) Indigenous peoples and the marketing of the rainforest. The Ecologist 20: 223–227.

Gray A, Parellada A, and Newing H (1998) From principles to practice: Indigenous peoples and biodiversity conservation in Latin America. Copenhagen: International Work Group for Indigenous Affairs.

Green C, Joekes S, and Leach M (1998) Questionable links: Approaches to gender in environmental research and policy. In C Jackson and R Pearson (eds.), Feminist visions of development. London: Routledge, pp. 259–283.

Green MJB, and Paine J (1997) State of the world's protected areas at the end of the twentieth century. Cambridge: World Conservation Monitoring Centre.

Grifo F, Newman D, Fairfield AS, Bhattacharya B, and Grupenhoff JT (1997) The origins of prescription drugs. In F Grifo and J Rosenthal (eds.), Biodiversity and human health. Washington, DC: Island Press, pp. 131–163.

Grifo F, and Rosenthal J (1997) Biodiversity and human health. Washington, DC: Island Press.

Grootenhuis JG, and Prins HHT (2000) Wildlife utilization: A justified option for sustainable land use in African savannas. In HHT Prins, JG Grootenhuis, and TT Dolan (eds.), Wildlife conservation by sustainable utilization. Boston: Kluwer Academic Publishers, pp. 469–482.

Grove RH (1992) Origins of Western environmentalism. Scientific American 267:42–47.

Grove RH (1995) Green imperialism: Colonial expansion, tropical island edens, and the origins of environmentalism, 1600–1860. New York: Cambridge University Press.

Groves CR, Jensen DB, Valutis LL, Redford KH, Shaffer ML, Scott JM, Baumgartner JV, Higgins JV, Beck MW, and Anderson MG (2002) Planning for biodiversity conservation: Putting conservation science into practice. Bioscience 52:499–512.

Grundmann R (1998) The strange success of the Montreal protocol: Why reductionist accounts fail. International Environmental Affairs 10:197–220.

Guha R (1989a) The unquiet woods. Berkeley: University of California Press.

Guha R (1989b) Radical American environmentalism and wilderness preservation: A third world critique. Environmental Ethics 11:71–83.

Guha R, ed. (1994) Social ecology. New York: Oxford University Press.

Guha R (1997) The authoritarian biologist and the arrogance of anti-humanism. The Ecologist 27:14–20.

Gullison RE, and Losos EC (1993) The role of foreign debt in deforestation in Latin America. Conservation Biology 7:140–147.

Gullison RE, Rice RE, and Blundell AG (2000) 'Marketing' species conservation. Nature *404*: 923–924.

Gurung CP (1995) People and their participation: New approaches to resolving conflicts and promoting cooperation. In JA McNeely (ed.), Expanding partnerships in conservation. Washington, DC: Island Press, pp. 223–233.

Guruswamy LD (1999) The convention on biological diversity: Exposing the flawed foundations. Environmental Conservation *26*:79–82.

Guyer J, and Richards P (1996) The invention of biodiversity: Social perspectives on the management of biological variety in Africa. Africa *66*:1–13.

Hackel JD (1999) Community conservation and the future of Africa's wildlife. Conservation Biology *13*:726–734.

Haines F (1970) The buffalo: The story of American bison and their hunters from prehistoric times to the present. New York: Thomas Y. Crowell.

Hales D (1989) Changing concepts of national parks. In D Western and M Pearl (eds.), Conservation for the twenty first century. New York: Oxford University Press, pp. 139–144.

Hames R (1987) Game conservation or efficient hunting. In BJ McCay and JM Acheson (eds.), The question of the commons: The culture and ecology of communal resources. Tucson: University of Arizona Press, pp. 92–107.

Hames R (1991) Wildlife conservation in tribal societies. In ML Oldfield and JB Alcorn (eds.), Biodiversity: Culture, conservation and ecodevelopment. Boulder: Westview Press, pp. 172–199.

Hamilton WD (1964) The genetical evolution of social behaviour. I & II. Journal of Theoretical Biology *7*:1–52.

Hanna S (1998) Managing for human ecological context in the Maine soft shell clam fishery. In F Berkes and C Folke (eds.), Linking social and ecological systems. Cambridge: Cambridge University Press, pp. 190–211.

Hanna S, Folke C, and Maler K-G, eds. (1996a) Rights to nature: Ecological, economic, cultural, and political principles of institutions for the environment. Washington, DC: Island Press.

Hanna S, Folke C, and Maler K-G (1996b) Property rights and natural environment. In S

Hanna, C Folke, and K-G Maler (eds.), Rights to nature: Ecological, economic, cultural, and political principles of institutions for the environment. Washington, DC: Island Press, pp. 1–10.

Hannah L, Rakotosamimanana B, Ganzhorn J, Mittermeier RA, Olivieri S, Iyer L, Rajaobelina S, Hough J, Andriamialisoa F, Bowles I, and Tilkin G (1998) Participatory planning, scientific priorities, and landscape conservation in Madagascar. Environmental Conservation *25*:30+.

Harcourt AH (1986) Gorilla conservation: Anatomy of a campaign. In K Berischke (ed.), Primates: The road to self-sustaining populations. New York: Springer-Verlag, pp. 32–46.

Harcourt AH (1995) Population viability estimates: Theory and practice for a wild gorilla population. Conservation Biology *9*:134–142.

Hardin G (1968) The tragedy of the commons. Science *162*:1243–1248.

Hardin G (1993) Living within limits: Ecology, economics, and population taboos. New York: Oxford University Press.

Hardner J, and Rice R (2002) Rethinking green consumerism. Scientific American *286*:89–95.

Harmon D (1991) National park residency in developed countries: The example of Great Britain. In PC West and SR Brechin (eds.), Resident peoples and national parks. Tucson: University of Arizona Press, pp. 33–39.

Harmon D (1996) Losing species, losing languages: Connections between biological and linguistic diversity. Southwest Journal of Linguistics *15*:89–108.

Haugerud A (1989) Land tenure and agrarian change in Kenya. Africa *59*:61–90.

Hawken P, Lovins AB, and Lovins LH (1999) Natural capitalism: Creating the next industrial revolution. Boston: Little, Brown and Co.

Hayes DJ, Sader SA, and Schwartz NB (2002) Analyzing a forest conversion history database to explore the spatial and temporal characteristics of land cover change in Guatemala's Maya Biosphere Reserve. Landscape Ecology *17*:299–314.

Headrick DR (1981) The tools of empire: Technology and European imperialism in the nineteenth century. New York: Oxford University Press.

Heang KB (1997) Sustainable development: A southeast Asian perspective. In F Smith (ed.),

Environmental sustainability: Practical global implications. Boca Raton, FL: St. Lucie Press, pp. 67–83.

Hearne J, and McKenzie M (2000) Compelling reasons for game ranching in Maputaland. In HHT Prins, JG Grootenhuis, and TT Dolan (eds.), Wildlife conservation by sustainable utilization. Boston: Kluwer Academic Publishers, pp. 417–438.

Hecht SB (1985) Environment, development and politics: Capital accumulation and the livestock sector in eastern Amazonia. World Development 13:663–684.

Hecht SB (1992) Valuing land uses in Amazonia: Colonist agriculture, cattle, and petty extraction in comparative perspective. In KH Redford and C Padoch (eds.), Conservation of neotropical forests. New York: Columbia University Press, pp. 379–399.

Hecht SB, and Cockburn A (1990) The fate of the forest: Developers, destroyers and defenders of the Amazon. New York: Harper Perennial.

Heerwagen JH, and Orians GH (1993) Humans, habitats, and aesthetics. In SR Kellert and EO Wilson (eds.), The biophilia hypothesis. Washington, DC: Island Press, pp. 138–172.

Heil MT, and Wodon QT (2000) Future inequality in CO_2 emissions and the impact of abatement proposals. Environmental & Resource Economics 17:163–181.

Heinen JT (1993) Park-people relations in Kosi Tappu wildlife reserve, Nepal: A socio-economic analysis. Environmental Conservation 20:25–34.

Heinen JT, and Low BS (1992) Human behavioral ecology and environmental conservation. Environmental Conservation 19:105–116.

Hemming J (1978) Red gold: The conquest of the Brazilian Indians. London: Macmillan.

Henderson N, and Sutherland WJ (1996) Two truths about discounting and their environmental consequences. Trends in Ecology & Evolution 11:527–528.

Henrich J, Boyd R, Bowles S, Camerer C, Fehr E, Gintis H, and McElreath R (2001) In search of Homo economicus: Behavioral experiments in 15 small-scale societies. American Economic Review 91:73–78.

Hill K (1995) Comment on Alvard. Current Anthropology 36:789–818.

Hill K (1996) The Mbaracayú reserve and the Aché of Paraguay. In KH Redford and JA Mansour (eds.), Traditional peoples and biodiversity conservation in large tropical landscapes. Arlington, VA: The Nature Conservancy, pp. 159–195.

Hill K, Padwe J, Bejyvagi C, Bepurangi A, Jakugi F, Tykuarangi R, and Tykuarangi T (1997) Impact of hunting on large vertebrates in the Mbaracayu reserve, Paraguay. Conservation Biology 11:1339–1353.

Hill MA, and Press AJ (1994) Kakadu National Park: An Australian experience in comanagement. In D Western and RM Wright (eds.), Natural connections: Perspectives on community-based conservation. Washington, DC: Island Press, pp. 135–158.

Hitchcock RK (1999) Resource rights and resettlement among the San of Bostwana. Cultural Survival Quarterly 22:51–55.

Hitchcock RK (2000) Traditional African wildlife utilization: Subsistence hunting, poaching and sustainable use. In HHT Prins, JG Grootenhuis, and TT Dolan (eds.), Wildlife conservation by sustainable use. Boston: Kluwer Academic Publishers, pp. 389–415.

Hitchcock RK, and Holm JD (1993) Bureaucratic domination of hunter-gatherer societies: A study of the San in Botswana. Development and Change 24:305–338.

Hobbelink H (1989) Biotechnology and the future of world agriculture. London; Atlantic Highlands: Zed Books.

Hobbes T (1996) Leviathan. Cambridge: Cambridge University Press.

Hobley M (1991) Gender, class and use of forest resources: Nepal. In A Rodd (ed.), Women and the environment. London: Zed Books.

Hodgson DL (2002) Precarious alliances: The cultural politics and structural predicaments of the indigenous rights movement in Tanzania. American Anthropologist 104:1086–1097.

Hodgson DL, and Schroeder RA (2002) Dilemmas of counter-mapping community resources in Tanzania. Development and Change 33:79–100.

Hogg R (1990) The politics of changing property rights among Isiolo Boran pastoralists in northern Kenya. In PTW Baxter (ed.), Property, poverty, and people. Manchester: Department of Social Anthropology and International Development Centre, University of Manchester, pp. 20–31.

Holdaway RN, and Jacomb C (2000) Rapid extinction of the moas (Aves: Dinorinthiformes): Model, test, and implications. Science 287:2250–2254.

Holdgate M, and Munro DA (1993) Limits to caring: A response. Conservation Biology 7: 938–940.

Holling CS (1986) Resilience of ecosystem: Local surprise and global change. In EC Clark and RE Munn (eds.), Sustainable development of the biosphere. Cambridge: Cambridge University Press, pp. 292–317.

Holling CS, Gunderson LH, and Ludwig D (2002) In search of a theory of adaptive change. In LH Gunderson and CS Holling (eds.), Panarchy: Understanding transformations in human and natural systems. Washington, DC: Island Press.

Holling CS, and Meffe GK (1996) Command and control and the pathology of natural resource management. Conservation Biology 10:328–337.

Holm J (1994) Attitudes to nature. London: Pinter Publishers.

Holmes CM (2003a) Assessing the perceived utility of wood resources in a protected area of western Tanzania. Biodiversity and Conservation 111:179–189.

Holmes CM (2003b) The influence of protected area outreach on conservation attitudes and resource use patterns: A case study from western Tanzania. Oryx 37:305–315.

Homewood K, Lambin EF, Coast E, Kariuki A, Kikula I, Kivelia J, Said M, Serneels S, and Thompson M (2001) Long-term changes in Serengeti-Mara wildebeest and land cover: Pastoralism, population, or policies? Proceedings of the National Academy of Sciences of the United States of America 98:12544–12549.

Homewood KM, and Rodgers WA (1984) Pastoralism and conservation. Human Ecology 12:431–441.

Homewood KM, and Rodgers WA (1987) Pastoralism, conservation and the overgrazing controversy. In D Anderson and R Grove (eds.), Conservation in Africa: People, policies and practice. Cambridge: Cambridge University Press.

Homewood KM, and Rodgers WA (1991) Maasailand ecology. Cambridge: Cambridge University Press.

Honey M (1999) Ecotourism and sustainable development: Who owns paradise? Washington, DC: Island Press.

Horowitz LS (1998) Integrating indigenous resource management with wildlife conservation: A case study of Batang Ai National Park, Sarawak, Malaysia. Human Ecology 26:371–403.

Hough JL, and Sherpa MN (1989) Bottom up vs basic needs: Integrating conservation and development in the Annapurna and Michiru Mountain conservation areas of Nepal and Malawi. Ambio 18:434–441.

Howard PC, Viskanic P, Davenport TRB, Kigenyi FW, Baltzer M, Dickinson CJ, Lwanga JS, Matthews RA, and Balmford A (1998) Complementarity and the use of indicator groups for reserve selection in Uganda. Nature 394:472–475.

Hulme D, and Murphree M (1999) Communities, wildlife, and the "new conservation" in Africa. Journal of International Development 11:277–285.

Hulme D, and Murphree M, eds. (2001) African wildlife and African livelihoods: The promise and performance of community conservation. Oxford: James Currey.

Hunn E (1982) Mobility as a factor limiting resource use in the Columbian Plateau of North America. In N Williams and E Hunn (eds.), Resource managers: North American and Australian hunter-gatherers. Boulder: Westview Press, pp. 17–43.

Hunter ML (1991) Coping with ignorance: The coarse filter strategy for maintaining biodiversity. In KA Kohm (ed.), Balancing on the brink of extinction: The Endangered Species Act and lessons for the future. Washington, DC: Island Press.

Hurrell A (1994) A crisis of ecological viability: Global environmental change and the nation state. Political Studies 42:146–165.

Hyden G (1990) Reciprocity and governance in Africa. In JS Winsch and D Olowu (eds.), The failure of centralized state: Institutions and self governance in Africa. Boulder: Westview Press.

Hyndman D (1994) Conservation through self-determination: Promoting the interdependence of cultural and biological diversity. Human Organization 53:296–302.

IIED (1994) Whose Eden? An overview of community approaches to wildlife management. London: International Institute for Ecology and Environment.

Illius AW, and O'Connor TG (1999) On the relevance of nonequilibrium concepts to arid and semiarid grazing systems. Ecological Applications 9:798–813.

Inamdar A, de Jode H, Lindsay K, and Cobb S (1999) Conservation: Capitalizing on nature: Protected area management. Science 283: 1856–1857.

Irwin A (1995) Citizen science: A study of people, expertise, and sustainable development. London: Routledge.

Ise J (1961) Our national park policy: A critical history. Baltimore: Johns Hopkins University Press.

IUCN (1994) Guidelines for protected area management categories. Gland, Switzerland.

IUCN/UNEP/WWF (1980) World conservation strategy: Living resource conservation for sustainable development. Gland, Switzerland.

IUCN/UNEP/WWF (1991) Caring for the earth: A strategy for sustainable living. Gland, Switzerland.

IWGIA (1992) Declaration by the indigenous peoples. International Work Group for Indigenous Affairs Yearbook:157–163.

IWGIA (1998) From principles to practice: Indigenous peoples and biodiversity conservation in Latin America. Copenhagen: International Work Group for Indigenous Affairs.

Iwu MM (1996) Resource utilization and conservation in Africa. In MJ Balick, E Elisabetsky, and SA Laird (eds.), Medicinal resources of the tropical forest. New York: Columbia University Press, pp. 233–250.

Jackson C (1993) Doing what comes naturally? Women and environment in development. World Development 21:1947–1963.

Jackson JBC (2001) What was natural in the coastal oceans? Proceedings of the National Academy of Sciences of the United States of America 98:5411–5418.

Jackson JBC, Kirby MX, Berger WH, Bjorndal KA, Botsford LW, Bourque BJ, Bradbury RH, Cooke R, Erlandson J, Estes JA, Hughes TP, Kidwell S, Lange CB, Lenihan HS, Pandolfi JM, Peterson CH, Steneck RS, Tegner MJ, and Warner RR (2001) Historical overfishing and the recent collapse of coastal ecosystems. Science 293:629–638.

Jacobson SK (1995) Conserving wildlife: International education and communication approaches. New York: Columbia University Press.

Jacobson SK, and McDuff MD (1998) Conservation education. In WJ Sutherland (ed.), Conservation science and action. Oxford: Blackwell Science, pp. 237–255.

James AN, Gaston KJ, and Balmford A (1999) Balancing the earth's accounts. Nature 401: 323–324.

Janzen DH (1987) How to grow a tropical national park: Basic philosophy for Guanacaste National Park, northwestern Costa Rica. Experientia 43:1037–1038.

Janzen DH (1994) Wildland biodiversity management in the tropics: Where are we now and where are we going? Vida Silvestre Neotropical 3:3–15.

Janzen DH (2000) Costa Rica's Area de Conservación Guanacaste: A long march to survival through non-damaging biodiversity and ecosystem development. Biodiversity 1:7–20.

Jochim MA (1981) Strategies for survival. New York: Academic Press.

Johannes RE (1978) Traditional marine conservation methods in Oceania and their demise. Annual Review of Ecological Systematics 9: 349–364.

Johannes RE (1998) Government-supported village-based management of marine resources in Vanuatu. Ocean and Coastal Management 40:165–186.

Johns ND (1999) Conservation in Brazil's chocolate forest: The unlikely persistence of the traditional cocoa agroecosystem. Environmental Management 23:31–47.

Johnson A (1989) How the Machiguenga manage resources: Conservation or exploitation of nature. In DA Posey and W Balée (eds.), Resource management in Amazonia: Indigenous and folk strategies. New York: The New York Botanical Garden, pp. 213–222.

Johnson C (2001) Community formation and fisheries conservation in southern Thailand. Development and Change 32:951–974.

Johnston BR (1997) Life and death matters: Human rights and the environment at the end of the millennium. Walnut Creek, CA: AltaMira Press.

Johnston BR (2001) Considering the power and potential of the anthropology of Trouble. In E Messer and M Lambek (eds.): Ecology and the sacred: Engaging the anthropology of Roy A. Rappaport. Ann Arbor: University of Michigan Press, pp. 99–121.

Judd RW (1997) Common lands, common people: The origins of conservation in northern New England. Cambridge, MA: Harvard University Press.

Juma C (1989) The gene hunters: Biotechnology and the scramble for seeds. Princeton, NJ: Princeton University Press.

Kacelnik A (1997) Normative and descriptive models of decision making: Time discounting and risk sensitivity. In G Bock and G Cardew (eds.), Characterizing human psychological adaptations. London: Wiley.

Kahn JR, and McDonald JA (1995) Third-world debt and tropical deforestation. Ecological Economics 12:107–123.

Kaplan H, and Hill K (1992) The evolutionary ecology of food acquisition. In EA Smith and B Winterhalder (eds.), Evolutionary ecology and human behavior. New York: Aldine de Gruyter, pp. 167–201.

Kaplan H, and Kopishke K (1992) Resource use, traditional technology, and change among native peoples of lowland South America. In KH Redford and C Padoch (eds.), Conservation of neotropical forests. New York: Columbia University Press, pp. 83–107.

Karanth KU (2002) Nagarahole: Limits and opportunities in wildlife conservation. In J Terborgh, C van Schaik, L Davenport, and M Rao (eds.), Making parks work: Strategies for preserving tropical nature. Washington, DC: Island Press, pp. 189–202.

Karanth KU, Bhargav P, and Kumar NS (2001) Karnataka tiger conservation project. New York: Wildlife Conservation Society.

Katon BM, Pomeroy RS, Garces LR, and Salamanca AM (1999) Fisheries management of San Salvador Island, Philippines: A shared responsibility. Society and Natural Resources 12:777–795.

Kaus A (1993) Environmental perceptions and social relations in the Mapimi Biosphere Reserve. Conservation Biology 7:398–406.

Kellert SR, and Wilson EO (1993) The biophilia hypothesis. Washington, DC: Island Press.

Kemf E, ed. (1993) Indigenous peoples and protected areas. London: Earthscan.

Kenchington RA (1989) Tourism in the Galapagos Islands: The dilemma of conservation. Environmental Conservation 16:227–232.

Keohane RO, and Ostrom E, eds. (1995) Local commons and global interdependence. London: Sage Publications.

King SR (1996) Conservation and tropical medicinal plant research. In MJ Balick, E Elisabetsky, and SA Laird (eds.), Medicinal resources of the tropical forest. New York: Columbia University Press, pp. 62–74.

Kirch PV (1984) The evolution of the Polynesian chiefdoms. New York: Cambridge University Press.

Kirch PV (2000) On the road of the winds: An archaeological history of the Pacific islands before European contact. Berkeley: University of California Press.

Kirch PV, and Hunt TL (1997) Historical ecology in the Pacific Islands: Prehistoric environmental and landscape change. New Haven: Yale University Press.

Kiss A, Castro G, and Newcombe K (2002) The role of multilateral institutions. Philosophical Transactions of the Royal Society, London, A. 360:1641–1652.

Kiss A, and World Bank (1990) Living with wildlife: Wildlife resource management with local participation in Africa. Washington, DC: World Bank.

Kjekshus H (1977) Ecology control and economic development in East African history. London: Heinemann.

Knight R (1965) A re-examination of hunting, trapping, and territoriality among the northeastern Algonkian Indians. In A Leeds and AP Vayda (eds.), Man, culture, and animals. Washington, DC: American Association for the Advancement of Science, pp. 27–42.

Koch E (1997) Ecotourism and rural reconstruction in South Africa: Reality or rhetoric. In KB Ghimire and MP Pimbert (eds.), Social change and conservation. London: Earthscan, pp. 213–238.

Koester F (2001) El hombre y la biosfera en Yasuni. Nuestra Ciencia 3:23–29.

Konner M (1990) Why the reckless survive: And other secrets of human nature. New York: Viking.

Kottak CP (1999) The new ecological anthropology. American Anthropologist 101:23–35.

Kramer RA, van Schaik C, and Johnson J (1997) Last stand: Protected areas and the defense of tropical biodiversity. New York: Oxford University Press.

Kramer RA, and Sharma N (1997) Tropical forest biodiversity protection: Who pays and why. In RA Kramer, Cv Schaik, and J Johnson (eds.), Last stand: Protected areas and the defense of tropical biodiversity. New York: Oxford University Press, pp. 162–186.

Krebs JR, and Davies NB (1993) An introduction to behavioural ecology. Oxford; Cambridge, MA: Blackwell Scientific Publications.

Kremen C, Merelender AM, and Murphy DD (1994) Ecological monitoring: A vital need for an integrated conservation and development programs in the tropics. Conservation Biology 8:388–397.

Kremen C, Niles JO, Dalton MG, Daily GC, Ehrlich PR, Fay JP, Grewal D, and Guillery RP (2000) Economic incentives for rain forest conservation across scales. Science 288:1828–1832.

Kremen C, Raymond I, and Lance K (1998) An interdisciplinary tool for monitoring conservation impacts in Madagascar. Conservation Biology 12:549–563.

Krishnaswamy A (1995) Sustainable development and community forest management in Bihar, India. Society and Natural Resources 8:339–350.

Kumar S (1993) Taiwan accuses princess of smuggling rhino horn. New Scientist 140:11.

Lam WF (1996) Improving the performance of small-scale irrigation systems: The effects of technological investments and governance structure on irrigation performance in Nepal. World Development 24:1301–1315.

Lambeck RJ (1997) Focal species: A multispecies umbrella for nature conservation. Conservation Biology 11:849–856.

Lamprey HF (1983) Pastoralism yesterday and today: The overgrazing problem. In F Bourliere (ed.), Tropical savannas. Amsterdam: Elsevier, pp. 643–666.

Lamprey HF, and Yussuf H (1981) Pastoralism and desert encroachment in Northern Kenya. Ambio 10:131–134.

Lamprey R, and Waller R (1990) The Loita-Mara region in historical times: Patterns of subsistence, settlement and ecological change. In P Robertshaw (ed.), Early pastoralists of South-western Kenya. Nairobi: British Institute in Eastern Africa, pp. 16–35.

Lane C (1996) Pastures lost: Barabaig economy, resource tenure, and the alienation of their land in Tanzania. Nairobi, Kenya: Initiatives Publishers.

Lane CR (1998) Introduction: Custodians of the commons: Pastoral land tenure in East and West Africa. London: Earthscan, pp. 1–25.

Langholz J (2002) Privately owned parks. In J Terborgh, C van Schaik, L Davenport, and M Rao (eds.), Making parks work: Strategies for preserving tropical nature. Washington, DC: Island Press, pp. 172–188.

Lansing JS (1987) Balinese "water temples" and the management of irrigation. American Anthropologist 89:326–341.

Lansing JS, and Kremer JN (1993) Emergent properties of Balinese water temple networks: Coadaptation on a rugged fitness landscape. American Anthropologist 95:97–114.

Lappé FM, and Schurman R (1989) Taking population seriously. London: Earthscan.

Lasserre P (1999) Broadening horizons. Sources 109:4–9.

Lavigne DM, Callaghan CJ, and Smith RJ (1996) Sustainable utilization: The lessons of history. In VJ Taylor and N Dunstone (eds.), The exploitation of mammal populations. London: Chapman and Hall, pp. 250–265.

Lawton JH, Bignell DE, Bolton B, Bloemers GF, Eggleton P, Hammond PM, Hodda M, Holt RD, Larsen TB, Mawdsley NA, Stork NE, Srivastava DS, and Watt AD (1998) Biodiversity inventories, indicator taxa and effects of habitat modification in tropical forest. Nature 391:72–76.

Le Maitre DC, Van Wilgen BW, Chapman RA, and McKelly DH (1996) Invasive plants and water resources in the Western Cape Province, South Africa: Modelling the consequences of a lack of management. Journal of Applied Ecology 33:161–172.

Leach ER (1972) Anthropological aspects: Conclusion. In PR Cox and J Peel (eds.), Population and pollution. New York: Academic Press, pp. 37–40.

Leach M (1996) West African forestry history. In TS Driver and GP Chapman (eds.), Time scales and environmental change. London: Routledge.

Leach M, and Fairhead J (2002) Manners of contestation: "Citizen science" and "indigenous knowledge" in West Africa and the Caribbean. International Social Science Journal 54:299–311.

Leach M, Mearns R, and Scoones I (1999) Environmental entitlements: Dynamics and institutions in community-based natural resource management. World Development 27: 225–247.

Leacock E (1954) The Montagnais "hunting territory and the fur trade." American Anthropological Association Memoir No. 78 56:1–59.

Leader-Williams N, and Albon SD (1988) Allocation of resources for conservation. Nature 336:533–535.

Leader-Williams N, and Dublin H (2000) Charismatic megafauna as "flagship species." In A Entwistle and N Dunstone (eds.), Priorities for the conservation of mammalian diversity: Has the panda had its day? Cambridge: Cambridge University Press, pp. 53–81.

Leader-Williams N, and Milner-Gulland EJ (1993) Policies for the enforcement of wildlife laws: The balance between detection and penalties in Luangwa Valley, Zambia. Conservation Biology 7:611–617.

Leakey RE, and Lewin R (1995) The sixth extinction: Patterns of life and the future of humankind. New York: Doubleday.

Lee RB, and Daly RH (1999) The Cambridge encyclopedia of hunters and gatherers. New York: Cambridge University Press.

Leimar O, and Hammerstein P (2001) Evolution of cooperation through indirect reciprocity. Proceedings of the Royal Society of London Series B-Biological Sciences 268: 745–753.

Lele SM (1991) Sustainable development: A critical view. World Development 19:607–621.

Leopold A (1933) Game management. New York: Scribner & Sons.

Leopold A (1970) A Sand County almanac, with essays on conservation from Round River. New York: Ballantine Books.

Leopold A (1997) The land ethic. In JR Des Jardins (ed.), Environmental ethics: An introduction to environmental philosophy. Belmont, CA: Wadsworth, pp. 170–197.

Levins R (1969) Some demographic and genetic consequences of environmental heterogeneity for biological control. Bulletin of the Entomological Society of America 15:237–240.

Lewis DM (1995) Importance of GIS to community-based management of wildlife: Lessons from Zambia. Ecological Applications 5: 861–871.

Lewis DM, Kaweche GB, and Mwenya A (1990) Wildlife conservation outside protected areas: Lessons from an experiment in Zambia. Conservation Biology 4:171–180.

Lewis HT (1989) Ecological and technical knowledge of fire: Aborigines versus park rangers in Northern Australia. American Anthropologist 91:940–961.

Lidicker WZ (1995) Landscape approaches in mammalian ecology and conservation. Minneapolis: University of Minnesota Press.

Lima DM (1999) Equity, sustainable development, and biodiversity preservation: Some questions about ecological partnerships in the Brazilian Amazon. In C Padoch, JM Ayres, M Pinedo-Vasquez, and A Henderson (eds.), Várzea: Diversity, conservation and development in Amazonia's whitewater floodplains. New York: New York Botanical Garden Press, pp. 247–263.

Lindsay WK (1987) Integrating parks and pastoralists: Some lessons from Amboseli. In D Anderson and R Grove (eds.), Conservation in Africa: People, policies and practice. Cambridge: Cambridge University Press, pp. 149–167.

Little PD (1994) The link between local participation and improved conservation: A review of issues and experiences. In D Western and RM Wright (eds.), Natural connections. Washington, DC: Island Press, pp. 347–372.

Loucks CJ, Lu Z, Dinerstein E, Wang H, Olson DM, Zhu CQ, and Wang DJ (2001) Ecology: Giant pandas in a changing landscape. Science 294:1465.

Lovelock J (1979) Gaia: A new look at life on earth. New York: Oxford University Press.

Low BS (1996) Behavioral ecology of conservation in traditional societies. Human Nature 7: 353–379.

Low BS, and Heinen JT (1993) Population, resources and environment. Population and Environment 15:7–41.

Lu F (1999) Changes in subsistence patterns and resource use of the Huaorani Indians in the Ecuadorian Amazon. PhD dissertation, Chapel Hill, NC.

Lubchenco J, and Menge BA (1978) Community development and persistence in a low rocky intertidal zone. Ecological Monographs 59:67–94.

Ludwig D (2000) Limitations of economic valuation of ecosystems. Ecosystems 3:31–35.

Ludwig D, Hilborn R, and Walters CJ (1993) Uncertainty, resource exploitation, and conservation: Lessons from history. Science 260: 17–18.

Lynge F (1992) Arctic wars, animal rights, endangered peoples. Hanover, NH: Dartmouth College : University Press of New England.

Maass A, and Anderson RL (1978) . . . And the desert shall rejoice: Conflict, growth, and justice in arid environments. Cambridge, MA: MIT Press.

MacArthur R (1958) Population ecology of some warblers of northeastern coniferous forests. Ecology 39:599–619.

MacArthur RH, and Wilson EO (1967) The theory of island biogeography. Princeton, NJ: Princeton University Press.

Mace GM, Balmford A, Boitani L, Cowlishaw G, Dobson AP, Faith DP, Gaston KJ, Humphries CJ, Vane-Wright RI, Williams PH, Lawton JH, Margules CR, May RM, Nicholls AO, Possingham HP, Rahbek C, and van Jaarsveld AS (2000) It's time to work together and stop duplicating conservation efforts. Nature 405:393.

Mace GM, and Lande R (1991) Assessing extinction threats: Toward a reevaluation of IUCN threatened species categories. Conservation Biology 5:148–157.

Mace R (1991) Overgrazing overstated. Nature 349:280–281.

MacKenzie JM (1987) Chivalry, social Darwinism and ritualised killing: The hunting ethos in Central Africa up to 1914. In D Anderson and R Grove (eds.), Conservation in Africa: People, policies and practice. Cambridge: Cambridge University Press, pp. 41–61.

MacKenzie JM (1988) The empire of nature: Hunting, conservation, and British imperialism. Manchester: Manchester University Press.

Maffi L (2001) On the interdependence of biological and cultural diversity. In L Maffi (ed.), On biocultural diversity: Linking language, knowledge, and the environment. Washington, DC: Smithsonian Institution Press, pp. 1–50.

Maffi L, Oviedo G, and Larsen PB (2000) Indigenous and traditional peoples of the world and ecoregion conservation. Gland, Switzerland: WWF.

Mahapatra A, and Mitchell CP (1997) Sustainable development of non-timber forest products: Implication for forest management in India. Forest Ecology and Management 94: 15–29.

Mahony R (1992) Debt-for-nature swaps: Who really benefits. The Ecologist 22:97–103.

Malthus TR (1959) Population: The first essay. Ann Arbor: University of Michigan Press.

Manwood J (1615) The treatise of the Lawes of the forest. London: Societie of Stationers.

Marcus RR (2001) Seeing the forest for the trees: Integrated conservation and development projects and local perceptions of conservation in Madagascar. Human Ecology 29:381–397.

Margoluis R, Margoluis C, Brandon K, and Safalsky N (2000) In good company: Effective alliances for conservation. Washington, DC: Biodiversity Support Program.

Margules CR, and Pressey RL (2000) Systematic conservation planning. Nature 405:243–253.

Martin C (1993) Introduction. In E Kemf (ed.), Indigenous peoples and protected areas. London: Earthscan, pp. xv–xix.

Mason P (1997) Tourism codes of conduct in the Arctic and Sub-Arctic region. Journal of Sustainable Development 5:151–164.

Mauro F, and Hardison PD (2000) Traditional knowledge of indigenous and local communities: International debate and policy initiatives. Ecological Applications 10:1263–1269.

Mawdsley E (1998) After Chipko: From environment to region in Uttaranchal. Journal of Peasant Studies 25:36–54.

May PH (1992) Common property resources in the neotropics: Theory, management progress, and an action agenda. In KH Redford and C Padoch (eds.), Conservation of neotropical forests. New York: Columbia University Press, pp. 359–378.

May R (1999) Unanswered questions in ecology. Philosophical Transactions of the Royal Society, London B. 354:1951–1959.

Maybury-Lewis D (1991) Becoming Indian in lowland South America. In G Urban and J Sherzer (eds.), Nation states and Indians in Latin America. Austin: University of Texas Press, pp. 207–235.

Maybury-Lewis D (1997) Indigenous peoples, ethnic groups and the state. Boston: Allyn and Bacon.

Maynard Smith J (1964) Group selection and kin selection. Nature 201:1145–1147.

McAfee K (1999) Selling nature to save it? Biodiversity and green developmentalism. Environment and Planning D: Society & Space 17:133–154.

McCabe JT (1990) Turkana pastoralism: A case against the tragedy of the commons. Human Ecology 18:81–103.

McCay BJ, and Acheson JM, eds. (1987) The question of the commons: The culture and ecology of communal resources. Tucson: University of Arizona Press.

McClanahan TR, Glaesel H, Rubens J, and Kiambo R (1997) The effects of traditional fisheries management on fisheries yields and the coral-reef ecosystems of southern Kenya. Environmental Conservation 24:105–120.

McDonald DR (1977) Food taboos: A primitive environmental protection agency (South America). Anthropos 72:734–748.

McDuff M, and Jacobson S (2000) Impacts and future directions of youth conservation organizations: Wildlife clubs in Africa. Wildlife Society Bulletin 28:414–425.

McGinn AP (1998) Promoting sustainable fisheries. In LR Brown, C Flavin, and H French (eds.), State of the world 1998. New York: Norton, pp. 59–78.

McGinnis MV (1999) Bioregionalism. New York: Routledge.

McIntosh IS (2001) Plan A and Plan B partnerships for cultural survival. Cultural Survival Quarterly 25:4–6.

McIntosh RP (1985) The background of ecology: Concept and theory. Cambridge: Cambridge University Press.

McKean MA (1992a) Management of traditional common lands (Iriaichi) in Japan. In DW Bromley and D Feeny (eds.), Making the commons work: Theory, practice, and policy. San Francisco: ICS Press, pp. 63–98.

McKean MA (1992b) Success on the commons: A comparative examination of institutions for common property resource management. Journal of Theoretical Politics 4:243–281.

McMichael AJ, Bolin B, Costanza R, Daily GC, Folke C, Lindahl-Kiessling K, Lindgren E, and Niklasson B (1999) Globalization and the sustainability of human health: An ecological perspective. Bioscience 49:205–210.

McNeely JA (1990) Conserving the world's biological diversity. Gland, Switzerland: IUCN.

McNeely JA (1995) Expanding partnerships in conservation. Washington, DC: Island Press.

McNeely JA (1999) The convention on biological diversity: A solid foundation for effective action. Environmental Conservation 26:250–251.

McNeely JA (2000) How countries with limited resources are dealing with biodiversity problems. In PH Raven (ed.), Nature and human society: The quest for a sustainable world. Washington, DC: National Academy Press, pp. 557–572.

McNeely JA, and Weatherly WP (1996) Innovative funding to support biodiversity conservation. International Journal of Social Economics 23:98–124.

Meadows DH, Meadows DI, Randers J, and Behrens WW III (1972) The limits to growth. New York: Universe Books.

Medley KE (1993) Extractive forest resources of the Tana River National Primate Reserve, Kenya. Economic Botany 47:171–183.

Meffe GK, and Carroll CR (1997) Principles of conservation biology. Sunderland, MA: Sinauer.

Meffe GK, Ehrlich AH, and Ehrenfeld D (1993) Human population control: The missing agenda. Conservation Biology 7:1–3.

Mehta JN, and Heinen JT (2001) Does community-based conservation shape favorable attitudes among locals? An empirical study from Nepal. Environmental Management 28:165–177.

Mehta JN, and Kellert SR (1998) Local attitudes toward community-based conservation policy and programmes in Nepal: A case study in the Makalu-Barun Conservation Area. Environmental Conservation 25:320–333.

Meine C (1988) Aldo Leopold: His life and work. Madison: University of Wisconsin Press.

Meinertzhagen R (1983) Kenya diary (1902–1906). London: Eland Books.

Meinzen-Dick R, Raju KV, and Gulati A (2002) What affects organization and collective action from managing resources? Evidence from canal irrigation systems in India. World Development 30:649–666.

Menzel P (1994) Material world: A global family portrait. San Francisco: Sierra Club Books.

Merchant C (1982) The death of nature: Women, ecology, and the scientific revolution. New York: Harper & Row.

Merson J (2000) Bio-prospecting or bio-piracy: Intellectual property rights and biodiversity in a colonial and postcolonial context. *Osiris* 15:282–296.

Metcalfe S (1994) The Zimbabwe communal areas management programme for indigenous resources (CAMPFIRE). In D Western and M Wright (eds.), Natural connections: Perspectives in community-based conservation. Washington, DC: Island Press, pp. 161–192.

Micklebugh S (2000) CITES: What role for science? *Oryx* 34:241–242.

Milner-Gulland EJ (2002) Is bushmeat just another conservation bandwagon? *Oryx* 32:1–2.

Milner-Gulland EJ, and Mace R, eds. (1998) Conservation of biological resources. Malden, MA: Blackwell Science.

Moegenburg SM, and Levey DJ (2002) Prospects for conserving biodiversity in Amazonian extractive reserves. *Ecology Letters* 5:320–324.

Moore JL, Manne L, Brooks T, Burgess ND, Davies R, Rahbek C, Williams P, and Balmford A (2002) The distribution of cultural and biological diversity in Africa. Proceedings of the Royal Society of London Series B-Biological Sciences 269:1645–1653.

Moorehead R (1991) Structural chaos: Community and state management of common property in Mali. PhD dissertation, University of Sussex, UK.

Moran D (1994) Contingent valuation and biodiversity: Measuring the user surplus of Kenyan protected areas. *Biodiversity and Conservation* 3:663–684.

Moran K (2001) Lessons from bioprospecting in India and Nigeria. *Cultural Survival Quarterly* 24:25–27.

Moran K, King SR, and Carlson TJ (2001) Biodiversity prospecting: Lessons and prospects. *Annual Review of Anthropology* 30:505–526.

Mosimann JE, and Martin PS (1975) Simulating overkill by Paleoindians. *American Scientist* 63:304–313.

Mukerjee R (1942) Social ecology. New York; Bombay: Longmans Green.

Muller S (2003) Toward decolonisation of Australia's protected area management: The Nantawarrina Indigenous Protected Area experience. *Australian Geographical Studies* 41:29–43.

Muniz-Miret N, Vamos R, Hiraoka M, Montagnini F, and Mendelsohn Robert O (1996) The economic value of managing the acai palm (Euterpe oleracea Mar) in the floodplains of the Amazon estuary, Para, Brazil. *Forest Ecology & Management* 87:163–173.

Murombedzi J (1991) Decentralizing common property resources management: A case study of the Nyaminyami District Council of Zimbabwe's wildlife management programme. London: IIED.

Murombedzi J (2001) Natural resource stewardship and community benefits in Zimbabwe's CAMPFIRE programme. In D Hulme and M Murphree (eds.), African wildlife and livelihoods: The promise and performance of community conservation. Oxford: James Currey, pp. 244–255.

Murphree M (1994) The role of institutions in community-based conservation. In D Western and RM Wright (eds.), Natural connections: Perspectives on community-based conservation. Washington, DC: Island Press, pp. 403–427.

Murphree M (2001) A case study in ecotourism development in Mahenye, Zimbabwe. In D Hulme and M Murphree (eds.), African wildlife and livelihoods: The promise and performance of community conservation. Oxford: James Currey, pp. 177–194.

Murray W (1995) Lessons from 35 years of private preserve management in the USA. In JA McNeely (ed.), Expanding partnerships in conservation. Washington, DC: Island Press, pp. 198–205.

Myers N (1987) The impending extinction spasm: Synergisms at work. *Conservation Biology* 14:15–22.

Myers N (1988) Tropical forests: Much more than stocks of wood. *Journal of Tropical Ecology* 4:209–221.

Myers N (1998) Lifting the veil on perverse subsidies. *Nature* 392:327–329.

Myers N, Mittermeier RA, Mittermeier CG, da Fonseca GAB, and Kent J (2000) Biodiversity hotspots for conservation priorities. *Nature* 403:853–858.

Nabhan GP (1997) Cultures of habitat: On nature, culture, and story. Washington, DC: Counterpoint.

Næss A, and Rothenberg D (1989) Ecology, community, and lifestyle: Outline of an ecosophy. New York: Cambridge University Press.

Nagami T (1995) Protection of the earth's environment and corporate ethics for the future. In JA McNeely (ed.), Expanding partnerships in conservation. Washington, DC: Island Press, pp. 24–29.

Nagendra H (2001) Incorporating landscape transformation into local conservation prioritization: A case study in the Western Ghats, India. Biodiversity and Conservation 10:353–365.

Nash RF (1989) The rights of nature: A history of environmental ethics. Madison: University of Wisconsin Press.

Naughton-Treves L (1999) Whose animals? A history of property rights to wildlife in Toro, western Uganda. Land Degradation & Development 10:311–328.

Nelson JG, and Serafin R (1992) Assessing biodiversity: A human ecological approach. Ambio 21:212–218.

Nepal SK (2002) Involving indigenous peoples in protected area management: Comparative perspectives from Nepal, Thailand, and China. Environmental Management 30:748–763.

Netting RM (1981) Balancing on an Alp: Ecological change and continuity in a Swiss mountain community. New York: Cambridge University Press.

Nettle D (1996) Language diversity in West Africa: An ecological approach. Journal of Anthropological Archaeology 15:403–438.

Nettle D (1998) Explaining global patterns of language diversity. Journal of Anthropological Archaeology 17:354–374.

Neumann RP (1996) Dukes, Earls, and Ersatz Edens: Aristocratic nature preservationists in Colonial Africa. Environment and Planning D-Society & Space 14:79–98.

Neumann RP (1997) Primitive ideas: Protected area buffer zones and the politics of land in Africa. Development and Change 28:559–582.

Neumann RP (1998) Imposing wilderness: Struggles over livelihood and nature preservation in Africa. Berkeley: University of California Press.

Newmark WD, and Hough JL (2000) Conserving wildlife in Africa: Integrated conservation and development projects and beyond. Bioscience 50:585–592.

Ngoile MN, Linden O, and Coughenour CA (1995) Coastal zone management in Eastern Africa including the island states: A review of issues and initiatives. Ambio 24:448–457.

Niamir-Fuller M (1998) The resilience of pastoral herding in Sahelian Africa. In F Berkes and C Folke (eds.), Linking social and ecological systems. Cambridge: Cambridge University Press, pp. 250–284.

Nietschmann BQ (1987) Militarization and indigenous people. Cultural Survival Quarterly 11:1–16.

Nietschmann BQ (1991a) Miskito Coast protected area. Research & Exploration 7:232–234.

Nietschmann BQ (1991b) Conservation by self-determination. Research & Exploration: 372–373.

Nigh R (2002) Maya medicine in the biological gaze: Bioprospecting research as herbal fetishism. Current Anthropology 43:451–477.

Niles JO, Brown S, Pretty J, Ball A, and Fay J (2002) Potential carbon mitigation and income in developing countries from changes in use and management of agricultural and forest lands. Philosophical Transactions of the Royal Society, London, A. 360:1621–1639.

Njiforti HL, and Tchamba NM (1993) Conflict in Cameroon: Parks for or against people. In E Kemf (ed.), Indigenous peoples and protected areas. London: Earthscan, pp. 173–178.

Nollman J (1990) Spiritual ecology: A guide for reconnecting with nature. New York: Bantam Books.

Norton BG (2000) Biodiversity and environmental values: In search of a universal earth ethic. Biodiversity and Conservation 9:1029–1044.

Norton DA (2000) Conservation biology and private land: Shifting the focus. Conservation Biology 14:1221–1223.

Norton-Griffiths M (1998) The economics of wildlife conservation policy in Kenya. In EJ Milner-Gulland and R Mace (eds.), Conservation of biological resources. Malden, MA: Blackwell Science, pp. 279–293.

Norton-Griffiths M, and Southey C (1995) The opportunity costs of biodiversity conservation in Kenya. Ecological Economics 12:125–139.

Noss AJ (1997) Challenges to nature conservation with community development in central African forests. Oryx 31:180–188.

Noss RF, Carroll C, Vance-Borland K, and Wuerthner G (2002) A multicriteria assessment of the irreplaceability and vulnerability of sites in the Greater Yellowstone Ecosystem. Conservation Biology 16:895–908.

Noss RF, O'Connell MA, and Murphy DD (1997) The science of conservation planning: Habitat conservation under the Endangered Species Act. Washington, DC: Island Press.

Noss RF, and Cooperrider A (1994) Saving nature's legacy. Washington, DC: Island Press.

Notzke C (1995) A new perspective in Aboriginal natural resource management: Co-management. Geoforum 26:187–209.

Novaro AJ (1995) Sustainability of harvest of culpeo foxes in Patagonia. Oryx 29:18–22.

Novaro AJ, Redford KH, and Bodmer RE (2000) Effect of hunting in source-sink systems in the neotropics. Conservation Biology 14:713–721.

Nowak MA, and Sigmund K (1993) A strategy of win-stay, lose-shift that outperforms tit-for-tat in the prisoner's dilemma game. Nature 364:56–58.

Nowak MA, and Sigmund K (1998) Evolution of indirect reciprocity by image scoring. Nature 393:573–577.

Nyerges AE, and Green GM (2000) The ethnography of landscape: GIS and remote sensing in the study of forest change in West African guinea savanna. American Anthropologist 102:271–289.

Oakerson RJ (1992) Analyzing the commons: A framework. In DW Bromley and D Feeny (eds.), Making the commons work: Theory, practice, and policy. San Francisco: ICS Press, pp. 41–59.

Oates JF (1995) The dangers of conservation by rural development: A case study from the forests of Nigeria. Oryx 29:115–122.

Oates JF (1999) Myth and reality in the rain forest: How conservation strategies are failing in West Africa. Berkeley: University of California Press.

O'Brien TG, and Kinnaird MF (2003) Caffeine and conservation. Science 300:587.

O'Brien TG, and Kinnaird MF (2000) Differential vulnerability of large birds and mammals to hunting in north Sulawesi, Indonesia and the outlook for the future. In J Robinson and E Bennett (eds.), Hunting for sustainability in tropical forests. New York: Columbia University Press, pp. 199–213.

Ochoa J (1998) Preliminary assessment of the effects of logging on the composition and structure of forests in the Venezuelan Guayana Region. Interciencia 23:197–207.

Oldfield TEE, Smith RJ, Harrop SR, and Leader-Williams N (2003) Field sports and conservation in the United Kingdom. Nature 423:531–533.

Olson DM, and Dinerstein E (1998) The global 200: A representation approach to conserving the Earth's most biologically valuable ecoregions. Conservation Biology 12:502–515.

Olson M (1965) The logic of collective action: Public goods and the theory of groups. Cambridge, MA: Harvard University Press.

Olsson P, and Folke C (2001) Local ecological knowledge and institutional dynamics for ecosystem management: A study of Lake Racken Watershed, Sweden. Ecosystems 4: 85–104.

O'Neill BC, MacKellar FL, Lutz W, and International Institute for Applied Systems Analysis (2001) Population and climate change. Cambridge: Cambridge University Press.

Orgeas J, and Andersen AN (2001) Fire and biodiversity: Responses of grass-layer beetles to experimental fire regimes in an Australian tropical savanna. Journal of Applied Ecology 38:49–62.

Orians GH (1998) Human behavioral ecology: 140 years without Darwin is too long. Bulletin of the Ecological Society of America 79: 15–28.

Orians GH, and Heerwagen JH (1992) Evolved responses to landscape. In JH Barkow, L Cosmide, and J Tooby (eds.), The adapted mind. New York: Oxford, pp. 555–579.

Orlove BS, and Brush SB (1996) Anthropology and the conservation of biodiversity. Annual Review of Anthropology 25: 329–352.

Ornstein RE, and Ehrlich PR (1989) New world new mind: Moving toward conscious evolution. New York: Doubleday.

Ostrom E (1990) Governing the commons: The evolution of institutions for collective action. Cambridge: Cambridge University Press.

Ostrom E (1998) A behavioral approach to the rational choice theory of collective action. American Political Science Review 92:1–22.

Ostrom E, Burger J, Field CB, Norgaard RB, and Policansky D (1999) Revisiting the commons: Local lessons, global challenges. Science 284:278–282.

Ostrom E, Dietz T, Dolšak N, Stern P, Stronich

S, and Weber E, eds. (2002) The drama of the commons. Washington, DC: National Academy Press.

Ostrom E, and Walker J (1997) Neither markets nor states: Linking transformation processes in collective action arenas. In DC Mueller (ed.), Perspectives on public choice: A handbook. Cambridge: Cambridge University Press.

Ostrom E, Walker J, and Gardner R, eds. (1994) Rules, games and common pool resources. Ann Arbor: University of Michigan Press.

OTA (1987) Technologies to maintain biological diversity. Washington, DC: U.S. Government Printing Office, Office of Technological Assessment, U.S. Congress.

Owen-Smith N (1987) Pleistocene extinctions: The pivotal role of megaherbivores. Paleobiology 13:351–362.

Paehlke RC (1989) Environmentalism and the future of progressive politics. New Haven, CT: Yale University Press.

Painter M, and Durham WH, eds. (1996) The social causes of environmental destruction in Latin America. Ann Arbor: University of Michigan Press.

Palmer CT (1991) Kin-selection, reciprocal altruism, and information sharing among Maine lobstermen. Ethology and Sociobiology 12:221–235.

Panayotou T, and Ashton PS (1992) Not by timber alone: Economics and ecology for sustaining tropical forests. Washington, DC: Island Press.

Parker E (1992) Forest islands and Kayapo resource management in Amazonia: A reappraisal of the Apete. American Anthropologist 94:406–428.

Parker E (1993) Fact and fiction in Amazonia: The case of the Apete. American Anthropologist 95:715–723.

Parkes M, and Panelli R (2001) Integrating catchment ecosystems and community health: The value of participatory action research. Ecosystem Health 7:85–106.

Parks SA, Harcourt AH, Tobler W, Deichmann U, Gottsegen J, and Maloy K (2002) Reserve size, local human density, and mammalian extinctions in US protected areas. Conservation Biology 16:800–808.

Parras DA (2001) Coastal resource management in the Philippines: A case study in the Central Visayas region. Journal of Environment and Development 10:80–103.

Parson EA, and Fisher-Vanden K (1999) Joint implementation of greenhouse gas abatement under the Kyoto protocol's "clean development mechanism": Its scope and limits. Policy Sciences 32:207–224.

Pearce DW (1993) Economic values and the natural world. Cambridge, MA: MIT Press.

Pelkey NW, Stoner CJ, and Caro TM (2000) Vegetation in Tanzania: Assessing long term trends and effects of protection using satellite imagery. Biological Conservation 94:297–309.

Peluso NL (1992) Rich forests, poor people: Resource control and resistance in Java. Berkeley: University of California Press.

Peluso NL (1995) Whose woods are these? Counter-mapping forest territories in Kalimantan, Indonesia. Antipode 27:383–406.

Peng JJ, Jiang ZG, and Hu JC (2001) Status and conservation of giant panda (Ailuropoda melanoleuca): A review. Folia Zoologica 50:81–88.

Penn D (2003) The evolutionary roots of our environmental problems: Towards a Darwinian ecology. Quarterly Review of Biology 78:275–301.

Peres CA (2000) Evaluating the impact and sustainability of subsistence hunting at multiple Amazonian forest sites. In JG Robinson and EL Bennett (eds.), Hunting for sustainability in tropical forests. New York: Columbia University Press, pp. 31–56.

Peres CA (2001) Synergistic effects of subsistence hunting and habitat fragmentation on Amazonian forest vertebrates. Conservation Biology 15:1490–1505.

Peres CA, Baider C, Zuidema PA, Wadt LHO, Kainer KA, Gomes-Silva DAP, Salomao RP, Simoes LL, Franciosi ERN, Valverde FC, Gribel R, Shepard GH, Kanashiro M, Coventry P, Yu DW, Watkinson AR, and Freckleton RP (2003) Demographic threats to the sustainability of Brazil nut exploitation. Science 302:2112–2114.

Peres CA, and Terborgh JW (1995) Amazonian nature reserves: An analysis of the defensibility status of existing conservation units and design criteria for the future. Conservation Biology 9:34–46.

Pereyra M (1998) Paraguay: The Mbaracayu

case. In IWGIA (ed.), From principles to practice: Indigenous peoples and biodiversity conservation in Latin America. Copenhagen: International Work Group for Indigenous Affairs, pp. 199–212.

Peters CM, Gentry AH, and Mendelsohn RO (1989) Valuation of an Amazonian rainforest. Nature 339:655–656.

Peters J (1998) Transforming the integrated conservation and development project (ICDP) approach: Observations from the Ranomafana National Park Project, Madagascar. Journal of Agricultural & Environmental Ethics 11:17–47.

Peters PE (1987) Embedded systems and rooted models: The grazing lands of Botswana and the commons debate. In BJ McCay and JM Acheson (eds.), The question of the commons. Tucson: University of Arizona Press, pp. 171–194.

Peterson G (2000) Political ecology and ecological resilience: An integration of human and ecological dynamics. Ecological Economics 35:323–336.

Pickup G, Bastin GN, and Chewings VH (1994) Remote-sensing-based condition assessment for nonequilibrium rangelands under large-scale commercial grazing. Ecological Applications 4:497–517.

Pimbert MP, and Pretty J (1997) Parks, people and professionals: Putting "participation" into protected area management. In KB Ghimire and MP Pimbert (eds.), Social change and conservation: Environmental politics and impacts of national parks and protected areas. London: Earthscan, pp. 297–330.

Pimentel D, Harman R, Pacenza M, Pecarsky J, and Pimentel M (1994) Natural resources and an optimum human population. Population and Environment 15:347–369.

Pimentel D, Wilson C, McCullum C, Huang R, Dwen P, Flack J, Tran Q, Saltman T, and Cliff B (1997) Economic and environmental benefits of biodiversity. Bioscience 47:747–757.

Pimm SL, Russell GJ, Gittleman JL, and Brooks TM (1995) The future of biodiversity. Science 269:347–350.

Pinedo-Vasquez M, Zarin D, and Jipp P (1992) Economic returns from forest conversion in the peruvian Amazon. Ecological Economics 6:163–173.

Place R, Roth M, and Hazell P (1994) Land tenure security and agricultural performance in Africa: Overview of research methodology. In J Bruce and SE Migot-Adholla (eds.), Searching for land tenure security in Africa. Dubuque, IO: Kendall/Hunt Publishing Company, pp. 15–39.

Pleumarom A (1994) The political economy of tourism. The Ecologist 24:142–148.

Pleumarom A (1998) Political ecology of tourism. Annals of Tourism Research 25:25–54.

Poffenberger M (1994) The resurgence of community forest management in Eastern India. In D Western and RM Wright (eds.), Natural connections. Washington, DC: Island Press.

Pomeroy RS, Katon BM, and Harkes I (2001) Conditions affecting the success of fisheries co-management: Lessons from Asia. Marine Policy 25:197–208.

Pomeroy RS, and Berkes F (1997) Two to tango: The role of government in fisheries co-management. Marine Policy 21:465–480.

Poole P, ed. (1995) Geomatics: Who needs it? Cultural Survival Quarterly 18:1–77.

Posey DA (2001) Biological and cultural diversity: The inextricable, linked by language and politics. In L Maffi (ed.), On biocultural diversity: Linking language, knowledge, and the environment. Washington, DC: Smithsonian Institution Press, pp. 379–396.

Posey DA, and Balée W, eds. (1989) Resource management in Amazonia: Indigenous and folk strategies. The Bronx: New York Botanical Gardens.

Prescott-Allen R, and Prescott-Allen C (1990) How many plants feed the world. Conservation Biology 4:365–374.

Pressey RL, and Cowling RM (2001) Reserve selection algorithms and the real world. Conservation Biology 15:275–277.

Pressey RL, Humphries CJ, Margules CR, Vanewright RI, and Williams PH (1993) Beyond opportunism: Key principles for systematic reserve selection. Trends in Ecology & Evolution 8:124–128.

Primack RB (2000) A primer of conservation biology. Sunderland, MA: Sinauer Associates.

Principe PP (1996) Monetarizing the pharmacological benefits of plants. In MJ Balick, E Elisabetsky, and SA Laird (eds.), Medicinal resources of the tropical forest. New York: Columbia University Press, pp. 191–218.

Prins HHT (1992) The pastoral road to extinction: Competition between wildlife and traditional pastoralism in East Africa. Environmental Conservation *19*:117–123.

Prins HHT, Grootenhuis JG, and Dolan TT (2000) Wildlife conservation by sustainable use. Boston: Kluwer Academic Publishers.

Puri RK (1995) Comment on Alvard. Current Anthropology *36*:789–818.

Purvis A, Gittleman JL, Cowlishaw G, and Mace GM (2000) Predicting extinction risk in declining species. Proceedings of the Royal Society of London Series B-Biological Sciences *267*:1947–1952.

Quammen D (1996) The song of the dodo: Island biogeography in an age of extinctions. New York: Scribner.

Rackham O (1986) The history of the countryside. London: J.M. Dent.

Rahel FJ (1990) The hierarchical nature of community persistence: A problem of scale. American Naturalist *136*:328–344.

Rangan H (2000) Of myths and movements: Rewriting Chipko into Himalayan history. London; New York: Verso.

Rao M, and McGowan PJK (2002) Wild meat use, food security, livelihoods, and conservation. Conservation Biology *16*:580–583.

Redclift MR (1987) Sustainable development: Exploring the contradictions. London: Methuen.

Redford KH (1990) The ecologically noble savage. Orion Nature Quarterly *9*:25–29.

Redford KH (1992) The empty forest. Bioscience *42*:412–422.

Redford KH, Brandon K, and Sanderson SE (1998) Holding ground. In K Brandon, KH Redford, and SE Sanderson (eds.), Parks in peril: People, politics, and protected areas. Washington, DC: Island Press, pp. 455–463.

Redford KH, Coppolillo P, Sanderson EW, da Fonseca GAB, Dinerstein E, Groves C, Mace G, Maginnis S, Mittermeier RA, Noss R, Olson D, Robinson JG, Vedder A, and Wright M (2003) Mapping the conservation landscape. *Conservation Biology* 17:116–131.

Redford KH, and Feinsinger P (2001) The half-empty forest: Sustainable use and the ecology of interactions. In J Reynolds, GM Mace, KH Redford, and JG Robinson (eds.), Conservation of exploited species. London: Cambridge University Press.

Redford KH, and Mansour JA, eds. (1996) Traditional peoples and biodiversity conservation in large tropical landscapes. Arlington, VA: The Nature Conservancy.

Redford KH, and Padoch C (1992) Conservation of neotropical forests: Working from traditional resource use. New York: Columbia University Press.

Redford KH, and Richter BD (1999) Conservation of biodiversity in a world of use. Conservation Biology *13*:1246–1256.

Redford KH, and Stearman AM (1993) Forest-dwelling native Amazonians and the conservation of biodiversity: Interests in common or in collision? Conservation Biology 7:248–255.

Redford KH, and Taber A (2000) Writing the wrongs: Developing a safe-fail culture in conservation. Conservation Biology *14*:1567–1568.

Reice SR (1994) Nonequilibrium determinants of biological community structure. American Scientist *82*:424–435.

Reichel-Dokmatoff G (1971) Amazonian cosmos: The sexual and religious symbolism of the Tukano Indians. Chicago: University of Chicago Press.

Reid WV (1997) Opportunities for collaboration between biomedical and conservation communities. In F Grifo and J Rosenthal (eds.), Biodiversity and human health. Washington, DC: Island Press, pp. 334–351.

Reid WV, Laird SA, Meyer CA, Gamez R, Sittenfeld A, Janzen DH, Gollin MA, and Juma C (1996) Biodiversity prospecting. In MJ Balick, E Elisabetsky, and SA Laird (eds.), Medicinal resources of the tropical forest. New York: Columbia University Press, pp. 142–173.

Repetto RC, and Gillis M (1988) Public policies and the misuse of forest resources. Cambridge: Cambridge University Press.

Resor JP (1997) Debt-for-nature swaps: A decade of experience and new directions for the future. Unasylva *188*:15–22.

Ribot JC (1996) Participation without representation: Chiefs, councils, and forestry law in the West African Sahel. Cultural Survival Quarterly *20*:40–44.

Ribot JC (2002) Democratic decentralization of natural resources: Institutionalizing popular participation. Washington, DC: World Resources Institute.

Rice RA, and Greenberg R (2000) Cacao cultivation and the conservation of biological diversity. Ambio *29*:167–173.

Richards M (2000) Can sustainable tropical

forestry be made profitable? The potential and limitations of innovative incentive mechanisms. World Development 28:1001–1016.

Richards P (1992) Saving the rain forest? Contested futures in conservation. In S Wallman (ed.), Contemporary futures: Perspectives from social anthropology. London: Routledge.

Ridley M (1997) The origins of virtue. Harmondsworth, England: Viking.

Ridley M, and Low BS (1993) Can selfishness save the environment? The Atlantic Monthly September:76–86.

Rietbergen-McCracken J, and Abaza H (2000) Environmental valuation: A worldwide compendium of case studies. London: Earthscan.

Riley M (2001) The Traditional Medicine Research Center (TMRC): A potential tool for protecting traditional tribal medicinal knowledge in Laos. Cultural Survival Quarterly 24: 21–24.

Roberts C, and Hawkins JP (2000) Fully-protected marine reserves: A guide. Washington, DC: WWF Endangered Seas Campaign.

Roberts RG, Flannery TF, Ayliffe LK, Yoshida H, Olley JM, Prideaux GJ, Laslett GM, Baynes A, Smith MA, Jones R, and Smith BL (2001) New ages for the last Australian megafauna: Continent-wide extinction about 46,000 years ago. Science 292:1888–1892.

Robertson DP, and Hull B (2001) Beyond biology: Toward a more public ecology for conservation. Conservation Biology 15:970–979.

Robinson JG (1993) The limits to caring: Sustainable living and the loss of biodiversity. Conservation Biology 7:20–28.

Robinson JG, and Bennett EL, eds. (2000) Hunting for sustainability in tropical forests. New York: Columbia University Press.

Robinson JG, and Redford KH (1991) Sustainable harvest of neotropical forest mammals. In JG Robinson and KH Redford (eds.), Neotropical wildlife use and conservation. Chicago: University of Chicago Press, pp. 415–429.

Robinson JG, and Redford KH (2004) Jack of all trades, master of none: Inherent contradictions among ICDP approaches. In T McShane and M Wells (eds.), Getting Biodiversity Projects to Work: Towards More Effective Conservation and Development. New York: Columbia University Press.

Rochleau D, Thomas-Slayter B, and Wangari E (1996) Gender and environment: A feminist political ecology perspective. In D Rochleau, B Thomas-Slayter, and E Wangari (eds.), Feminist political ecology: Global issues and local experiences. New York: Routledge, pp. 3–23.

Rogers A (1994) Evolution of time preference by natural selection. The American Economic Review 84:460–481.

Röper M (2000) On the way to a better state? The role of NGOs in the planning and implementation of protected areas in Brazil. GeoJournal 52:61–69.

Ross A, and Pickering K (2002) The politics of reintegrating Australian Aboriginal and American Indian indigenous knowledge into resource management. Human Ecology 30: 187–213.

Roth AE (1995) Bargaining experiments. In JH Kagel and AE Roth (eds.), Handbook of experimental economics. Princeton, NJ: Princeton University Press, pp. 253–248.

Rozzi R, Silander J, Armesto JJ, Feinsinger P, and Massardo F (2000) Three levels of integrating ecology with the conservation of South American temperate forests: The initiative of the Institute of Ecological Research Chiloe, Chile. Biodiversity and Conservation 9:1199–1217.

Runge CF (1981) Common property externalities: Isolation, assurance, and resource depletion in a traditional grazing context. American Journal of Agricultural Economics 63: 595–606.

Runge CF (1986) Common property and collective action in economic development. World Development 14:623–635.

Ruttan LM (1998) Closing the commons: Cooperation for gain or restraint? Human Ecology 26:43–66.

Ruttan LM, and Borgerhoff Mulder M (1999) Are East African pastoralists truly conservationists? Current Anthropology 40:621–652.

Ryle J (1998) Lord of all he surveys. Outside 29:59+.

Safina C (1995) The world's imperiled fish. Scientific American 273:46–53.

Sagoff M (1988) The economy of the earth: Philosophy, law, and the environment. Cambridge: Cambridge University Press.

Sahlins M (1968) Notes on the original affluent society. In R Lee and I DeVore (eds.), Man the hunter. Chicago: Aldine, pp. 85–89.

Salafsky N, Cauley H, Balachander G, Cordes

B, Parks J, Margoluis C, Bhatt S, Encarnacion C, Russell D, and Margoluis R (2001) A systematic test of an enterprise strategy for community-based biodiversity conservation. Conservation Biology 15:1585–1595.

Salafsky N, Dugelby BL, and Terborgh JW (1993) Can extractive reserves save the rain forest: An ecological and socioeconomic comparison of nontimber forest product extraction systems in Peten, Guatemala, and West Kalimantan, Indonesia. Conservation Biology 7:39–52.

Salafsky N, and Margoluis R (1999) Threat reduction assessment: A practical and cost-effective approach to evaluating conservation and development projects. Conservation Biology 13:830–841.

Salafsky N, Margolius R, Redford KH, and Robinson JG (2002) Improving the practice of conservation: A conceptual framework and research agenda for conservation science. Conservation Biology 16:1469–1479.

Sally D (1995) Conversation and cooperation in social dilemmas: A meta-analysis of experiments from 1958–1992. Rationality and Society 7:58–92.

Sandalow DB, and Bowles IA (2001) Climate change: Fundamentals of treaty-making on climate change. Science 292:1839–1840.

Sandbrook R (1997) UNGASS has run out of steam. International Affairs 73:641–654.

Sanderson EW, Jaiteh M, Levy MA, Redford KH, and Wannebo AV (2002) The human footprint and the last of the wild. Bioscience 52:891–904.

Sandford S (1983) Management of pastoral development in the third world. London: John Wiley and Sons.

Sanjayan MA, Shen S, and Jansen M (1997) Experiences with integrated-conservation development projects in Asia. Washington, DC: World Bank.

Sarkar S (1990) Accommodating industrialism: A third world view of the West German ecological movement. The Ecologist 20:147–152.

Sarkar S (1998) Restoring wilderness or reclaiming forests? Terra Nova 3:35–51.

Sarkar S (1999) Wilderness preservation and biodiversity conservation: Keeping divergent goals distinct. Bioscience 49:405–412.

Sayre R, Mansour J, Li X, Boucher T, Sheppard S, and Redford K (1998) The parks in peril network: An ecogeographic perspective. In K Brandon, KH Redford, and SE Sanderson (eds.), Parks in peril: People, politics, and protected areas. Washington, DC: Island Press, pp. 37–61.

Schaller GB (1993) The last panda. Chicago: University of Chicago Press.

Schelhas J (2001) The USA national parks in international perspective: Have we learned the wrong lesson? Environmental Conservation 28:300–304.

Schelling T (1960) The strategy of conflict. Cambridge, MA: Harvard University Press.

Schroeder RA (1999) Shady practices: Agroforestry and gender politics in the Gambia. Berkeley: University of California Press.

Schroeder RA (2000) Beyond distributive justice: Resource extraction and environmental justice in the tropics. In C Zerner (ed.), People, plants and justice: The politics of nature conservation. New York: Columbia University Press, pp. 52–66.

Schroeder RA (2002) Debating distributive environmental justice: The politics of sharing wildlife wealth with rural communities in Tanzania.

Schuster BG, Jackson JE, Obijiofor CN, Okunji CO, Milbous W, Losos E, Agyafor JF, and Iwu MM (1999) Drug development and conservation of biodiversity in West and Central Africa: A model for collaboration with indigenous people. Pharmaceutical Biology 37:84–99.

Schwartz MW (1999) Choosing the appropriate scale of reserves for conservation. Annual Review of Ecology and Systematics 30:83–108.

Schwartz MW, Jurjavcic NL, and O'Brien JM (2002) Conservation's disenfranchised urban poor. BioScience 52:601–606.

Schwartzman S, Moreira A, and Nepstad D (2000a) Rethinking tropical forest conservation: Perils in parks. Conservation Biology 14:1351–1357.

Schwartzman S, Nepstad D, and Moreira A (2000b) Arguing tropical forest conservation: People versus parks. Conservation Biology 14:1370–1374.

Scoones I (1999) New ecology and the social sciences: What prospects for a fruitful engagement? Annual Review of Anthropology 28:479–507.

Seidensticker J, Christie S, and Jackson P (1999) Riding the tiger: Tiger conservation in human-dominated landscapes. Cambridge: Cambridge University Press.

Sen AK (1970) Collective choice and social welfare. San Francisco: Holden-Day.

Sen AK (1977) Rational fools: A critique of the behavioural foundations of economic theory. Philosophy and Public Affairs 6:317–344.

Sessions G, and Devall B (1990) Deep ecology. In R Nash (ed.), American environmentalism: Readings in conservation history. New York: McGraw-Hill, pp. 309–315.

Shanley P, and Luz L (2003) The impacts of forest degradation on medicinal plant use and implications for health care in Eastern Amazonia. BioScience 53:573–584.

Shiva V (1990) Biodiversity, biotechnology and profit. The Ecologist 20:44–47.

Shogren JF, Tschirhart J, Anderson T, Ando AW, Beissinger SR, Brookshire D, Brown GM, Coursey D, Innes R, Meyer SM, and Polasky S (1999) Why economics matters for endangered species protection. Conservation Biology 13:1257–1261.

Sillitoe P (1998) The development of indigenous knowledge. Current Anthropology 39:223–252.

Simberloff D (1998a) Small and declining populations. In WJ Sutherland (ed.), Conservation science and action. Oxford: Blackwell Science, pp. 114–134.

Simberloff D (1998b) Flagships, umbrellas, and keystones: Is single-species management passé in the landscape era? Biological Conservation 83:247–257.

Simbotwe MP (1993) African realities and western expectations. In D Lewis and N Carter (eds.), Voices from Africa: Local perspectives on conservation. Washington, DC: World Wildlife Fund, pp. 15–22.

Simon H (1990) A mechanism for the social selection of successful altruism. Science 250:1665–1668.

Simon JL (1990) Population matters: People, resources, environment, and immigration. New Brunswick, NJ: Transaction Publishers.

Slattersfield AJ, Crosby MJ, Long AJ, and Wege DC (1998) Endemic bird areas of the world: Priorities for biodiversity conservation. Cambridge: BirdLife International.

Smil V (1984) The bad earth: Environmental degradation in China. London: Zed Press.

Smith EA (1983) Anthropological applications of optimal foraging theory: A critical review. Current Anthropology 24:625–651.

Smith EA (1995) Comment on Alvard. Current Anthropology 36:789–818.

Smith EA (2001) On the coevolution of cultural, linguistic and biological diversity. In L Maffi (ed.), On biocultural diversity: Linking language, knowledge and the environment. Washington, DC: Smithsonian Institution Press, pp. 95–117.

Smith EA, and Winterhalder B (1992) Natural selection and decision making: Some fundamental principles. In EA Smith and B Winterhalder (eds.), Evolutionary ecology and human behavior. New York: Aldine de Gruyter, pp. 25–60.

Smith EA, and Wishnie M (2000) Conservation and subsistence in small-scale societies. Annual Review of Anthropology 29:493–524.

Sneath D (1998) Ecology: State policy and pasture degradation in inner Asia. Science 281:1147–1148.

Snelson D (1995) Neighbors as partners of protected areas. In JA McNeely (ed.), Expanding partnerships in conservation. Washington, DC: Island Press, pp. 280–290.

Songorwa AN (1999) Community-based wildlife management (CWM) in Tanzania: Are the communities interested? World Development 27:2061–2079.

Soulé ME (1985) What is conservation biology. BioScience 35:727–734.

Soulé ME, ed. (1986) Conservation biology: The science of scarcity and diversity. Sunderland, MA: Sinauer.

Soulé ME (2000) Does sustainable development help nature? Wild Earth Winter:57–64.

Soulé ME, and Sanjayan MA (1998) Ecology—conservation targets: Do they help? Science 279:2060–2061.

Sousa WP (1979) Experimental investigations of disturbance and ecological succession in a rocky intertidal algal community. Ecological Monographs 49:227–254.

Southwick EE, and Southwick L (1992) Estimating the economic value of honey bees (Hymenoptera, Apidae) as agricultural pollinators in the United-States. Journal of Economic Entomology 85:621–633.

Spaeder JJ (2003) Co-management in a landscape of resistance: The political ecology of wildlife management in Western Alaska. Anthropologica (in press).

Spergel B (2002) Financing protected areas. In J Terborgh, C van Schaik, L Davenport, and M Rao (eds.), Making parks work: Strategies for preserving tropical nature. Washington, DC: Island Press, pp. 364–382.

Spinage C (1998) Social change and conservation misrepresentation in Africa. Oryx 32: 265–276.

Sponsel LE, Bailey RC, and Headland TN. (1996) Anthropological perspectives on the causes, consequences, and solutions of deforestation. In LE Sponsel, TN Headland, and RC Bailey (eds.), Tropical deforestation: The human dimension. New York: Columbia University Press, pp. 3–52.

Sporrong U (1998) Dalecarlia in central Sweden before 1800: A society of social stability and ecological resilience. In F Berkes and C Folke (eds.), Linking social and ecological systems. Cambridge: Cambridge University Press, pp. 67–94.

Stapp WB, Cromsell MM, and Wals Arjen W (1995) The Global Rivers Environmental Education Network. In SK Jacobson (ed.), Conserving wildlife: International education and communication approaches. New York: Columbia University Press, pp. 177–197.

Steadman DW (1995) Prehistorical extinctions of Pacific island birds: Biodiversity meets zooarchaeology. Science 267:1123–1130.

Stearman AM (1994) Only slaves climb trees. Human Nature 5:339–357.

Steele P (1995) Ecotourism: An economic analysis. Journal of Sustainable Tourism 3:29–44.

Steinman AD, Havens KE, Aumen NG, James RT, Jin KR, Zhang J, and Rosen B (1999) Phosphorus inlake Okeechobee: Sources, sinks and strategies. In KR Reddy, GA O'Connor, and CL Schelske (eds.), Phosphorus biogeochemistry of subtropical ecosystems: Florida as a case example. New York: CRC/Lewis Publishers.

Stevens S (1997a) The legacy of Yellowstone. In S Stevens (ed.), Conservation through cultural survival: Indigenous peoples and protected areas. Washington, DC: Island Press, pp. 13–32.

Stevens S, ed. (1997b) Conservation through cultural survival. Washington, DC: Island Press.

Stevens S (1997c) Consultation, co-management, and conflict in Sagarmatha (Mount Everest) National park, Nepal. In S Stevens (ed.), Conservation through cultural survival: Indigenous peoples and protected areas. Washington, DC: Island Press, pp. 63–97.

Stevenson, GG (1991). Common property economics: A general theory and land use appli-

cations. Camabridge: Cambridge University Press.

Steward JH (1946) Handbook of South American Indians. Washington, DC: U.S. Government Printing Office: for sale by the Superintendent of Documents.

Steward JH (1955) Theory of culture change. Urbana: University of Illinois Press.

Stocking M, Perkin S, and Brown K (1995) Coexisting with nature in a developing world. In S Morse and P Stocking (eds.), People and environment. London: University College London Press, pp. 155–185.

Stocks A (1983) Cocamilla fishing: Patch modification and environmental buffering in the Amazon Varzea. In RB Hames and WT Vickers (eds.), New York: Academic Press, pp. 239–267.

Stonich SC (1998) Political ecology of tourism. Annals of Tourism Research 25:25–54.

Stronza A (2001) Anthropology of tourism: Forging new ground for ecotourism and other alternatives. Annual Review of Anthropology 30:261–283.

Sturgeon N (1997) Ecofeminist natures: Race, gender, feminist theory, and political action. New York: Routledge.

Sutherland WJ (2000) The conservation handbook: Research, management and policy. Oxford; Malden, MA: Blackwell Science.

Sutherland WJ (2003) Parallel extinction risk and global distribution of languages and species. Nature 423:276–279.

Swanson T (1995) The international regulation of biodiversity decline: Optimal policy and evolutionary product. In C Perrings, K-G Mäler, and C Folke (eds.), Biodiversity loss: Economic and ecological issues. Cambridge: Cambridge University Press, pp. 225–259.

Swanson T (1999a) Why is there a biodiversity convention? The international interest in centralized development planning. International Affairs 75:307–331.

Swanson T (1999b) Conserving global biological diversity by encouraging alternative development paths: Can development coexist with diversity? Biodiversity and Conservation 8: 29–44.

Tang SY (1992) Institutions and collective action: Self-governance in irrigation. San Francisco: Institute for Contemporary Studies Press.

Taylor B (1991) The religion and politics of Earth First! The Ecologist 21:258–266.

Ten Kate K, and Laird SA (2000) Biodiversity and business: Coming to terms with the 'grand bargain'. International Affairs 76: 241–264,U4,U5.

Terborgh J (1992) Maintenance of diversity in tropical forests. Biotropica 24:283–292.

Terborgh J (1999) Requiem for nature. Washington, DC: Island Press.

Terborgh J (2000) The fate of tropical forests: A matter of stewardship. Conservation Biology 14:1358–1361.

Terborgh J, Estes JA, Paquet P, Ralls K, Boyd-Heger D, Miller BJ, and Noss RF (1999) The role of top carnivores in regulating terrestrial ecosystems. In ME Soule and J Terborgh (eds.): Continental conservation: Scientific foundations of regional reserve networks. Washington, DC: Island Press, pp. 39–64.

Terborgh J, van Schaik C, Davenport L, and Rao M (2002) Making parks work: Strategies for preserving tropical nature. Washington, DC: Island Press.

Thaler RH (1992) The winner's curse: Paradoxes and anomalies of economic life. Princeton, NJ: Princeton University Press.

Thapa B (1998) Debt-for-nature swaps: An overview. International Journal of Sustainable Development and World Ecology 5: 249–262.

Ticktin TP, Nantel P, Ramirez F, and Johns T (2002) Effects of variation on harvest limits for nontimber forest species in Mexico. Conservation Biology 16:691–705.

Tiffen M, and Mortimore MJ (1994) Environment, population growth and productivity in Kenya: A case study of Machakos District. London: International Institute for Environment and Development.

Tilman D, Fargione J, Wolff B, D'Antonio C, Dobson A, Howarth R, Schindler D, Schlesinger WH, Simberloff D, and Swackhamer D (2001) Forecasting agriculturally driven global environmental change. Science 292: 281–284.

Tisdell C (1994) Conservation, protected areas and the global economic system: How debt, trade, exchange rates, inflation and macroeconomic policy affect biological diversity. Biodiversity and Conservation 3:419–436.

Tisdell CA (1995) Issues in biodiversity conservation including the role of local communities. Environmental Conservation 22:216–222, 228.

Trillmich F (1992) Conservation problems on Galapagos: The showcase of evolution in danger. Naturwissenschaften 79:1–6.

Trivers RL (1971) The evolution of reciprocal altruism. Quarterly Review of Biology 46: 189–226.

Turner BL, Villar SC, Foster D, Geoghegan J, Keys E, Klepeis P, Lawrence D, Mendoza PM, Manson S, Ogneva-Himmelberger Y, Plotkin AB, Salicrup DP, Chowdhury RR, Savitsky B, Schneider L, Schmook B, and Vance C (2001) Deforestation in the southern Yucatan peninsular region: An integrative approach. Forest Ecology and Management 154:353–370.

Turner T (1992) Defiant images: The Kayapo appropriation of video. Anthropology Today 8:5–16.

Turner T (1995) An indigenous peoples' struggle for socially equitable and ecologically sustainable production: The Kayapo revolt against extractivism. Journal of Latin American Anthropology 1:98–121.

UNEP (1999) Global environmental outlook—2000. Nairobi, Kenya: United Nations Environment Program.

van Jaarsveld AS, Freitag S, Chown SL, Muller C, Koch S, Hull H, Bellamy C, Kruger M, EndrodyYounga S, Mansell MW, and Scholtz CH (1998) Biodiversity assessment and conservation strategies. Science 279:2106–2108.

Vandergeest P (1996) Property rights in protected areas: Obstacles to community involvement as a solution in Thailand. Environmental Conservation 23:259–268.

Vasquez R, and Gentry AH (1989) Use and misuse of forest-harvested fruits in the Iquitos Area. Conservation Biology 3:350–361.

Vayda AP (1983) Progressive contextualization: Methods for research in human ecology. Human Ecology 11:265–281.

Vayda AP, and Walters BB (1999) Against political ecology. Human Ecology 27:167–179.

Veblen T (1899) The theory of the leisure class; an economic study in the evolution of institutions. New York: Macmillan.

Ventocilla J, Nunez V, Herrera H, Herrera F, and Chapin M (1997) The Kuna Indians and conservation. In KH Redford and JA Mancour (eds.), Traditional peoples and biodiversity conservation in large tropical landscapes. Arlington, VA: The Nature Conservancy, pp. 33–56.

Verissimo A, Barreto P, Tarifa R, and Uhl C (1995) Extraction of a high-value natural resource in Amazonia: The case of mahogany. Forest Ecology and Management 72:39–60.

Vickers WT (1991) Hunting yields and game composition over ten years in an Amazon Indian territory. In JG Robinson and KH Redford (eds.), Neotropical wildlife use and conservation. Chicago: University of Chicago Press, pp. 55–81.

Vickers WT (1994) From opportunism to nascent conservation: The case of the Siona-Secoya. Human Nature 5:307–337.

Vietmeyer N (1986) Lesser-known plants of potential use in agriculture and forestry. Science 232:1379–1384.

Visser DR, and Mendoza GA (1994) Debt-for-nature swaps in Latin-America. Journal of Forestry 92:13–16.

Vitousek PM, Mooney HA, Lubchenco J, and Melillo JM (1997) Human domination of Earth's ecosystems. Science 277:494–499.

Wallace A, and Naughton-Treves L (1998) Belize: Rio Bravo conservation and management area. In K Brandon, KH Redford, and SE Sanderson (eds.), Parks in peril: People, politics, and protected areas. Washington, DC: Island Press, pp. 217–247.

Walpole MJ, and Goodwin HJ (2001) Local attitudes towards conservation and tourism around Komodo National Park, Indonesia. Environmental Conservation 28:160–166.

Walpole MJ, and Leader-Williams N (2001) Masai Mara tourism reveals partnership benefits. Nature 413:771.

Walters BB, Cadelina A, Cardano A, and Visitacion E (1999) Community history and rural development: Why some farmers participate more readily than others. Agricultural Systems 59:193–214.

Ward D, Ngairorue BT, Kathena J, Samuels R, and Ofran Y (1998) Land degradation is not a necessary outcome of communal pastoralism in arid Namibia. Journal of Arid Environments 40:357–371.

Warren DM, and Pinkston J (1998) Indigenous African resource management of a tropical rainforest ecosystem: A case study of the Yoruba of Ara, Nigeria. In F Berkes and C Folke (eds.), Linking social and ecological systems. Cambridge: Cambridge University Press, pp. 158–189.

Warren KJ (1999) Environmental justice: Some ecofeminist worries about a distributive model. Environmental Ethics 21:151–161.

Warren S (2001) Landmark ruling favors Sarawak's indigenous communities. Cultural Survival Quarterly 25:55.

Watson R, and Pauly D (2001) Systematic distortions in world fisheries catch trends. Nature 414:534–536.

Watts M (1983) Silent violence: Food, famine, & peasantry in northern Nigeria. Berkeley: University of California Press.

Wearing S, and Neil J (1999) Ecotourism: Impacts, potentials, and possibilities. Oxford: Butterworth-Heinemann.

Weaver DB (1999) Magnitude of ecotourism in Costa Rica and Kenya. Annals of Tourism Research 26:792–816.

Weber B, and Vedder A (2001) Afterword: Mountain gorillas at the turn of the century. In MR Robbins, P Sicotte, and KJ Stewart (eds.), Mountain gorillas. Cambridge: Cambridge University Press, pp. 414–423.

Weinberg A, Bellows S, and Ekster D (2002) Sustaining ecotourism: Insights and implications from two successful case studies. Society & Natural Resources 15:371–380.

Wells AC (2001) Caribbean pride. Living Bird Autumn:10–16.

Wells M (1992) Biodiversity conservation, affluence and poverty: Mismatched costs and benefits and efforts to remedy them. Ambio 21:237–243.

Wells M (1993) Neglect of biological riches: The economics of nature tourism in Nepal. Biodiversity and Conservation 2:445–464.

Wells, M (1994) A profile and interim assessment of the Annapurna Conservation Area Project, Nepal. In D Western and RM Wright (eds.), Natural connections: Perspectives in community-based conservation. Washington, DC: Island Press, pp. 261–281.

Wells M, Brandon K, and Hannah L (1992) Parks and people: Linking protected area management with local communities. Washington, DC: World Bank.

Wells M, Guggenheim S, Khan A, Wardojo W, and Jepson P (1999) Investing in biodiversity: A review of Indonesia's integrated conservation and development projects. Washington, DC: World Bank.

West PC, and Brechin SR, eds. (1991) Resident peoples and national parks. Tucson: University of Arizona Press.

Western D (1994) Ecosystem conservation and rural development: The case of Amboseli. In D Western and RM Wright (eds.), Natural connections: Perspectives in community-based conservation. Washington, DC: Island Press, pp. 15–52.

Western D, and Wright M, eds. (1994) Natural connections: Perspectives in community-based conservation. Washington, DC: Island Press.

Westoby JC (1987) The purpose of forests: Follies of development. New York: B. Blackwell.

White LW (1967) The historical roots of our ecologic crisis. Science 155:1203–1207.

Wilcox BA, and Duin KN (1995) Indigenous cultural and biological diversity: Overlapping values of Latin American ecoregions. Cultural Survival Quarterly 18:49–53.

Wilhere GF (2002) Adaptive management in habitat conservation plans. Conservation Biology 16:20–29.

Wilkie DS, Carpenter JF, and Zhang QF (2001) The under-financing of protected areas in the Congo Basin: So many parks and so little willingness-to-pay. Biodiversity and Conservation 10:691–709.

Wilkie DS, and Godoy RA (2001) Income and price elasticities of bushmeat demand in lowland Amerindian societies. Conservation Biology 15:761–769.

Williams GC (1966) Adaptation and natural selection: A critique of some current evolutionary thought. Princeton, NJ: Princeton University Press.

Williams N, and Baines G, eds. (1993) Traditional ecological knowledge: Wisdom for sustainable development. Canberra: Australian National University.

Wilson A (1993) Sacred forests and the elders. In E Kemf (ed.), Indigenous peoples and protected areas. London: Earthscan.

Wilson EO (1984) Biophilia. Cambridge, MA: Harvard University Press.

Wilson EO (1988) The current state of biological diversity. In EO Wilson and FM Peter (eds.), Biodiversity. Washington, DC: National Academy Press, pp. 3–20.

Wilson EO (1992) The diversity of life. Cambridge, MA: Belknap Press of Harvard University Press.

Wilson EO, and Peter FM (1988) Biodiversity. Washington, DC: National Academy Press.

Wilson EO (1998) Consilience: The unity of knowledge. New York: Knopf.

Wilson GM (1953) The Tatoga of Tanganyika (Part II). Tanganyika notes and records 34: 35–56.

Wilson M, Daly M, and Gordon S (1998) The evolved psychological apparatus of human decision-making is one source of environmental problems. In T Caro (ed.), Behavioral ecology and conservation biology. New York: Oxford University Press, pp. 501–523.

Winterhalder B (1981) Foraging strategies in the boreal forest: An analysis of Cree hunting and gathering. In B Winterhalder and EA Smith (eds.), Hunter-gatherer foraging strategies. Chicago: University of Chicago Press, pp. 66–98.

Winterhalder B, and Lu F (1997) A forager-resource population ecology model and implications for indigenous conservation. Conservation Biology 11:1354–1364.

Winterhalder BW, Baillargeon W, Cappelletto F, Daniel I, and Prescott C (1988) The population ecology of hunter-gatherers and their prey. Journal of Anthropological Archaeology Research 7:289–328.

Woinarski JCZ, Milne DJ, and Wanganeen G (2001) Changes in mammal populations in relatively intact landscapes of Kakadu National Park, Northern Territory, Australia. Austral Ecology 26:360–370.

Wood D (1995) Conserved to death: Are tropical forests being over-protected from people. Land Use Policy 12:115–135.

Woodroffe R, and Ginsberg JR (1998) Edge effects and the extinction of populations inside protected areas. Science 280:2126–2128.

World Bank (1998) Assessing aid: What works, what doesn't, and why. New York: Oxford University Press.

World Commission on Environment and Development (1987) Our common future. New York: Oxford University Press.

World Conservation Monitoring Centre (1992) Global biodiversity: Status of the earth's living resources. London: Chapman and Hall.

WRI (1992) Global biodiversity strategy: Guidelines for action to save, study, and use the earth's biotic wealth sustainably and equitably. Washington, DC: World Resources Institute.

Wright PC (1992) Primate ecology, rainforest ecology, and economic development: Building a national park in Madagascar. Evolutionary Anthropology 1:25–33.

Wright PC (1997) The future of biodiversity in Madagascar: A view from Ranomafana National Park. In SM Goodman and BD Patterson (eds.): Natural change and human impact in Madagascar. Washington, DC: Smithsonian Institution Press, pp. 381–405.

Wright PC, and Andriamihaja B (2002) Making a rain forest national park work in Madagascar: Ranomafana National Park and its long-term research commitment. In J Terborgh, van Schaik C, L Davenport, and M Rao (eds.), Making parks work: Strategies for preserving tropical nature. Washington, DC: Island Press, pp. 112–136.

Wright RM (1988) Anthropological presuppositions of indigenous advocacy. Annual Reviews in Anthropology 17:365–390.

Wright SJ, Zeballos H, Dominguez I, Gallardo MM, Moreno MC, and Ibanez R (2000) Poachers alter mammal abundance, seed dispersal, and seed predation in a neotropical forest. Conservation Biology 14:227–239.

Wunder S (2001) Poverty alleviation and tropical forests: What scope for synergies? World Development 29:1817–1833.

Wurm S (1991) Language death and disappearance: Causes and circumstances. In RH Robins and E Uhlenbeck (eds.): Endangered languages. New York: Berg.

Wyckoff-Baird B, Klaus A, Christen C, and Keck M (2000) Shifting the power: Decentralization and biodiversity conservation. Washington, DC: Biodiversity Support Program.

Yibarbuk D, Whitehead PJ, Russell-Smith J, Jackson D, Godjuwa C, Fisher A, Cooke P, Choquenot D, and Bowman DMJS (2001) Fire ecology and Aboriginal land management in central Arnhem Land, northern Australia: A tradition of ecosystem management. Journal of Biogeography 28:325–343.

Young TP (1993) Development and conservation: More on caring. Conservation Biology 7:750–751.

Yu DW, Sutherland WJ, and Clark C (2002) Trade versus environment. Trends in Ecology and Evolution 17:341–344.

Zhi L, Wenshi P, Xiaojian Z, and Dajun W (2000) What has the panda taught us? In A Entwistle and N Dunstone (eds.), Priorities for the conservation of mammalian diversity: Has the panda had its day? Cambridge: Cambridge University Press, pp. 325–334.

Zimmerman B, Peres CA, Malcolm JR, and Turner T (2001) Conservation and development alliances with the Kayapo of south-eastern Amazonia, a tropical forest indigenous people. Environmental Conservation 28:10–22.

Index